Advanced Polarimetry and Polarimetric Imaging

Advanced Polarimetry and Polarimetric Imaging

Xiaobo Li
Fei Liu
Jian Liang

MDPI

Basel • Beijing • Wuhan • Barcelona • Belgrade • Novi Sad • Cluj • Manchester

Xiaobo Li
School of Marine Science and Technology
Tianjin University
Tianjin
China

Fei Liu
School of Optoelectronic Engineering
Xidian University
Xi'an
China

Jian Liang
School of Physics and Information Technology
Shaanxi Normal University
Xi'an
China

Editorial Office
MDPI AG
Grosspeteranlage 5
4052 Basel, Switzerland

This is a reprint of articles from the Special Issue published online in the open access journal *Photonics* (ISSN 2304-6732) (available at: www.mdpi.com/journal/photonics/special_issues/154N325O38).

For citation purposes, cite each article independently as indicated on the article page online and using the guide below:

Lastname, A.A.; Lastname, B.B. Article Title. *Journal Name* **Year**, *Volume Number*, Page Range.

ISBN 978-3-7258-1732-0 (Hbk)
ISBN 978-3-7258-1731-3 (PDF)
https://doi.org/10.3390/books978-3-7258-1731-3

Contents

photonics

Editorial

Advanced Polarimetry and Polarimetric Imaging

Xiaobo Li [1,*], Fei Liu [2] and Jian Liang [3]

[1] School of Marine Science and Technology, Tianjin University, Tianjin 300072, China
[2] School of Optoelectronic Engineering, Xidian University, Xi'an 710071, China; feiliu@xidian.edu.cn
[3] School of Physics and Information Technology, Shaanxi Normal University, Xi'an 710119, China
* Correspondence: lixiaobo@tju.edu.cn

Polarization, a core attribute of light waves, offers insights into light's physical properties and its interactions with materials [1,2]. This unique aspect of polarization paves the way for its application across a broad range of fields, from object detection and biomedical imaging to remote sensing and beyond [3–6]. This Editorial is part of the Special Issue "Advanced Polarimetry and Polarimetric Imaging", which highlights new theories in and applications of advanced polarimeters and polarimetric imaging. Seventeen manuscripts were submitted to this Special Issue, and all of them were subject to a rigorous review process. In total, fourteen papers were finally accepted for publication and included in this Special Issue (twelve articles and two reviews). These contributions are listed below.

Precision in polarization measurements is foundational to achieving superior polarimetric imaging [7–9], and there are four contributions related to advanced polarization measurements in this issue (1, 2, 3, 4). The Stokes–Mueller formalism models how a light beam's polarization state is altered through its linear interactions with materials [1]. In Contribution 1, Gil et al. introduced a novel, synthesized Mueller polarimetry imaging approach, creating new avenues for the enhancement of Mueller polarimetry's application. Liu et al. (Contribution 2) engineered a vertically aligned, high-speed Mueller matrix ellipsometer, achieving a remarkable, dynamic monitoring of rapidly changing processes, with a measurement resolution in the 10 μs range and unparalleled accuracy. The calibration challenges arising from waveplate field of view effects, often caused by manufacturing or installation errors, can significantly impact optical systems' precision. In Contribution 3, Jiang et al. responded to this challenge with a calibration strategy that mitigates these effects, improving systems' accuracy. Taking cues from nature, Liu et al. (Contribution 4) developed a polarization sensor inspired by insects' sky-based navigation, integrating an image chip with nanograting via nanoimprint lithography for autonomous navigation, with impressive accuracy in polarization angle measurements.

Polarized light research extends into the realms of metasurfaces and metamaterials [10]. In Contribution 5, Shang et al. demonstrated dynamic, tunable structural coloration through a polarization-sensitive metasurface, showcasing its full-color image display and switching capabilities—a promising tool for virtual reality and high-density data storage. Li et al. (Contribution 6) crafted a fully dielectric chiral metasurface, achieving a full-color display with anti-counterfeiting features. Liu et al. (Contribution 7) delved into perfect absorber metamaterials, exploring the principles of impedance matching and coherent perfect absorption. Fundamental research, such as Smolkin et al.'s numerical method for wave propagation constant calculations (Contribution 8) and Li et al.'s limb boundary extraction technique for EUV solar images (Contribution 9), further broadens the scope of this Special Issue.

Polarimetric imaging excels in enhancing the image quality of images captured in scattering environments, leveraging the partial polarization of light scattered by microparticles [11–15]. In Contribution 10, Zhang et al. proposed an active polarization imaging technique tailored to turbid waters, dramatically improving image quality. Deep learning, as utilized by Lin et al. in Contribution 11, augments the capabilities of polarimetric

Citation: Li, X.; Liu, F.; Liang, J. Advanced Polarimetry and Polarimetric Imaging. *Photonics* **2024**, *11*, 317. https://doi.org/10.3390/photonics11040317

Received: 19 March 2024
Accepted: 28 March 2024
Published: 29 March 2024

imaging, achieving high-performance imaging reconstructions. Polarimetric technology can also advance 3D reconstruction efforts, with applications ranging from tunnel crack detection to passive 3D face reconstruction, as demonstrated by Zhang et al. (Contribution 12) and Han et al. (Contribution 13), respectively. Polarization lidar, or P-lidar, expands our detection capabilities by harnessing polarization's physical properties, thus enriching the information obtained from targets. In Contribution 14, Liu et al. provided a comprehensive overview of P-lidar's principles and applications, highlighting its potential use in atmospheric, oceanic, and terrestrial observations.

This Special Issue on polarization technology illustrates the field's notable progress and potential. It features research articles that introduce innovative solutions and tackle key challenges in polarimetric image restoration, 3D reconstruction, high-speed Mueller ellipsometry, and P-lidar. These promising applications and novel approaches in polarimetry and imaging technology herald a promising future.

Funding: This research received no external funding.

Acknowledgments: As Guest Editors of the Special Issue "Advanced Polarimetry and Polarimetric Imaging", we would like to express our deep appreciation to all authors whose valuable work was published in this issue and who thus contributed to its success.

Conflicts of Interest: The authors declare no conflicts of interest.

List of Contributions:

1. Gil, J.J.; San José, I. Synthetic Mueller Imaging Polarimetry. *Photonics* **2023**, *10*, 969.
2. Liu, J.; Zhang, S.; Deng, B.; Li, L.; Gu, H.; Zhu, J.; Jiang, H.; Liu, S. Development and Calibration of a Vertical High-Speed Mueller Matrix Ellipsometer. *Photonics* **2023**, *10*, 1064.
3. Jiang, Z.; Zhang, S.; Jiang, H.; Liu, S. Calibration of Waveplate Retardance Fluctuation Due to Field-of-View Effect in Mueller Matrix Ellipsometer. *Photonics* **2023**, *10*, 1038.
4. Liu, Z.; Chu, J.; Zhang, R.; Guan, C.; Fan, Y. Preparation of an Integrated Polarization Navigation Sensor via a Nanoimprint Photolithography Process. *Photonics* **2022**, *9*, 806.
5. Shang, X.; Niu, J.; Li, H.; Li, L.; Hu, H.; Lu, C.; Shi, L. Polarization-Sensitive Structural Colors Based on Anisotropic Silicon Metasurfaces. *Photonics* **2023**, *10*, 448.
6. Li, L.; Li, H.; Hu, H.; Shang, X.; Xue, H.; Hu, J.; Lu, C.; Zhao, S.; Niu, J.; Shi, L. Full-Color and Anti-Counterfeit Printings with All-Dielectric Chiral Metasurfaces. *Photonics* **2023**, *10*, 401.
7. Liu, X.; Xia, F.; Wang, M.; Liang, J.; Yun, M. Working Mechanism and Progress of Electromagnetic Metamaterial Perfect Absorber. *Photonics* **2023**, *10*, 205.
8. Smolkin, E.; Smirnov, Y. Numerical Study of the Spectrum of TE-Polarized Electromagnetic Waves of a Goubau Line Coated with Graphene. *Photonics* **2023**, *10*, 1297.
9. Li, S.; Zhang, J.; Liu, B.; Jiang, C.; Ren, L.; Xue, J.; Song, Y. An Algorithm to Extract the Boundary and Center of EUV Solar Image Based on Sobel Operator and FLICM. *Photonics* **2022**, *9*, 889.
10. Zhang, H.; Gong, J.; Ren, M.; Zhou, N.; Wang, H.; Meng, Q.; Zhang, Y. Active Polarization Imaging for Cross-Linear Image Histogram Equalization and Noise Suppression in Highly Turbid Water. *Photonics* **2023**, *10*, 145.
11. Lin, B.; Fan, X.; Li, D.; Guo, Z. High-Performance Polarization Imaging Reconstruction in Scattering System under Natural Light Conditions with an Improved U-Net. *Photonics* **2023**, *10*, 204.
12. Zhang, Y.; Zhang, X.; Su, Y.; Li, X.; Ma, S.; Zhang, S.; Ren, W.; Li, K. Tunnel Lining Crack Detection Method Based on Polarization 3D Imaging. *Photonics* **2023**, *10*, 1085.
13. Han, P.; Li, X.; Liu, F.; Cai, Y.; Yang, K.; Yan, M.; Sun, S.; Liu, Y.; Shao, X. Accurate Passive 3D Polarization Face Reconstruction under Complex Conditions Assisted with Deep Learning. *Photonics* **2022**, *9*, 924.
14. Liu, X.; Zhang, L.; Zhai, X.; Li, L.; Zhou, Q.; Chen, X.; Li, X. Polarization Lidar: Principles and Applications. *Photonics* **2023**, *10*, 1118.

References

1. Goldstein, D.H. *Polarized Light*; CRC Press: Boca Raton, FL, USA, 2017.
2. Pérez, J.J.G.; Ossikovski, R. *Polarized Light and the Mueller Matrix Approach*; CRC Press: Boca Raton, FL, USA, 2017.

3. Sun, W.; Liu, Z.; Videen, G.; Fu, Q.; Muinonen, K.; Winker, D.M.; Lukashin, C.; Jin, Z.; Lin, B.; Huang, J. For the depolarization of linearly polarized light by smoke particles. *J. Quant. Spectrosc. Radiat. Transf.* **2013**, *122*, 233–237. [CrossRef]
4. Kong, Z.; Ma, T.; Zheng, K.; Cheng, Y.; Gong, Z.; Hua, D.; Mei, L. Development of an all-day portable polarization Lidar system based on the division-of-focal-plane scheme for atmospheric polarization measurements. *Opt. Express* **2021**, *29*, 38512–38526. [CrossRef] [PubMed]
5. Li, X.; Hu, H.; Zhao, L.; Wang, H.; Yu, Y.; Wu, L.; Liu, T. Polarimetric image recovery method combining histogram stretching for underwater imaging. *Sci. Rep.* **2018**, *8*, 12430. [CrossRef] [PubMed]
6. Shibata, Y. Particle polarization Lidar for precipitation particle classification. *Appl. Opt.* **2022**, *61*, 1856–1862. [CrossRef] [PubMed]
7. Li, X.; Liu, T.; Huang, B.; Song, Z.; Hu, H. Optimal distribution of integration time for intensity measurements in Stokes polarimetry. *Opt. Express* **2015**, *23*, 27690–27699. [CrossRef] [PubMed]
8. Morio, J.; Goudail, F. Influence of the order of diattenuator, retarder, and polarizer in polar decomposition of Mueller matrices. *Opt. Lett.* **2004**, *29*, 2234–2236. [CrossRef] [PubMed]
9. Chu, J.; Zhao, K.; Zhang, Q.; Wang, T. Construction and performance test of a novel polarization sensor for navigation. *Sens. Actuators A Phys.* **2008**, *148*, 75–82. [CrossRef]
10. Kim, G.; Kim, Y.; Yun, J.; Moon, S.W.; Kim, S.; Kim, J.; Park, J.; Badloe, T.; Kim, I.; Rho, J. Metasurface-driven full-space structured light for three-dimensional imaging. *Nat. Commun.* **2022**, *13*, 5920. [CrossRef] [PubMed]
11. Qi, P.; Li, X.; Han, Y.; Zhang, L.; Xu, J.; Cheng, Z.; Liu, T.; Zhai, J.; Hu, H. U2R-pGAN: Unpaired underwater-image recovery with polarimetric generative adversarial network. *Opt. Lasers Eng.* **2022**, *157*, 107112. [CrossRef]
12. Wei, Y.; Han, P.; Liu, F.; Shao, X. Enhancement of underwater vision by fully exploiting the polarization information from the Stokes vector. *Opt. Express* **2021**, *29*, 22275–22287. [CrossRef] [PubMed]
13. Li, H.; Zhu, J.; Deng, J.; Guo, F.; Zhang, N.; Sun, J.; Hou, X. Underwater active polarization descattering based on a single polarized image. *Opt. Express* **2023**, *31*, 21988–22000. [CrossRef] [PubMed]
14. Liang, J.; Ren, L.; Qu, E.; Hu, B.; Wang, Y. Method for enhancing visibility of hazy images based on polarimetric imaging. *Photonics Res.* **2014**, *2*, 38–44. [CrossRef]
15. Liu, F.; Han, P.; Wei, Y.; Yang, K.; Huang, S.; Li, X.; Zhang, G.; Bai, L.; Shao, X. Deeply seeing through highly turbid water by active polarization imaging. *Opt. Lett.* **2018**, *43*, 4903–4906. [CrossRef] [PubMed]

photonics

MDPI

Review

Polarization Lidar: Principles and Applications

Xudong Liu [1,2], Liping Zhang [3], Xiaoyu Zhai [4], Liye Li [5], Qingji Zhou [1], Xue Chen [6,*] and Xiaobo Li [1,2,*]

1 School of Marine Science and Technology, Tianjin University, Tianjin 300054, China; sunny777@tju.edu.cn (X.L.); zqj@tju.edu.cn (Q.Z.)
2 College of Mechanical Engineering and Automation, Fuzhou University, Fuzhou 350108, China
3 Department of Imaging and Interventional Radiology, The Chinese University of Hong Kong, Hong Kong 999077, China; lpzhang@cuhk.edu.hk
4 National Ocean Technology Center, Tianjin 300112, China; cynthiachai@126.com
5 China Ship Scientific Research Center, Wuxi 214082, China; lly850189@163.com
6 Law School, Tianjin University, Tianjin 300054, China
* Correspondence: xuechen@tju.edu.cn (X.C.); lixiaobo@tju.edu.cn (X.L.)

Abstract: Traditional lidar techniques mainly rely on the backscattering/echo light intensity and spectrum as information sources. In contrast, polarization lidar (P-lidar) expands the dimensions of detection by utilizing the physical property of polarization. By incorporating parameters such as polarization degree, polarization angle, and ellipticity, P-lidar enhances the richness of physical information obtained from target objects, providing advantages for subsequent information analysis. Over the past five decades, the application fields of P-lidar have rapidly expanded, starting from its early use in atmospheric remote sensing to later applications in oceanic remote sensing. This review first provides a brief introduction to the basic principles of both polarization and P-lidar, along with typical systems. It then explores the applications of P-lidar in various remote sensing fields, including atmospheric, oceanic, and terrestrial domains. Additionally, we propose potential research directions based on current cutting-edge applications, with the aims of providing critical insights to researchers in the fields of polarization and lidar and inspiring further exciting ideas.

Keywords: polarization; lidar; remote sensing; P-lidar

Citation: Liu, X.; Zhang, L.; Zhai, X.; Li, L.; Zhou, Q.; Chen, X.; Li, X. Polarization Lidar: Principles and Applications. *Photonics* **2023**, *10*, 1118. https://doi.org/10.3390/photonics10101118

Received: 19 September 2023
Revised: 1 October 2023
Accepted: 2 October 2023
Published: 4 October 2023

1. Introduction

Polarization, akin to parameters such as frequency, phase, and amplitude/intensity, represents a fundamental physical property of light [1–3]. The earliest records of polarized light can be traced back to 1669, when Bartholin discovered the phenomenon of double-refraction in a piece of Iceland spar (calcite), which paved the way for human exploration of polarized optics [2]. Subsequently, in 1678, Huygens proposed the wave theory of light, providing a satisfactory explanation for the polarization characteristics of light [2]. Therefore, Huygens is recognized as the first scientist in the history of physics to discover the properties of polarized light.

Polarized light is prevalent in the natural world, primarily resulting from reflection and scattering processes. This is because reflection and scattering often induce varying optical efficiencies and/or phase changes for the orthogonal polarization components of the incident light [3,4]. Consequently, differences in the surface structure and texture of an object can influence the polarization state of reflected and/or scattered light [5,6]. By measuring the polarization characteristics of reflected or scattered light, the analysis of surface morphology information becomes possible, leading to extensive applications of polarized light in remote sensing [7,8]. For example, in passive remote sensing, the polarization characteristics of solar spectral lines serve as important carriers for navigation [9,10]; when sunlight interacts with water vapor, ice crystals, dust, sand, smoke, and other substances, polarized light can be generated, proving the physical properties of such materials [11,12]. Furthermore, polarized light plays a significant role in areas including resource exploration, vegetation and soil classification, research on the sea surface, and global atmospheric aerosol studies [5,13]. In active remote sensing, polarized light is indispensable for detecting aerosols' shape, identifying cloud phases, and determining particles' orientation [14–16].

In 1916, Einstein proposed the theory of light's stimulated emission (LSE) [17], which guided humanity to recognize such light with a precise single color and wavelength, which is also named light amplification by stimulated emissions of radiation (laser) [18]. However, it was not until 1960 that the first laser generator was developed, marking the beginning of laser utilization [19]. Actually, the idea of using a laser in radar systems, i.e., light detection and ranging (Lidar), was immediately considered after the laser's invention [20]. For instance, in 1969, an American team installed a retro-reflector device similar to a mirror on the lunar surface. Laser beams were directed towards this device from Earth, enabling accurate measurement of the Earth–Moon distance [21,22]. However, lasers quickly expanded beyond this application and found widespread utility in various fields, including surveying, atmospheric remote sensing, and oceanic remote sensing.

Lidar systems consist of three main components: transmitter, receiver, and processing. The transmitter emits a laser beam into a target scene, initiating an interaction between the laser and the target object (such as suspended particles in the atmosphere or seawater), which causes changes in various aspects of the optical signal, including propagation direction, intensity, frequency, polarization, and phase [23]. By detecting these changes and combining transmission models, information related to the target's physical properties, such as its position, velocity, and composition, can be inferred [24,25]. Figure 1 presents the basic elements of lidar: the laser sources, ranging principles, beam modulations, and detectors. It should be noted that all the used abbreviations and the related full definitions are listed in Appendix A.

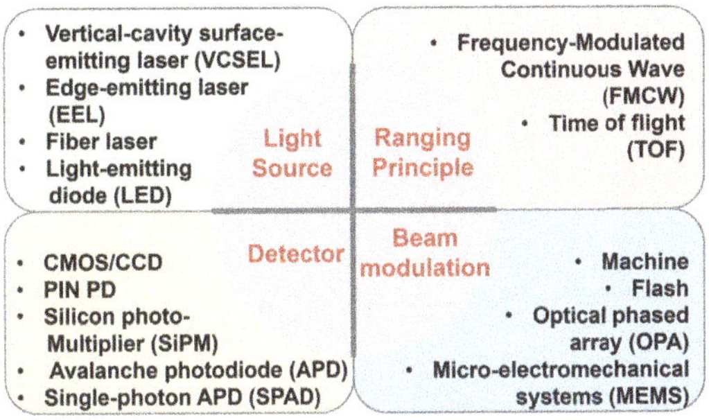

Figure 1. Basic components of lidar.

Lidar is a remote sensing technology similar to radar, which uses laser pulses as the radiation source instead of continuous electromagnetic waves. Commonly used laser sources include vertical-cavity surface-emitting lasers (VCSELs) [26], edge-emitting lasers (EELs) [27], fiber lasers, and light-emitting diodes (LEDs) [28]. It is worth noting that both pulsed and continuous-wave lasers are employed in lidar systems, where the pulsed systems measure the round-trip time of a short light pulse from the laser to the target and back to the receiver, i.e., the ranging principle is named time of flight (TOF) [29]; the continuous wave systems range by measuring the phase difference between the transmitted and received signals, for example the frequency-modulated continuous wave (FMCW) [30]. In lidar systems, three different wavelength regions are used: near-infrared (NIR) excitation at 1064 nm using diode-flashlamp-pumped solid-state or thulium-doped fiber lasers, visible light (VIS) excitation at 532 nm generated through frequency doubling of the 1064 nm laser or ultraviolet (UV) at 355 nm though frequency tripling, and short-wave infrared (SWIR) excitation at 1550 nm using erbium-doped fiber lasers. The advantages and disadvantages of these choices depend on the target reflectance and absorption, background radiation, atmospheric transmission, and eye safety considerations. According to the different ways the laser beam is emitted towards the targets, lidar can be roughly divided into flash lidar and scanning lidar. Flash lidar systems observe the complete field of view (FOV) simultaneously and often employ a charge-coupled device (CCD) or complementary metal–oxide semiconductor (CMOS) sensors as detection devices [31]. Conversely, scanning lidar systems focus on a subset of the FOV before moving on to the next subset, covering the entire FOV sequentially. Scanning lidar systems have the ability to detect objects at greater area compared to flash lidar systems since the laser beam is concentrated on a subset of the FOV at a time [32]. However, the laser beam of the scanner necessitates redirection from one subset to another to cover the entire FOV. Current systems typically use mechanical-beam-steering principles, achieved either by rotating the entire sensor head (rotational lidar) or using internal micro-electromechanical systems (MEMS) within the sensor for beam steering. In this case, the detector may be a single-point detector, such as the PIN photodiode (PD) [33], silicon photo-multiplier (SiPM) [34], avalanche photodiode (APD), or single-photon APD (SPAD) [35]. More information about lidar's basic components can be found in previous reviews [32,36–38].

Initially, lidar only detected intensity (or power) changes of backscattering or echo light signals; later, wavelength and/or frequency modulation was added; then, polarization modulation was introduced, forming polarization lidar, i.e., P-lidar. Researchers [39] from New York University initially borrowed the P-lidar technique from the microwave radar methods, which were developed in about the 1950s (before the laser was invented). They found that laser depolarization is considerably stronger compared to microwave depolarization from non-spherical particles, suggesting that P-lidar is more promising for studying aerosols, particles in clouds, and precipitation (i.e., hydrometeors). Polarization has become a crucial aspect in contemporary radar systems, with most systems incorporating polarization components to enable either the partial or complete functionality of P-lidar. A concise chronological chart of P-lidar (including spaceborne P-lidar launched by different countries and various P-lidar technologies) is presented in Figure 2.

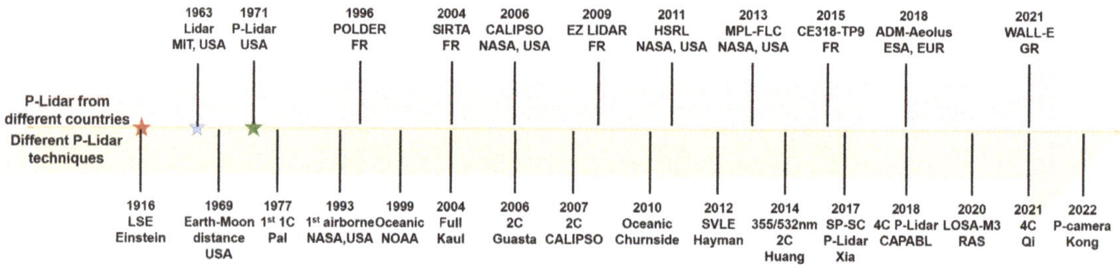

Figure 2. History of P-lidar. 1C: single-channel; 2C: dual-channel; 4C: four-channel; SP: single-photon; other abbreviations can be found in the main text. The corresponding references for the second line are [17,21,22,40–55].

In terms of applications, according to the Mie scattering theory [56], when polarized light is incident on spherical particles (such as water droplets), the backscattered light is in the same polarization state as the emitted light. However, when polarized light is incident on non-spherical particles (such as ice crystals), a portion of the backscattered light becomes depolarized, resulting in partially polarized light. By analyzing the changes in the echo signal's polarization state, P-lidar can distinguish between spherical and non-spherical particles in the atmosphere [15]. As a result, P-lidar was initially widely used for identifying liquid water clouds and ice clouds in the atmosphere. Furthermore, P-lidars have proven to be valuable for the remote sensing of the Earth's surface parameters, i.e., the surface texture and dielectric properties, which influence the multi-scattering processes and result in depolarization [41,57]. For example, P-lidars have the potential to provide information on soil, sand, and volcanic ash properties such as particle size and moisture content [58–60]; snow and ice characteristics such as age and types [61,62]; as well as plants and ground cover such as types and classification [41]. In oceanic remote sensing, polarimetry has been a well-known tool in ocean optics. For example, with the elastic Mie backscattering, seawater's optical properties can be estimated by retrieving the lidar attenuation coefficient over the laser penetrating depth [63]; with the depolarization effect of non-spherical particles, P-lidar can recognize ocean communities using the linear depolarization ratio [64,65]. Recent advancements have unveiled several promising applications in diverse fields, including the detection of oceanic scattering layers, fish schools, phytoplankton, seawater optical properties, and internal waves [66–70]. Figure 3 presents the triad of applications for P-lidar: (1) atmosphere remote sensing, which involves detecting clouds, precipitation, aerosol, and wind velocity; (2) oceanic remote sensing, such as phytoplankton layers' detection, turbulence measurement, and fish school detection; (3) the remote sensing of the Earth's surface, such as detecting and classifying vegetation, smoke, and urban objectives. From the figure, it is evident that P-lidar (as well as other lidar types) can be categorized into four groups: near-ground (terrestrial [71], ground-based [72], mobile [73], and UAV-based [74]), airborne [75], shipborne [76], and spaceborne/satellite-based lidar [77].

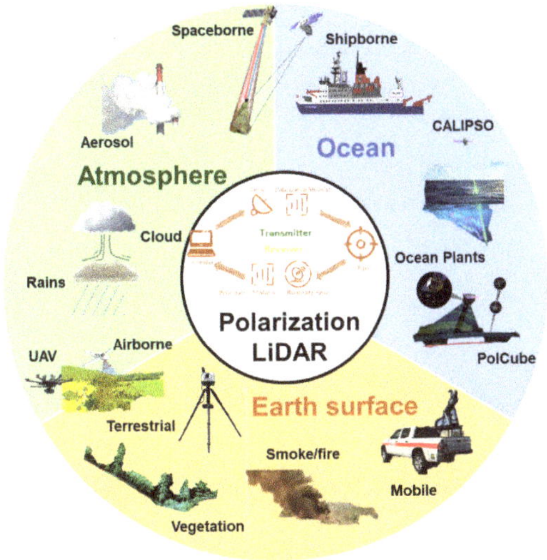

Figure 3. Triad of applications of P-lidar.

The rest of this review is organized as follows: Section 2 introduces the principles of polarization and lidar configurations. Section 3 surveys the representative applications of P-lidar in three fields, i.e., atmosphere, ocean, and the Earth's surface. Finally, conclusions, a critical summary, and an outlook toward future research are presented in Section 4.

2. Principles of Polarization and P-Lidar

2.1. Polarized Light and Its Description

Currently, there are three main mathematical representations for polarized optics: Jones vector, Stokes vector, and Mueller matrix [1,2]; the first two are used to describe the polarization state of light, while the last one describes the polarization characterizes of materials or targets [78,79]. According to Maxwell's electromagnetic theory, light propagates in space as a transverse wave, where the direction of the optical vector vibration is perpendicular to the direction of light propagation. Based on the theory of polarization, polarized light can be decomposed into two mutually orthogonal vectors, $\mathbf{E}_x$ and $\mathbf{E}_y$ [2].

The Jones vector (i.e., $\mathbf{J}$) describes both the amplitude and phase of the two orthogonal components, usually horizontal $\mathbf{E}_x$ and vertical $\mathbf{E}_y$, of the electric field, as:

$$\mathbf{J} = \begin{bmatrix} \mathbf{E}_x \\ \mathbf{E}_y \end{bmatrix} = \begin{bmatrix} E_{0x} e^{j\delta_x} \\ E_{0y} e^{j\delta_y} \end{bmatrix}, \tag{1}$$

where (E_{0x}, E_{0y}) and (δ_x, δ_y) denote the corresponding amplitudes and phases. If we use $\mathbf{J}_i$ and $\mathbf{J}_o$, respectively, to represent the Jones vectors of the incident light and the out-going light after reflecting or scattering from a target, we can use a 2×2 matrix $\mathbb{J}$ (named the Jones matrix) to describe how the target changes the polarization of incident light, as:

$$\mathbf{J}_o = \mathbb{J}\,\mathbf{J}_i. \tag{2}$$

It is necessary to point out that the Jones matrix $\mathbb{J}$ has four elements, all of which are complex numbers, which means it has a maximum of eight degrees of freedom [6].

Stokes first proposed a method in 1852 to represent arbitrary polarized light using four intensities, but he did not apply this method to the process of light scattering [80]. It was not until 1947 that Chandrasekhar utilized the Stokes vector to calculate the intensity vector of scattered light and represent the effects of polarization in the scattering process [2]. The Stokes vector, i.e., $\mathbf{S} = [s_0,\ s_1,\ s_2,\ s_3]^T$, can be obtained from only power or intensity measurements and is sufficient to characterize the magnitude and relative phase, i.e., the polarization of a monochromatic electromagnetic wave [2]. The Stokes vector can also be written as a function of the polarization ellipse parameters: orientation angle ϕ, ellipticity angle χ, and ellipse magnitude A, as:

$$\mathbf{S} = \begin{bmatrix} E_{0x}^2 + E_{0y}^2 \\ E_{0x}^2 - E_{0y}^2 \\ 2E_{0x}E_{0y}\cos\delta \\ 2E_{0x}E_{0y}\sin\delta \end{bmatrix} = \begin{bmatrix} A^2 \\ A^2\cos(2\phi)\cos(2\chi) \\ A^2\sin(2\phi)\cos(2\chi) \\ A^2\sin(2\chi) \end{bmatrix} \tag{3}$$

where $\delta = \delta_x - \delta_y$. From $\mathbf{S}$, one can obtain other polarization parameters. Three of them are the DoP (i.e., P), the degree of linear polarization (i.e., DoLP), and the degree of circle polarization (i.e., DoCP):

$$\mathrm{P} = \frac{\sqrt{s_1^2 + s_2^2 + s_3^2}}{s_0}, \quad \mathrm{DoLP} = \frac{\sqrt{s_1^2 + s_2^2}}{s_0}, \quad \mathrm{DoCP} = \frac{s_3}{s_0}. \tag{4}$$

Obviously, $P \in [0, 1]$, and the light is non-polarized when $P = 0$; completely polarized when $P = 1$; and partially polarized when $P \in (0, 1)$ [81]. It should be noted that one also defines the DoLP by the following Equation (5) when there only two orthogonal polarization components (i.e., $I_{\parallel}$ and $I_{\perp}$) [81,82]. Another important parameter in the field of P-lidar is the depolarization ratio (i.e., R), which is also determined by $I_{\parallel}$ and $I_{\perp}$ [83,84].

$$\mathrm{DoLP} = \frac{s_1}{s_0} = \frac{I_{\parallel} - I_{\perp}}{I_{\parallel} + I_{\perp}}, \quad R = \frac{I_{\perp}}{I_{\parallel}}. \tag{5}$$

Similar to the Jones vector and Jones matrix, the polarization characteristics of targeted objects can also be represented by corresponding matrices, connecting the incident and out-going light with Stokes vectors. The polarization states of the incident and out-going light are described by $\mathbf{S}_i$ and $\mathbf{S}_o$, respectively [79]. Then, the polarization characteristics of the object can be represented by a 4×4 Mueller matrix $\mathbb{M}$, as:

$$\mathbf{S}_o = \mathbb{M}\mathbf{S}_i. \tag{6}$$

Due to the fact that the elements of the Stokes vector are real numbers and have dimensions of intensity, the elements of the Mueller matrix $\mathbb{M}$ are also real numbers and dimensionless. This is in contrast to the Jones vector and Jones matrix, which exhibit significant differences as they involve complex numbers and consider both amplitude and phase information [1]. The Mueller matrix solely represents the polarization characteristics of an optical element in terms of intensity transformations, without incorporating phase information [1]. The Jones matrix and the corresponding Mueller matrix can be connected by:

$$\mathbb{M} = \mathbb{C}\left(J \bigotimes J^*\right)\mathbb{C}^{-1}. \tag{7}$$

where $\bigotimes$ denotes the Kronecker product, $*$ represents the conjugation, and $\mathbb{C}$ is

$$\mathbb{C} = \begin{pmatrix} 1 & 0 & 0 & 1 \\ 1 & 0 & 0 & -1 \\ 0 & 1 & 1 & 0 \\ 0 & i & -i & 0 \end{pmatrix} \tag{8}$$

Just as we claimed, both the Jones matrix and Mueller matrix are related to the ability of target objects to change the polarization state of incident light. In fact, when light interacts with targets, the polarization state can undergo the following changes: (1) alteration of the amplitude/intensity difference of the orthogonal polarization components; (2) modification of the phase difference between the two orthogonal polarization components; (3) changes in the direction of the two orthogonal optical fields. To characterize such changes, one can decompose, for example, the Mueller matrix into three different sub-matrices according to the Lu–Chipman decomposition method [85–87], as:

$$\mathbb{M} = \mathbb{M}_{dep} \cdot \mathbb{M}_{ret} \cdot \mathbb{M}_{dia}. \tag{9}$$

where the three sub-matrices represent the matrix related to depolarization, retardance, and diattenuation. The measurement of the Mueller matrix requires the establishment of at least 16 sets of intensity equations corresponding to different polarization states. This is because the Mueller matrix consists of 16 elements [1,88]. A typical Mueller measurement system includes two core modules: the polarization state generator (PSG) and the polarization state analyzer (PSA) [89,90]. The PSG and PSA are modulated four times each, resulting in a total of 16 intensity recordings, as:

$$\mathbb{I} = \mathbb{W}_{PSA} \cdot \mathbb{M} \cdot \mathbb{W}_{PSG}. \tag{10}$$

where $\mathbb{W}_{PSG}$ and $\mathbb{W}_{PSA}$ denote measurement matrices corresponding to the PSG and PSA, respectively. Then, by solving these equations, the Mueller matrix can be calculated, i.e.,

$$\mathbb{M} = \mathbb{W}_{PSA}^{-1} \cdot \mathbb{I} \cdot \mathbb{W}_{PSG}^{-1}. \tag{11}$$

More information about Mueller polarimeters can be found in previous publications [79,90].

2.2. Principle of P-Lidar

The lidar system is primarily divided into two parts: the signal-transmission system and the signal-receiving system [91–93]. At the transmission end of the lidar system, a pulse laser beam is emitted, interacting with the target object (such as particles in the air) through processes such as absorption, reflection, and scattering. The receiving system utilizes an optical telescope to receive the echo signals, which are then subjected to spectral analysis and opto-electronic conversion. Finally, through mathematical calculations and inversion, various parameters of the target object with different meanings can be obtained. The lidar equation at a height/distance z is given by [94]:

$$P(z) = P_0 Y(z) \frac{ct_p}{2} \frac{Ar}{z^2} \beta(z) \exp\left[-2\int_0^z \alpha(z)dz\right] \quad (12)$$

where $P(z)$ denotes the received power at height z; P_0 the peak power of the laser pulse; $Y(z)$ the geometric overlap factor; c the speed of light; t_p the pulse width; Ar the receiving telescope area; $\beta(z)$ the backscatter coefficient; and $\alpha(z)$ the extinction coefficient [95,96].

P-lidar emits a polarized modulated (or unmodulated) beam at the laser-transmission end. Then, the beam interacts with the target object and is recorded at the receiving end through different polarization modulation techniques, such as a polarization beam splitter (PBS) [15,97]. The power of the echo signal corresponding to a special polarization component, for example at the polarization direction of θ, can be obtained by:

$$P_\theta(z) = \frac{k_\theta P_0}{z^2} \beta_\theta(z) \exp\left[-2\int_0^z \alpha_\theta(z)dz\right] \quad (13)$$

where k_θ denotes the system parameter at the state of polarization direction θ, and can be considered as a constant. Based on the Mueller decomposition, Hayman et al. defined the Stokes vector lidar equation (SVLE) by linking the transmitted/received polarization states and the polarization properties of the target [47,48]. The equation is given by:

$$N(z) = \mathbf{o}^T \mathbb{M}_R \left[\left(G(z) \frac{A}{R^2} \Delta z \right) \mathbb{T}(\mathbf{k}_b, z) \mathbb{F}(\mathbf{k}_i, \mathbf{k}_b, z) \mathbb{T}(\mathbf{k}_o, z) \mathbb{M}_T \mathbf{S}_L + \mathbf{S}_B \right] \quad (14)$$

where $N(z)$ is the photon count vector for each polarization channel as a function of range z; $\mathbf{o}$ is the observation vector describing each polarization observation channel, each row of which is the polarization eigenvector related to each channel; and $\mathbb{M}_T$ and $\mathbb{M}_R$ are Mueller matrices related to the transmitter and receiver systems. G is the physical overlap function of the transmitter and receiver; A is the telescope area; R is the range resolution of the counting system; $\mathbb{T}(\mathbf{k}_o, z)$ and $\mathbb{T}(\mathbf{k}_b, z)$ are the out-going and back-going transmission Mueller matrices (wave direction vectors of $\mathbf{k}_i$ and $\mathbf{k}_b$); $\mathbb{F}$ is the phase matrix, which is a function of both transmitted and received wave vectors and range; $\mathbf{S}_L$ denotes the Stokes vector of laser; $\mathbf{S}_n$ denotes the Stokes vector of background noises.

Inspired by the formalism in Equation (10), we can simplify Equation (14) by using two generated measurement vectors $(\mathbf{O}, \mathbf{B})$, as:

$$N(z) = \mathbf{B}^T \mathbb{F}(\mathbf{k}_o, \mathbf{k}_b, z) \mathbf{O} \quad (15)$$

where $(\mathbf{O}, \mathbf{B})$, which are very similar to the measurement vectors related to the PSG and PSA part in a Mueller polarimeter, are determined by all the Mueller matrices on the way of out $(\mathbf{O})$- and back-going $(\mathbf{B})$:

$$\mathbf{O} = \mathbf{o}^T \mathbb{M}_R \mathbb{T}(\mathbf{k}_b, z), \quad \mathbf{B} = \mathbb{T}(\mathbf{k}_o, z) \mathbb{M}_T \mathbf{S}_L. \quad (16)$$

Based on the SVLE model, the basic principle of P-lidar is shown in Figure 4. All mathematical notations relevant to the model are, respectively, labeled at the transmitter and receiver locations to facilitate the readers' clear understanding of the model. Besides, in the configuration, the polarization modulation is optional, which depends on whether one needs a circular or linearly polarized beam in the P-lidar. The pulsed lasers generally used in lidar naturally produce linearly polarized light as the crystalline nature of the lasing media (e.g., a doped glass rod) and the method used in giant-pulsing, which typically relies on a polarization rotation device (e.g., a Pockels cell) to stop the cavity from lasing until the most-propitious instant. Sometimes, one may also use a cleaning polarizing sheet filter to ensure the output beam's polarization state [98]. Thus, basic P-lidar applications involve the transmission of a linearly polarized laser pulse [16].

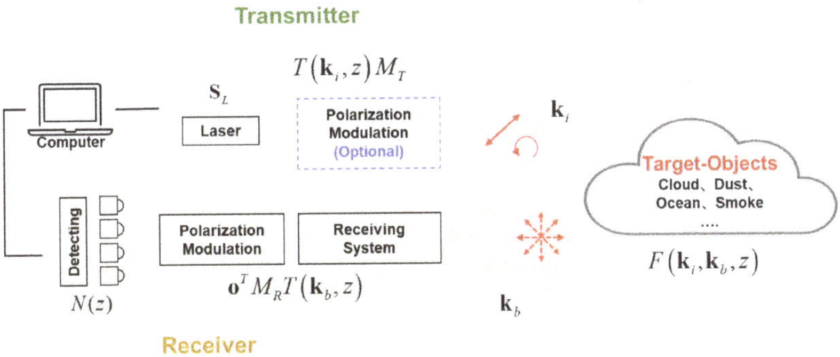

Figure 4. Principle of P-lidar.

According to the number of polarization modulation channels, which also is the number of detectors in the receiver, as shown in Figure 4, common P-lidars include dual-channel (2C), single-channel (1C), three-channel (3C), and four-channel (4C) configurations. Besides, if one wants to measure all elements in phase matrix $\mathbb{F}$, we need to obtain 16 measurements; in this case, we call it full P-lidar, and the receiver must have at least four channels, while the transmitter's polarization modulation must include circularly polarized elements.

In the following, we will introduce the four types of P-lidar. However, it is important to note that the initial design of these four types was for single-point detection and cannot achieve two-dimensional or three-dimensional imaging. By adding a scanning module or an array imaging module to the system, lidar imaging can be achieved. Therefore, in the following, we also included an introduction to relevant imaging P-lidar systems.

2.2.1. Dual-Channel P-Lidar

The dual-channel configuration is the most-commonly used type of P-lidar. It can simultaneously measure the vertical polarization signal (i.e., P_s) and the parallel polarization (i.e., P_p) signal of the echo signal, providing real-time information about the depolarization ratio of the scattering medium [15,16,99]. Guasta et al. [44] from Italy were the first to propose the dual-channel P-lidar system, as shown in Figure 5.

The Cloud-Aerosol lidar with Orthogonal Polarization (CALIOP), which is carried by the CALIPSO satellite launched by NASA [99,100], is a typical dual-channel P-lidar. CALIOP has been in orbit and operating stably, providing abundant data for global atmospheric observations [101].

Although dual-channel P-lidar can provide real-time measurement of the vertical and parallel channel components of the echo signal and obtain depolarization ratio information, it is prone to misjudgment due to the influence of gain coefficients and system errors from various sources. As a result, there are many challenges in system calibration [102,103], and we will discuss the calibration process in the following Section 2.2.6.

2.2.2. Single-Channel P-Lidar

The main idea of single-channel P-lidar is to use the same detector to receive both the parallel polarization component and the vertical polarization component of the echo signal, thereby avoiding the need for gain ratio calibration. Figure 6 presents an example configuration of single-channel P-lidar.

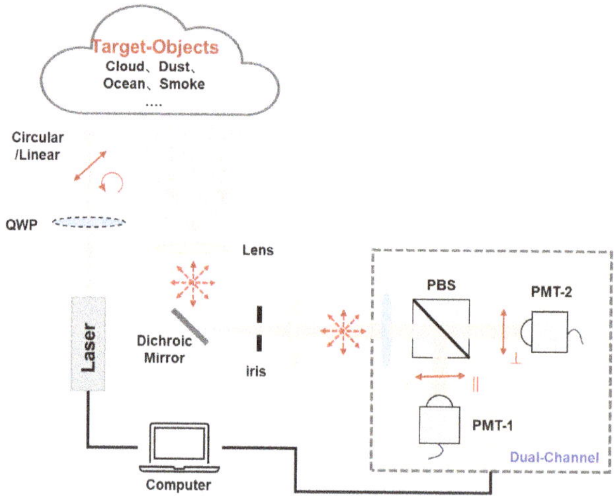

Figure 5. Examples of dual-channel P-lidar. PMT: photomultiplier, a detector.

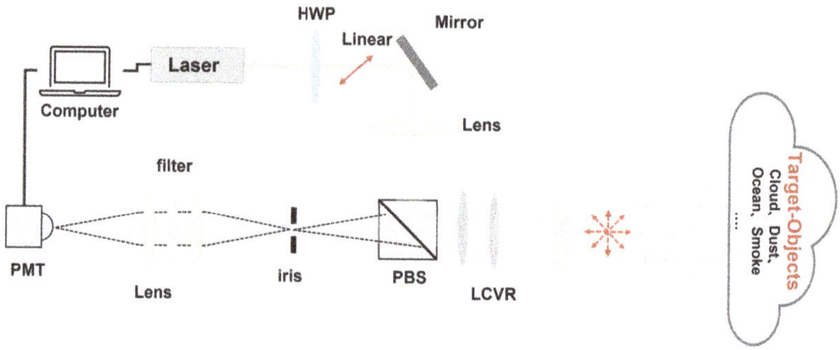

Figure 6. Examples of single-channel P-lidar.

In 1977, Plat et al. first proposed a single-channel P-lidar system, in which the transmitted laser polarization state remains unchanged, while the polarization direction of the analyzer in the receiving optical path rotates around the central optical axis by 90° with a rotation frequency consistent with the transmitted laser's pulse repetition frequency [40,104]. As a result, the parallel polarization component and the vertical polarization component in the echo signal are sequentially received by the same detector, resolving the issue of system errors caused by gain coefficient influence.

However, the single-channel solution also has inherent issues. For dual-channel P-lidar, the parallel and vertical polarization components of the echo signal at any altitude are always synchronously received by two detectors, resulting in a depolarization ratio profile for each laser pulse. However, in the single-channel one, the detector can only receive the parallel polarization component of each laser pulse's echo signal. It requires the combination of the echo signals from two consecutive laser pulses (with a time delay) to obtain a depolarization ratio profile, making the detection results highly susceptible to environment changes [105].

To address the real-time issue of single-channel solutions, Xia et al. in 2017 developed a new single-channel P-lidar based on a single-photon detector [50]. They utilized a polarization-maintaining fiber to receive the reflection signal from the PBS, reducing the time difference (40.6 μs) for the signal to reach the detectors in the reflection and transmission channels. This allowed them to obtain real-time depolarization ratio profiles using a single detector. However, this single-mode polarization-maintaining fiber is relatively long (8 km) and introduces an additional energy loss of approximately 30%.

2.2.3. Three/Four-Channel P-Lidar

The two types of P-lidar mentioned above are based on the parallel and vertical components of the echo signal. However, when the reflection or scattering from the target objects becomes more complex, errors can occur in depolarization analysis based on orthogonal polarization states. This can also be seen from the definition of the DoP, as shown in Equations (4) and (5), which state that the depolarization ratio (i.e., R) calculated based on orthogonal polarization components is just a simplified version of DoLP. Taking this into consideration and to further study the influence of multiple scattering on polarization changes, Pal et al. proposed a three-channel P-lidar [106]. At the receiving end, three telescopes are used to receive the signals, and polarization plates are used to modulate each beam channel with polarization directions of $\|$, $\perp$, and 45°, respectively, to obtain the first three parameters of the Stokes vector, i.e., s_0, s_1, and s_2, by

$$\begin{cases} s_0(z) = P_{\|}(z) + P_{\perp}(z) \\ s_1(z) = P_{\|}(z) - P_{\perp}(z) \\ s_2(z) = 2P_{45}(z) - s_0(z) \end{cases}, \tag{17}$$

Furthermore, based on Equation (3), angle ϕ and ellipticity angle χ can also be calculated, which provides more comprehensive information about the polarization characteristics of the measured object.

Three-channel P-lidar has primarily employed only a portion of the polarization information available by utilizing linear depolarization as the main observable. To extract more polarization information, the full Stokes vector has been studied, i.e., four independent measurements for s_0, s_1, s_2, and s_3. Houston et al. [107] designed a four-channel P-lidar, where the first receiver channel was used with its polarizer aligned parallel to the direction of linear polarization of the transmitted signal and the second with its polarizer aligned perpendicular to that of the transmitted signal. The third channel was set with its polarizer at an angle of 45° to this direction. In the fourth channel, the QWP reference axis was aligned with the transmitted polarization direction and the polarizer transmission axis at 45°. The recording four powers/intensities are denoted as $P_{\|}(z), P_{\perp}(z), P_{45}(z)$, and $P_{circle}(z)$, and the four Stokes parameters can be solved by:

$$\begin{cases} s_0(z) = P_{\|}(z) + P_{\perp}(z) \\ s_1(z) = P_{\|}(z) - P_{\perp}(z) \\ s_2(z) = 2P_{45}(z) - s_0(z) \\ s_3(z) = 2P_{circle}(z) - s_0(z) \end{cases}, \tag{18}$$

However, Houston et al. stated that the two-channel configuration for measuring R seems to be more efficient when compared to the four-channel configuration. This is because there is no significant additional information gained, but it does require additional instrumentation and data handling for the full four-channel measurements, at least for the examples presented in this work [107]. Of course, they also mentioned that further studies using a four-channel configuration are necessary to address certain issues and provide answers to specific questions.

In fact, the studies for three- and four-channel P-lidar have never stopped. The Clouds Aerosol Polarization and Backscatter Lidar (CAPABL) has been updated to include four on-orthogonal receiver polarization channels. These polarizations are all linear, i.e., 0°, 90°, 30°, and 110°. It should be noted that this design does not aim to measure the full Stokes vector, but to provide more inversion methods, granting flexibility to optimize the polarization measurements [51].

2.2.4. Full P-Lidar

As we mentioned in the above section, to measure the polarization characteristics, denoted as $\mathbb{F}$, of a target object (such as scattered particles in the atmosphere), at least 16 detection signals are required. Additionally, the out-going light needs four linearly independent polarization states. The telescope's receiving system needs to simultaneously measure **S** (which means at least four channels are necessary). In this configuration, both the transmitting end and the receiving end require a rotatable retarder. By continuously changing the polarization state of the transmitted laser and altering the polarization state of the received light path, multiple equations can be established to solve $\mathbb{F}$.

In 2004, Kaul et al. first proposed the use of a full-polarization lidar to detect and compute the atmospheric backscattering Mueller matrix [43]. In 2012, the Atmospheric Research Center in the United States retrofitted an existing high-spectral-resolution lidar, as shown in Figure 7 [48]. A QWP was rotated at the receiving end to generate circularly polarized light, obtaining the full Stokes vector and providing a comprehensive characterization of raindrop evaporation, condensation, and liquid water content.

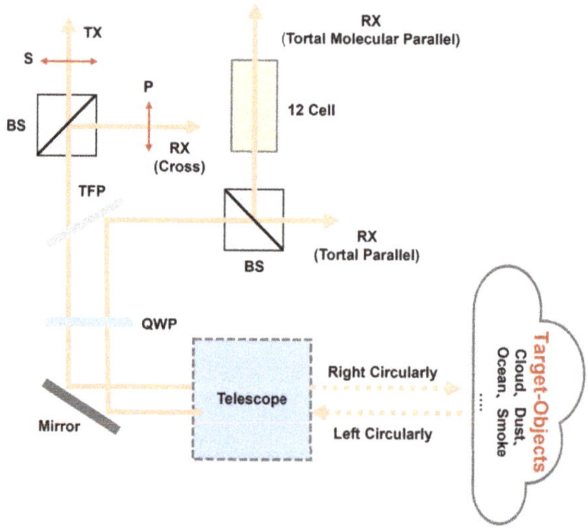

Figure 7. Examples of full P-lidar.

2.2.5. Scanning and Imaging P-Lidar Systems

The P-lidar systems introduced above do not incorporate scanning or mechanical motion devices, so they can only acquire point- or line-polarization information; we also call them non-scanning lidar or flash lidar [37]. In contrast, scanning P-lidar systems steer the laser beam and are more popular and mature as the captured data can reveal more diverse and complex polarization information of the target under test, such as information related to its direction or shape [108,109]. The simplest way to achieve this is to fix the transmitter or receiver ends of the above-mentioned system on a platform capable of mechanical motion or scanning [52,110]. Alternatively, scanning of the emitted beam can be achieved by adding a scanning mirror (or MEMS) at the transmitter end (for 2D scanning with a one-dimensional mirror or using a two-dimensional mirror for 3D scanning [111]). Generally, there are two classes of laser beam scanning: non-mechanical scanning and mechanical scanning. The former is also called solid-state beam scanning because it has no moving parts. We present these lidar types, i.e., non-scanning, optical-phased-array-scanner-based, motorized-spinning-scanner-based, and MEMS-/mirror-scanner-based, in Figure 8 to aid the readers' understanding, and more details can be found in [37,112].

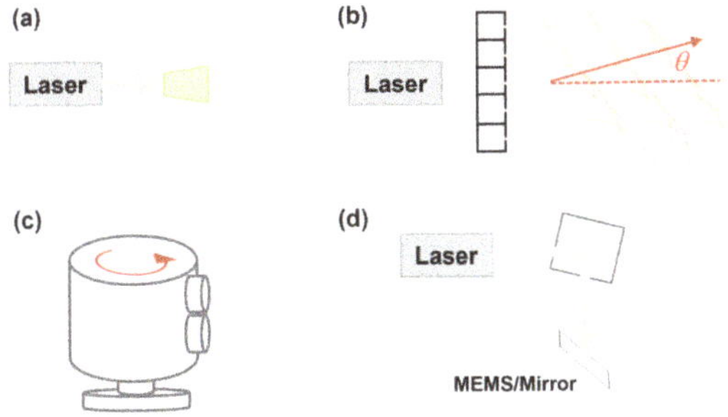

Figure 8. Different types of lidar. (**a**) Non-scanning; (**b**) optical phased array scanner; (**c**) motorized spinning scanner; (**d**) microelectromechanical systems (MEMS)/mirror laser scanner.

LOSA-M3, developed at the V. E. Zuev Institute of Atmospheric Optics, the Siberian Branch of the Russian Academy of Sciences, is a scanning P-lidar [52]. As depicted in Figure 9, it is equipped with an automatic scanning device that allows for changing the sensing direction in the upper hemisphere at speeds of up to 1.5° s^{-1} with an angle measurement accuracy setting of at least 1 arcmin. Bissonnette et al. chose to perform the scanning by rotating a mirror mounted at 45° to the laser axis, with a speed of 3.5° s^{-1}, and its rotation axis is the same as the transceiver axis. However, it should be noted that, for an accurate assessment of measurement errors, it is necessary to measure the refractive index of the applied mirror coating with good accuracy [113]. To some extent, the process of acquiring data using scanning P-lidar is imaging. One can also use imaging sensors (traditional imaging CCD, electron multiplying CCD, or polarization camera), instead of PMT in non-scanning P-lidar, to achieve polarization-sensitive imaging lidar [54,55,114].

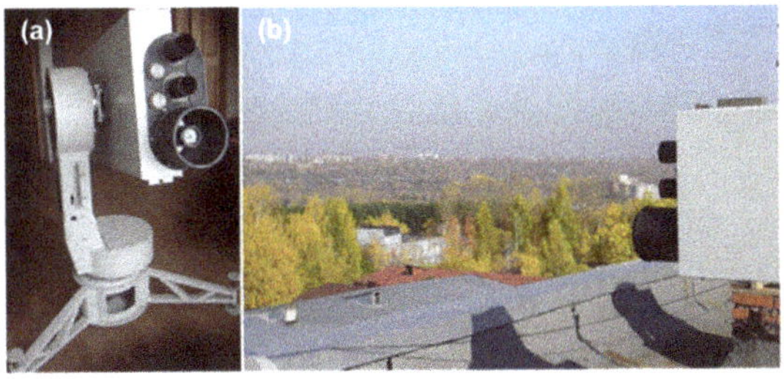

Figure 9. Photographs of the scanning polarizing lidar LOSA-M3 on a rotary column in the laboratory room (**a**) and on the roof of the institute's building (**b**) [52].

2.2.6. Calibration for P-Lidar Systems

In fact, achieving high-precision depolarization ratio measurements is crucial for subsequent applications and has been a focal point of P-lidar [115]. The main error sources for the depolarization ratio include the calibration error of the gain ratio, the error caused by impurities in the transmitted laser's polarization state, the alignment error between the laser polarization vector and the incident plane of the PBS, and the polarization crosstalk caused by the inability of the PBS to achieve 100% reflection and transmission [103,116,117]. Among these errors, the calibration error of the gain ratio (the first one) plays a decisive role in determining the final measurement accuracy; as a result, various methods for calibrating the gain ratio are continuously being proposed.

The most-common calibration methods are the +45° method, the Δ90° method, the Δ45° method, the rotation-fitting method, and the pseudo-depolarizer method. For example, CALIOP used the pseudo-depolarizer method to calibrate the gain ratio of the system [118], while the Multichannel Lidar System (MULIS) and Portable Lidar System (POLIS) used the Δ90° method [119]. Liu et al. from Zhejiang University systematically compared the accuracy and pros and cons of the four methods under different alignment angle conditions for gain ratio calibration [120]. Through theoretical and experimental comparisons, they provided the best choice for gain ratio calibration methods. The $+45°$ method's advantage lies in its simple operation, but its drawback is that it neglects the effects of alignment angle errors and polarization crosstalk, which may introduce errors [121]. The $\Delta 90°$ method is relatively easy to operate and provides higher accuracy, but it cannot eliminate alignment angle errors [98,122]. The $\Delta 45°$ method improves calibration speed while ensuring calibration accuracy, but its drawback is that it cannot exclude the influence of atmospheric state changes [116]. The rotation fitting method's advantage is that it can simultaneously invert three unknowns, namely the gain ratio, depolarization ratio, and initial alignment angle, without requiring prior knowledge. However, its drawbacks are that calibration takes a longer time, the operation is cumbersome, and it is only suitable for relatively stable atmospheric environments [103]. The depolarizer method is advantageous for its simplicity and real-time calibration, which eliminates the influence of atmospheric environment changes. However, the commercial depolarizer may still have difficulties generating completely depolarized light, introducing other errors [117,118].

3. Applications

3.1. Atmospheric Remote Sensing

Lidar and polarimeter technologies complement each other effectively. Lidar excels at detailing aerosol profiles and types, while polarimeters offer constraints on overall aerosol abundance, absorption, and microphysical properties [13,101]. Combining these methods enhances aerosol property observations; this is also why P-lidar has been widely employed in atmospheric remote sensing to study the optical properties of the atmosphere and the characteristics of particles such as aerosols, clouds, and water vapor. Specifically, P-lidar can measure the polarization scattering characteristics of aerosol particles, providing information about aerosol concentration, size distribution, shape, etc. This is crucial for understanding aerosol sources, transport, and interaction mechanisms in the atmosphere. By measuring polarization scattering properties using P-lidar, information such as aerosol types, cloud optical thickness, particle size, and shape distribution can be obtained.

This is significant for studying cloud formation and evolution and their impact on climate and weather. P-lidar can measure atmospheric transparency parameters such as atmospheric optical thickness and aerosol optical depth. This is important for monitoring atmospheric pollution and studying atmospheric composition [14,49].

In 1963, the Massachusetts Institute of Technology (MIT) established the first lidar based on ruby lasers, which was used to detect high-altitude aerosols in the troposphere and middle atmosphere [123]. This marked the beginning of the development and application of lidar for aerosol detection. Around the same time, mature polarization optical techniques were applied to lidar systems. In the 1960s, New York University began using polarization lidar, i.e., P-lidar, systems to observe atmospheric ice crystals and water droplets [39]. As shown in Figure 10, ice crystal detection hinges on measuring the volume depolarization ratio. While backscattering from spherical objects (such as liquid drops) at exactly $180°$ yields no depolarization, non-spherical crystal backscattering introduces notable depolarization through multiple internal reflections. Thus, the volume depolarization ratio effectively distinguishes between cloud layers with water drops and those where backscattering by ice crystals prevails [124]. In 1971, Schotland et al. conducted research on the depolarization ratio of water vapor condensates [39]. Understanding and quantifying the various forms of ice crystals in the atmosphere and precipitation are crucial for comprehending microphysical and radiative processes in different scenarios and improving regional and global climate models. In 2017, Sergey et al. used measurements from the U.S. Department of Energy's Atmospheric Radiation Measurement (ARM) program's cloud P-lidar to retrieve the nonsphericity of ice particles. The observed ice particles included irregularly shaped crystals and aggregates, with aspect ratios spanning from around 0.3 to 0.8 [125].

Using polarization information for particle shape measurement and properties' analysis is a well-established technique. In 2015, Wu et al. developed the Water Vapor, Cloud, and Aerosol Lidar (WACAL) for comprehensive atmospheric measurements, including the water vapor mixing ratio, depolarization ratio, backscatter and extinction coefficients, and cloud information [126]. The WACAL system, featuring Raman, polarization, and infrared channels, was installed at Qingdao Ocean University, enabling the assessment of aerosol and cloud optical properties and water vapor mixing ratios. In 2019, Tan et al. introduced a novel method to infer the phase state of submicron particles using linear depolarization ratios obtained from P-lidar [127]. This innovative approach demonstrated the feasibility of inferring aerosol phase distributions and established a parameterization scheme for deducing aerosol phase states from backscatter depolarization ratios, marking a significant advancement in real-time aerosol phase state profiling. In 2020, Jimenez et al. first introduced a novel cloud-retrieval technique using lidar observations of the volume linear depolarization ratio at two different receiver field of views (FOVs) to retrieve the micro-physical properties of liquid cloud layers [128] and then applied it to cloud measurements in pristine marine conditions at Punta Arenas in southern Chile [129]. Following Jimenez's work, in 2023, Zhang et al. introduced a dual-FOV high-spectral-resolution Lidar (HSRL) for simultaneous analysis of aerosol and water cloud properties, particularly the microphysical properties of liquid water clouds. This instrument allowed for continuous monitoring of aerosols and clouds and underwent validation through synchronous observations, including Monte Carlo simulations and other methods, to investigate the interplay between aerosol levels and the microphysical properties of liquid water clouds [130].

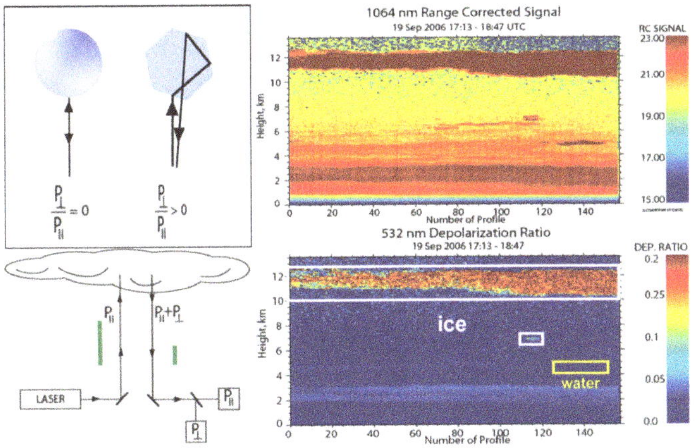

Figure 10. Demonstration of P-lidar for distinguishing liquid cloud droplets and ice crystals [124].

Another important application, aside from distinguishing various aerosols' shapes, is the identification of different types/altitudes of clouds. In 1977, Pal and Carswell utilized ruby lasers at 347 nm and 694 nm wavelengths as fundamental and second harmonic laser sources, respectively, to measure the depolarization ratio of falling snow, ice crystals, cumulonimbus clouds, and low-level rain clouds [61,131]. Their findings indicated a positive correlation between the depolarization ratio and cloud height, with clouds exhibiting a higher depolarization effect on 347 nm laser light compared to 694 nm laser light. In 1991, Sassen conducted research on the polarization characteristics of various cloud types using a P-lidar. He observed depolarization ratios of less than 0.15 for liquid water clouds, around 0.50 for cirrus clouds, and values between these two for mixed clouds [15,132]. P-lidars can be also applied to separate the dust and non-dust (e.g., the smoke) parts [133–135]. For example, Sugimoto et al. used the depolarization ratio and the volume backscatter coefficient (both at 532 nm and 1064 nm) to retrieve the dust and non-dust and the spectral dependence of the backscatter-related Ângström exponent [136,137]. By combining the multi-wavelength Raman lidar and P-lidar, Tesche et al. separated the optical properties of desert dust and biomass burning particles by using the depolarization ratio [138].

The advantages of P-lidar in detecting optical and microphysical properties of aerosol particles can also be extended to aerosol classification as different aerosols exhibit varying responses to different optical wavelengths. Among these different choices of optical wavelength, the 355 nm and 532 nm wavelengths are the most-widely used combinations. In 2014, Huang et al. developed a dual-polarization lidar system that simultaneously measured polarization at 355 nm and 532 nm wavelengths. During observations of dust and haze events in northern China, it was found that dust-dominated aerosols exhibited a higher depolarization ratio at 532 nm compared to 355 nm, whereas air pollutants showed relatively low depolarization ratios. This suggests that such multi-wavelength systems have the potential to enhance aerosol classification accuracy [49]. In 2021, Qi et al. developed a ground-based dual-polarization lidar system capable of simultaneously measuring polarization at 355 nm and 532 nm wavelengths to identify aerosol and cloud types [53]. Their findings revealed significant differences in volume depolarization ratios between typical aerosols and cloud layers at the two wavelengths. In addition to the above two wavelengths, the depolarization ratio measurement at 1064 nm is also useful. More details about its applications can be found in previous references [139–141]

In 2021, Kong et al. introduced a P-lidar system designed for precise all-weather retrieval of atmospheric depolarization ratios. This system simultaneously captured four-directional P-lidar signals, offering numerous possibilities for real-time field measurements of dust, clouds, and urban aerosols, directional particles (another typical work is [142]). Furthermore, the team employed laser diodes and a polarization camera to create a visible and near-infrared dual-polarization lidar technology for unattended atmospheric aerosol field measurements in all-weather conditions. Using one month of continuous atmospheric observation data, they analyzed and assessed the spectral features, including the aerosol extinction coefficient and the linear particle depolarization ratio (LPDR). Through this analysis, different types of aerosols were able to be classified [55].

There are three primary methods for detecting atmospheric aerosols: ground-based observations, airborne, and satellite remote sensing. Among these, using a lidar on moving platforms such as aircraft or satellites is the most-effective way to gather regional-scale aerosol data. In 2006, CALIPSO, a spaceborne P-lidar, successfully launched, equipped with a dual-wavelength (532/1064 nm) laser and a three-channel (532 nm P/S channels and 1064 nm) receiver. CALIPSO identifies clouds, measures particle content, and creates atmospheric profiles for research purposes. It is employed to detect the vertical distribution of aerosols and clouds, ascertain the cloud particle phase (via the signal ratio at 532 nm in parallel and perpendicular polarization channels), and classify aerosol sizes using the wavelength-dependent backscatter-related Ângström exponent [45,143].

Recently, MIT has been designing a satellite-based P-lidar system to categorize aerosols, as shown in Figure 11a. The system comprises two satellites: the primary satellite emits polarized laser pulses, while the second satellite is equipped with a lidar receiver that captures images of scattered polarization [144]. The proposed architecture enhances a satellite-based lidar system by introducing a second lidar receiver satellite, flying in formation with the transmitting satellite, to capture obliquely scattered light. The primary satellite emits polarized laser pulses, and the secondary satellite generates polarization-analyzed lidar images of the illuminated atmospheric column. Due to its oblique perspective, the secondary satellite observes changes in polarization that are not accessible to the primary satellite. These polarization variations provide critical information for aerosol classification.

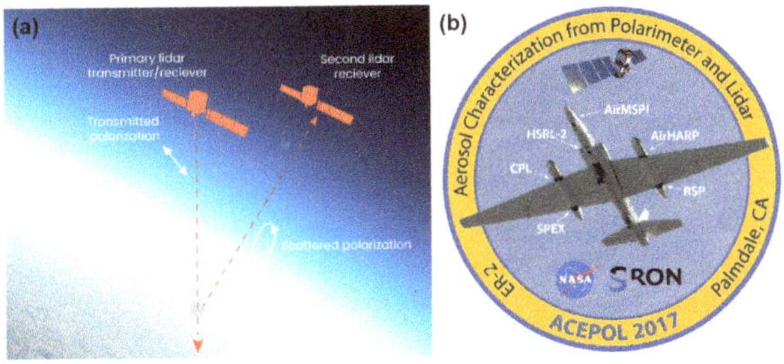

Figure 11. (**a**) P-lidar system with two satellites [144]. (**b**) The ACEPOL field campaign emblem [145].

Airborne lidar provides mobility and a high signal-to-noise ratio (SNR), serving as a valuable complement to spaceborne lidar. In 2003, Dulac et al. utilized the airborne P-lidar known as ALEX to study multi-layer aerosol structures in the Eastern Mediterranean [146]. In 2012, Bo et al. developed an all-weather atmospheric aerosol–water vapor lidar system for aircraft. This system simultaneously captures backscatter at 355 nm/532 nm wavelengths, 532 nm depolarization, and nitrogen and water vapor molecular Raman signals, facilitating long-term monitoring of aerosols and water vapor [147]. Another famous airborne system is the Aerosol Characterization from Polarimeter and Lidar (ACEPOL), which was conducted in the fall of 2017 by NASA. Figure 11b shows the ACEPOL emblem, which illustrates the locations of the remote sensing instruments on the aircraft, with two on the fuselage and two in each wing pod [145]. Additionally, there is a growing emphasis on developing portable P-lidar systems. For instance in 2021, Kong et al. introduced a portable P-lidar system using a focal-plane-splitting scheme. This system is designed in a T-shaped structure, featuring a sealed transmitter and a detachable large-aperture receiver. It is well-suited for cost-effective, low-maintenance outdoor unmanned measurements [13].

As detection instruments continue to progress, research algorithms are also undergoing gradual evolution. Many solutions to the lidar equation for elastic scattering (e.g., Fernald et al. [148], Klett [149], Davis [150], Sasano and Nakane [151], and Collis and Russell [152]) have been proposed. Among these solutions, the Fernald analysis method treats atmospheric molecules and aerosol particles separately, making it the current pinnacle of inversion methods under development [153]. Building upon Fernald's forward inversion technique, CALIOP employs the hybrid extinction retrieval algorithm (HERA), known for its flexibility and robustness as an iterative inversion method [99]. It utilizes hierarchical position data from the selective iterative boundary locator (SIBYL) and layer classification results from the scene classification algorithms (SCAs) to determine particle backscatter and extinction coefficients. It should be noted that Raman and HSRL, as the two typical systems to provide high-quality backscatter and extinction coefficients, have the potential for providing vertically resolved information about aerosol size/concentration [154,155]. For example, Thorsen et al. developed a comprehensive set of algorithms for processing the Raman lidar data to obtain the retrievals of aerosol extinction and feature detection. The details of the Raman algorithms for aerosol backscattering and extinction can be found in [156,157]. The HSRL data were processed in a manner akin to that of the Raman lidar data, ensuring alignment in terms of timestamps, altitudes, as well as temporal and vertical resolutions. While it is worth noting that the HSRL does not employ distinct low- and high-sensitivity channels, the processing and averaging approach for HSRL data is to mimic the processing of the Raman lidar low- and high-sensitivity channels for consistency. In 2017, based on the above algorithms, Ferrare et al. processed both the Raman lidar and HSRL to obtain a consistent set of profiles, with equivalent resolutions and averaging, across all wavelengths [158]. Recently, there have also been many researchers focused on developing advanced algorithms. For example, in 2022, the researchers in Italy employed a Bayesian parameter approach to infer the atmospheric particle size distribution [159]. In April of the same year, based on the spaceborne Aerosol and Cloud High-Spectral-Resolution Lidar (ACHSRL), capable of highly precise global aerosol and cloud detection, launched by China, Ju et al. proposed a retrieval algorithm for deriving aerosol and cloud optical properties from ACHSRL data, comparing it against end-to-end Monte Carlo simulations for validation. These efforts involved the use of the airborne prototype of ACHSRL [160].

3.2. Remote Sensing of Earth's Surface

In applications related to terrain characterization, vegetation remote sensing, and other surface environmental studies, microwave radar has been proven to possess certain advantages. However, due to its relatively long wavelengths, it cannot provide high-resolution information about targets. lidar, on the other hand, operates at shorter wavelengths, enabling sufficient resolution of surface environmental structures and, thus, enhancing the performance of relevant applications.

In vegetation remote sensing, traditional lidars are mainly used for measuring height and echo intensity, thus inferring the three-dimensional surface structure of vegetation. One example is the Scanning lidar Imager of Canopies by Echo Recovery (SLICER) scanning lidar system [161]. SLICER used a 1064 nm neodymium-doped yttrium aluminum garnet (Nd:YAG) laser with a beam divergence angle of approximately 2 mrad. Under normal operating conditions, SLICER had a footprint diameter of about 9 m, and it could obtain the canopy height, above-ground biomass, cross-sectional area, and other forest characteristics. Another example is the vegetation canopy lidar, a spaceborne lidar system with three 1064 nm Nd:YAG lasers, providing a 25 m field of view at an altitude of 400 km. It can characterize the three-dimensional structure of the Earth's surface on a global scale and offer improved global forest canopy height detection and biomass estimation [162]. However, all these lidars are non-polarimetric, and the ability to study the polarimetric properties of vegetation is of great significance.

In 1993, the NASA Goddard Space Flight Center made the first airborne laser polarization sensor (ALPS), i.e., a P-lidar, for the remote sensing of the Earth's surface (vegetation) [41]. ALPS has a linearly polarized Nd:YAG laser source at both 1064 and 532 nm, with a detector field of view of approximately 0.03 rad. It can measure desired parameters such as the total backscatter and the polarization state. Using the ALPS system, researchers were able to distinguish unique cross-polarization signatures for different tree species, such as broadleaf and coniferous trees [41]. This system also revealed a significant correlation between near-infrared depolarization and crop parameters, specifically nitrogen fertilization. The depolarization spectral difference index proved to be effective for estimating crop yields [163]. However, one limitation of the ALPS system was its inability to capture the lidar return waveform, which prevented obtaining detailed information about the vertical structure of the vegetation canopy. To handle this issue, the University of Nebraska has refurbished the ALPS system and developed it into a multi-wavelength airborne P-lidar system, named MAPL [164]. MAPL's receiver has four channels (dual-wavelength and dual-polarization detection) and records the entire lidar waveform. Therefore, it can study both vegetation canopy structure and the characterization of vegetation cross-polarization [165]. The same team also used MAPL to study the polarimetric reflectance from different tree species in the forest and proved its ability to detect different trees by analyzing the lidar waveform shapes, the depolarization ratios, and the reflectance percentages [166]. In 2018, Tian et al. proposed the measurement of co-polarized and cross-polarized components of maize leaves at a 532 nm laser wavelength [167]. They analyzed the depolarization differences under varying biochar contents and demonstrated that the laser depolarization ratio could serve as an indicator to monitor plant growth conditions.

P-lidars can also be applied for forest fire/smoke detection. By detecting and analyzing the backscattering process caused by the interaction between atmospheric particles and lasers, lidar can achieve precise measurements of smoke. However, traditional intensity-based lidar needs to find a commanding height for scanning, which is often hard to fulfill [168,169]. Moreover, due to the complexity of forest terrains, as they contain leaves and other obstacles, the lidar signals must pass through these leaves and obstacles, which makes it difficult to distinguish forest-fire smoke with a single-channel lidar. To solve this problem, Xian et al., in 2020, proposed a scanning P-lidar system to detect smoke from forest fires [108]. This system uses a 1064 nm pulsed laser and differentiates smoke from lidar signals contaminated by forest obstacles through the depolarization ratio.

A very important function of scanning or imaging lidar is its ability of 3D imaging in Urban remote sensing. The first polarization-modulated 3D lidar was proposed by Taboada and Tamburino in 1992 [170]. In order to improve the image quality, Chen et al. proposed an electron-multiplying charge-coupled-devices (EMCCDs)-based lidar system, in which the echo signal is separated into two orthogonal polarized components via a PBS. Such a polarization-modulated method improves the range accuracy of the objects of interest from 4.4 to 0.26 m with a gate opening range of 200 m [171]. In 2018, the same team found that the 3D imaging P-lidar had very promising performance in an FOV of 0.9 mrad [172]. The mentioned types of P-lidar, to some extent, belong to dual-channel P-lidar because they can only simultaneously acquire orthogonal polarization states. However, these systems require two cameras and a PBS to obtain orthogonal polarization states, making pixel-level alignment challenging [173]. In 2016, Jo et al. proposed a 3D flash P-lidar based on a micro-polarizer camera, as shown in Figure 12a, which can obtain the linear Stokes vector with a single shot [174], and achieved a spatial resolution and range precision of 0.12 mrad and 5.2 mm at 16 m, respectively [175].

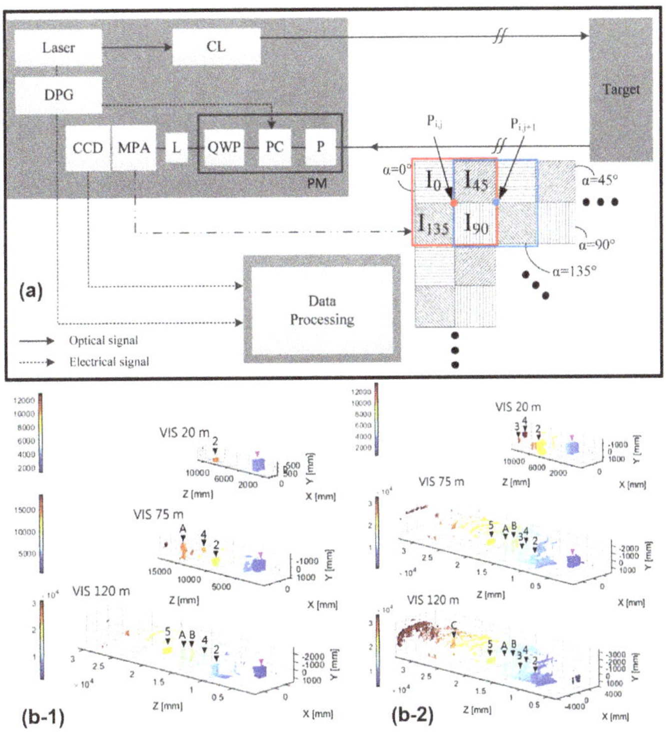

Figure 12. (**a**) P-lidar based on a polarization camera [175]. (**b**) Reconstructed point clouds for (**b-1**) co-polarization and (**b-2**) cross-polarization configurations at 20 m, 75 m, and 120 m of visibility [7].

Another important application is autonomous driving, as lidar sensors are one of the key supporting technologies for implementing autonomous driving. Nunes-Pereira et al. demonstrated through experiments that the reflection signals from metallic car paints have distinct polarization characteristics [176]. Therefore, by using a P-lidar, distance measurements can be supplemented, thereby aiding in target classification. In their experiments, they utilized a custom-built P-lidar system, which employed a pulsed, linearly polarized 785 nm laser, a pair of 2D scanning galvanometer mirrors, and a linear polarizer positioned in front of the collection objective. The polarizer was alternately set for co-linear and cross-linear polarized detection. Including polarization into the lidar can improve autonomous driving performance in a dense atmosphere. In 2021, Ronen et al. introduced a model that combines traditional lidar and Stokes–Mueller formulations and conducted experiments inside an aerosol chamber [177]. The results showed that the use of a polarized source together with a cross-polarized receiver can improve the target-signal-to-atmospheric-signal ratio in a dense aerosol medium for a lidar system. Therefore, implementing polarimetric imaging techniques in lidars can enhance the performance of autonomous vehicles in poor-visibility conditions. Actually, this characteristic of polarization information can be found in many applications of polarization imaging [5,6]. In 2022, Ballesta-Garcia et al. studied the performance of P-lidar in a macro-scale fog chamber under controlled fog conditions and demonstrated many interesting findings [7]. For example, a system based on circularly polarized incident light and cross-polarized configuration helps to reduce the SNR, and a cross-polarized configuration enables the detection of objects while allowing the filtering out of most of the fog response, as shown in Figure 12b.

3.3. Ocean Remote Sensing

Aerosol observations utilize various passive and active remote sensing techniques, which can be applied to the ocean to better characterize hydrosols and enhance the atmospheric-correction process. While spectral radiance provides sensitivity to the absorption and scattering properties of constituents within the water column, polarized light emerging from the Earth system carries a wealth of information about the atmosphere, ocean, and surface, which remains underutilized in ocean color remote sensing. Polarized light originating beneath the ocean surface contains valuable microphysical details about hydrosols, including their shape, composition, and attenuation. Retrieving such information is challenging, if not impossible, using traditional scalar remote sensing methods alone. Moreover, polarimetric measurements offer opportunities to enhance the characterization and removal of atmospheric and surface reflectance that can interfere with ocean color measurements.

Kattawar et al. [178] were pioneers in conducting vector radiative transfer simulations for a coupled atmosphere–ocean system. It was not until 30 years later, in 2006, that Chowdhary et al. [179] first introduced models specifically addressing the polarized contribution from the ocean for photopolarimetric remote sensing observations of aerosols above the ocean [180]. This marked the beginning of increased interest in ocean-related applications of polarimetry. In 2007, Chami demonstrated the potential advantages of utilizing polarimetry to understand the optical and microphysical properties of suspended oceanic particles (hydrosols) through radiative transfer (RT) simulations [181]. In aquatic environments characterized by the prevalence of phytoplankton, the polarized reflectance at the top of the atmosphere exhibits a high degree of insensitivity to fluctuations in chlorophyll concentration. In 2009, Tonizzo et al. [182] developed a hyperspectral, multiangular polarimeter designed to measure the polarized light field in the ocean, accompanied by an RT closure analysis, which validated the theoretical analysis. Additionally, in 2010, Voss and Souaidia successfully measured the upwelling hemispheric polarized radiance at various visible wavelengths, revealing the geometrical dependence of polarized light [183].

Oceanic lidars can penetrate seawater to acquire highly accurate vertical profiles of multiple parameters within the ocean. Since the introduction of the first bathymetry lidar in 1968, various types of ocean lidars have been developed to assess different oceanic parameters and constituents. In particular, P-lidar, widely employed in oceanic studies, offers significant advantages in providing multiple oceanic parameters. Analyzing the elastic Mie backscattering signal, seawater's optical properties can be estimated by retrieving the lidar attenuation coefficient within the laser's penetration depth. Additionally, it can identify oceanic communities through the linear depolarization ratio, owing to the depolarization effect of non-spherical particles on incident light. In other words, oceanic P-lidar has achieved successful applications in various domains such as the retrieval of depolarization optical products in the upper ocean [67], the detection of phytoplankton layers [184,185], turbulence measurement [186,187], and marine biological population detection [188,189].

Over the past decade, oceanic P-lidar has found numerous applications in oceanographic research. For example, Vasilkov et al. employed an airborne P-lidar to generate profiles of the scattering coefficient and identified subsurface layers with high scattering properties during their field experiments [75]. Furthermore, Churnside et al. developed a radiative transfer equation for airborne polarized lidar returns, facilitating the detection of scattering layers, fish schools, seawater optical properties, and internal waves [67]. To expand the scope of ocean observing efforts, Collister et al. designed a shipborne lidar to investigate the combined impacts of particle composition and seawater multiple scattering based on the lidar's linear depolarization ratio [190]. Behrenfeld et al. quantified phytoplankton biomass and diel vertical migration using the particulate backscattering coefficient and diffuse attenuation coefficient derived from the spaceborne P-lidar CALIOP [64]. Chen et al. observed the vertical distribution of subsurface phytoplankton layers in the South China Sea using a dual-wavelength airborne P-lidar [70]. Furthermore, Chen et al. introduced the planned "Guanlan" ocean remote sensing mission, featuring a near-nadir-pointing oceanic lidar and a dual-frequency interferometric altimetry system [70]. The oceanic lidar payload is expected to contribute significantly to our understanding of the marine food chain and ecosystem by providing data with a 10 m vertical resolution within the euphotic layer, advancing our knowledge of both the dynamic and bio-optical characteristics of the oceanic mixed layer.

Marine biological population detection is also a promising application for oceanic lidars. Likely, the initial development of the theory for using lidar to detect fish schools was pioneered by Murphree et al. [191]. The first experimental trials, conducted in a controlled environment, were carried out by Swedish scientists [192]. In 1999, Churnside from the National Oceanic and Atmospheric Administration (NOAA) developed an airborne lidar for marine fisheries. This lidar is indeed a single-channel P-lidar and uses a polarizer in front of the telescope system to select either the component of the return that is co-polarized with the laser or the cross-polarized component [42]. Their results showed that one can see fish from an airborne P-lidar. In clear water, one can see to depths of 40–50 m, and in turbid waters, this depth penetration is reduced. In 2010, Churnside took a case study in Chesapeake Bay, by using the cross-polarized component as its contrast between fish, and the background-scattering level was greater than that of the co-polarized return. They found that the average depth penetration of lidar was 12 m, and the average depth of detected schools was 3 m [46]. In 2018, Shamanaev proposed a method of P-lidar sensing of marine fish schools based on a comparison of the numerical values of the lidar return power and depolarization with their threshold levels determined by the sea water extinction index in the fishery region [193].

Another example is jellyfish detection, which facilitates the acquisition of data encompassing jellyfish taxonomy, population metrics, spatial dispersion, and related particulars. This process holds a pivotal role within the ambit of jellyfish prevention and management. The existing methodologies for jellyfish detection presently exhibit limitations in terms of detection efficiency, precision, and vertical distribution insights. P-lidar holds the potential to accomplish remote sensing of individual jellyfish organisms. This advanced technology offers an efficient, cost-effective, and accurate means of monitoring variations in jellyfish distribution and population dynamics. Currently, there is a relatively limited amount of research focused on utilizing marine lidar for jellyfish detection. In 2015, Churnside et al. [189] employed an airborne polarized marine lidar to observe the phenomenon of hollow aggregations within jellyfish populations. Their findings were corroborated by sonar detection results, thereby substantiating the viability of using polarized marine lidar for jellyfish population detection. In August 2017, a shipborne polarimetric marine lidar, independently developed by the research team at Zhejiang University, China, conducted experimental measurements in the Yellow Sea [194]. During these experiments, it observed a wealth of strong scattering signals. By combining these data with video monitoring information, the team was able to identify the source of these signals as jellyfish, demonstrating that jellyfish in the same area exhibited clustering patterns in their optical characteristics. Furthermore, jellyfish signals from different regions had similar contrast, but varied depolarization rates, suggesting that the optical properties of jellyfish are closely related to their local environmental conditions.

Besides, P-lidar can be used to detect sea surface oil spills. Oil spillage on the sea surface tends to create a film, impeding the exchange of gases between the sea and the atmosphere. This results in a reduction of dissolved oxygen in the seawater. Research has revealed that the complete oxidation of 1 kg of petroleum requires the consumption of approximately 400,000 L of dissolved oxygen from seawater. Consequently, oil spill pollution can lead to extensive areas of hypoxia [195]. During the 1990s, the University of Oldenburg (Germany) achieved a significant milestone by successfully developing a marine lidar system, which was subsequently employed for real-time monitoring of oil spill areas [196]. Furthermore, countries such as Canada, France, and Italy have also sequentially developed airborne lidar systems for oil spill monitoring [197]. In 2009, the marine fluorescence lidar system developed by Zhao et al. underwent multiple experiments in the vicinity of Qingdao. In 2021, the Aerospace Science and Technology Corporation's $9th$ Academy successfully completed emergency monitoring tasks for oil spill incidents in the Yellow Sea, utilizing unmanned aerial vehicles (UAVs) equipped with their self-developed dual-wavelength polarized marine-detection lidar, X-band target surveillance radar, and miniaturized synthetic aperture radar (MiniSAR) [198]. The results indicated that the P-lidar system exhibited reliable performance in monitoring oil spills on the sea surface. It could accurately differentiate the types of oil spills and assess the degree of oil pollution [199].

Previous studies predominantly used 532 nm lasers in oceanic P-lidar due to mature solid-state Nd:YAG lasers. Different seawater compositions across coastal and open seas result in varying optimal optical penetration wavelengths. Single-wavelength lidar cannot reveal particle size and absorption effects. Exploring diverse wavelengths in seawater is essential. Using the multi-wavelength volume linear depolarization ratio (VDR) and color ratio measurements from lidar, specific substances in seawater can be identified and distinguished. Meanwhile, in situ measurements are essential for validating the optical properties obtained from marine lidar measurements and for establishing simulation tools for future spaceborne marine lidar missions. In 2022, Liu et al. designed a shipborne variable-FOV, dual-wavelength oceanic P-lidar (as shown in Figure 13), named lidar for Ocean Optics Profiler (LOOP), to obtain the VDR, color ratio, and optical parameter profiles of seawater [63]. High-power pulse lasers and photon-counting detectors effectively increased the detectable seawater depth of the lidar in clean open seas. The system ensured the reliability of oceanic P-lidar by controlling the peak pulse intensity of backscattered signals and calibrating the after-pulse effects of the PMT to maintain the linear response of PMTs, providing seawater backscattered signal profiles. The consistency of the results demonstrated its ability to obtain the optical properties of seawater at certain depths, validating its reliability.

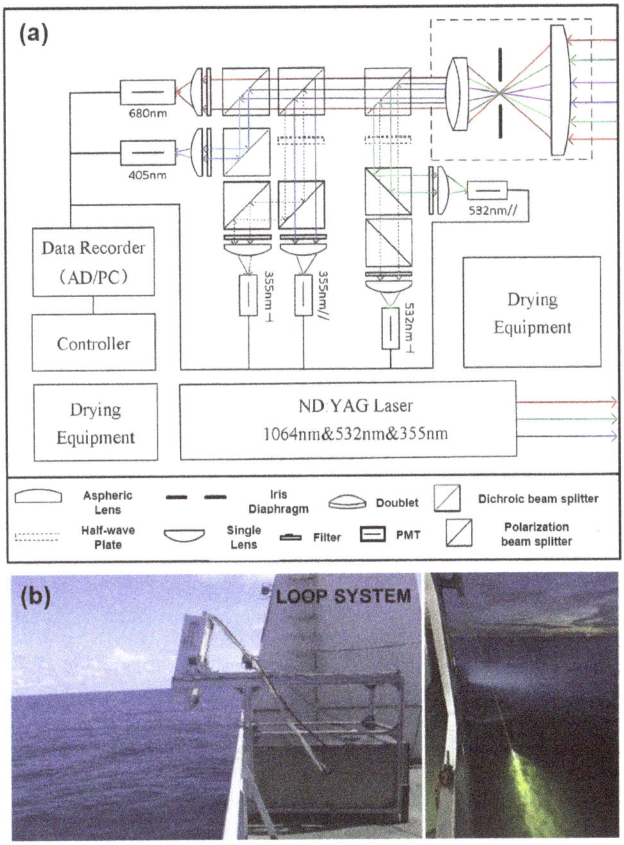

Figure 13. (**a**) The optical layout and (**b**) the photo of the LOOP system [63].

Compared to atmospheric or Earth remote sensing, oceanic P-lidar faces greater challenges due to strong seawater attenuation, limiting laser beam penetration. In nearshore seawater, the 532 nm laser energy attenuates by nearly five orders of magnitude when penetrating 30 m deep. Additionally, intense backscattered signals from the sea surface and shallow subsurface seawater can saturate the photodetectors, causing nonlinearity and after-pulse effects, leading to incorrect results. The complex marine environment also complicates accurate VDR. P-lidar for ocean applications, in conjunction with other oceanic measurements, has the potential to revolutionize the capacity for oceanic measurements.

4. Conclusions

This paper systematically reviewed the principles related to P-lidar and introduced the evolution of P-lidar systems from fixed to scanning configurations from an optical perspective. Moreover, we explored the applications of P-lidar technology in various fields, including atmospheric, oceanic, and terrestrial environments, with a special emphasis on intriguing areas such as aerosol detection and classification, ocean profiling, and unmanned driving. Table 1 provides a summary of the P-lidar systems related to various applications discussed in this paper; this also serves as a callback to Figure 3 presented in the Introduction.

Table 1. List of representative applications.

Scene	Application	Reference
Atmospheric	Precise all-weather retrieval of atmospheric depolarization ratios	[13]
	Research on the polarization characteristics of various cloud types	[15,132]
	Observe atmospheric ice crystals and water droplets	[39]
	Detect the vertical distribution of aerosols and clouds, ascertain cloud particle phase	[45,143]
	Identify aerosol and cloud types	[53]
	Classification of various aerosol types	[55]
	Measure the depolarization ratio of falling snow	[61,131]
	Detect high-altitude aerosols	[123]
	Retrieve the nonsphericity of ice particles	[125]
	Comprehensive atmospheric measurements	[126]
	Infer the phase state of submicron particles	[127]
	Retrieve micro-physical properties of liquid cloud layers	[128,129]
	Analysis of aerosol and water cloud properties	[130]
	Detect dust orientation	[142]
	Study multi-layer aerosol structures	[146]
Earth surface	Distinguish unique cross-polarization signatures for different tree species	[41]
	Detect smoke from forest fires	[108]
	Infer three-dimensional surface structure of vegetation	[161]
	Characterize the three-dimensional structure of the Earth's canopy height detection	[162]
	Study vegetation canopy structure and vegetation cross polarization characterization	[165]
	3D imaging in Urban remote sensing	[170]
	Autonomous driving	[176]
Oceanic	Obtain optical properties of seawater at certain depths	[63]
	Detection of scattering layers, fish schools, seawater properties, and internal waves	[67]
	Oceanographic research	[75]
	Observations of aerosols above the ocean	[180]
	Understand the optical and microphysical properties of suspended oceanic particles	[181]
	Measure the polarized light field in the ocean	[182]
	Detection of phytoplankton layers	[184,185]
	Turbulence measurement	[186,187]
	Jellyfish detection	[189,194]
	Retrieval of depolarization optical products in the upper ocean	[190]
	Marine biological population detection	[191]
	Sea surface oil spill detection	[196,197]

Thanks to priors of the polarization differences in target reflection and the polarization characteristics of medium-scattered light, the integration of polarization information acquisition and processing into lidar technology has demonstrated exceptional performance. Recent rapid advancements in both laser source and detector technologies have made it possible to realize multi-channel, multi-spectral (even full-spectral), and full-polarization lidar systems, thereby expanding the application fields of P-lidar. The foundational technologies of P-lidars (e.g., laser sources, polarization detectors, real-time processing algorithms, and system integration) are interdependent and mutually reinforcing.

The increasing demand for practical applications will undoubtedly drive the advancement of P-lidar systems and methodologies. A long-standing pursuit has been the development of a compact and cost-effective P-lidar. The advantage of P-lidar lies in its ability to integrate 3D imaging/detecting and optical material characterization, allowing for the measurement of spectral reflectance and polarimetric reflectance (i.e., represented using Mueller matrices) along with other sensing modalities. Moreover, the demand for miniaturization and cost-effectiveness necessitates novel approaches in designing P-lidar systems, while the need for enhanced target detection and recognition drives research into light–material interactions and new machine learning algorithms. Based on these insights, we propose several intriguing topics for future research.

Multi-modal integration: Integrating P-lidar with various other remote sensing technologies, including traditional lidar, radar, infrared imaging, and more, or merging systems that operate at different wavelengths offers the potential to gather a more-comprehensive set of environmental data. The integration of multiple modalities holds the promise of enhancing our capacity for comprehensive data analysis, enabling a deeper understanding of target recognition and environmental perception. As an illustrative example, Kong et al. designed a visible, near-infrared (Vis–NIR) dual-polarization imaging lidar [55]. Their research demonstrated the system's remarkable ability to characterize aerosol optical properties, distinguish between different aerosol types, and analyze the long-distance movement of aerosols. This exemplifies the practical advantages of multi-modal integration in environmental sensing. Nevertheless, it is essential to acknowledge that combining multiple sensor modalities also presents certain challenges. These include issues such as data fusion, calibration, synchronization, and increased computational complexity. To fully harness the advantages of integrating diverse sensors, it is imperative to develop appropriate data processing and integration techniques. These solutions will be crucial in overcoming the complexities associated with multi-modal data and ensuring that the combined data streams are effectively utilized.

Customized P-lidar tailored for oceanic remote sensing: P-lidar has found mature applications in atmospheric remote sensing. Many airborne or spaceborne P-lidars in use worldwide are primarily tailored for aerosol remote sensing. While it is possible to modify these instruments and systems for ocean color remote sensing, they were not originally designed for this purpose, mainly due to their lower spatial resolution (around 3–4 km). Additionally, ocean polarimetry requires measuring light at an off-principal viewing plane to reduce glint while staying reasonably close to the principal plane to maximize the polarized signal. This preference results in reduced ocean coverage. In ocean applications, it is crucial for the instrument to maintain a very high radiometric accuracy [200,201]. This accuracy is necessary to detect subtle variations in polarized light within the ocean, especially at small scales. Promising solutions include designing advanced sensors with greater sensitivity to polarization measurements and improved noise suppression capabilities. Additionally, the development of spaceborne high-spectral-resolution lidars at wavelengths of 355 and 532 nm is also a viable avenue [202,203].

Deep-learning-based P-lidar techniques: Deep learning technology, with its nonlinear convolution operations and powerful implicit correlation learning, leverages the advantages of data-driven approaches to enhance performance in various tasks related to polarization or lidar [204–209]. Compared to traditional intensity-based lidar, P-lidar offers additional information, including time-of-flight data and polarization information. Extracting meaningful features from P-lidar data for use by deep learning models or employing automatic feature selection to enhance model performance is essential. Huang et al. [210] compared distance-resolved aerosol volume concentration (VC) and effective radius (ER) retrieval methods based on partial least-squares regression and deep neural networks (DNNs). They found that the inversion results from DNN outperformed PLSR, particularly in cases with higher VC and ER values. Di Noia et al. (2015) discussed the use of DNN for retrieving the aerosol refractive index, size, and optical depth based on ground-based SPEX measurements [211]. In their subsequent work in 2017, they developed a DNN inversion approach for airborne MAP measurements over land using a research scanning polarimeter [212]. In both studies, the results obtained from DNN inversions were employed as initial values for iterative optimization, resulting in notable improvements in efficiency and retrieval accuracy. However, while using a DNN for direct inversion proves efficient, it is often perceived as a "black box", making it challenging to account for measurement uncertainties. The combination of DNN inversion with P-lidar's physical characterizations holds great promise. Besides, in the future, custom-designed deep learning strategies and network architectures hold the potential to enhance the efficiency of P-lidar in improving resolution, suppressing signal noise, and optimizing radiative transfer iteration.

Imaging snapshot P-lidar technology: With the rapid advancement of semiconductor processes and optoelectronic integration techniques, the development of compact, cost-effective, high-resolution, and user-friendly snapshot imaging P-lidar has become possible. In 2021, the California-based startup SiLC Technologies [213] introduced the first commercial chip-level FMCW lidar named "Eyeonic". Technologically more complex than traditional lidar systems, it has the capability to miniaturize the entire system to chip-level specifications, as shown in Figure 14. In terms of functionality, Eyeonic's visual sensor showcases an industry-leading detection range of over 1000 m, with three flyovers depicted in different modes, including distance, instantaneous velocity, and polarization intensity, all generating the same point cloud. Another promising direction is to build an imaging P-lidar based on division-of-focal-plane (DoFP) polarization cameras. This involves leveraging the partial polarization characteristics of backscattering light caused by scattering particles in challenging environments such as seawater, low light, or hazy weather [214–217]. This approach can include the development of polarization-suppression algorithms to enhance imaging quality in these complex conditions. Alternatively, constructing a polarimetric-gated lidar imaging device [171,218] can be explored to achieve high-quality imaging in such challenging environments.

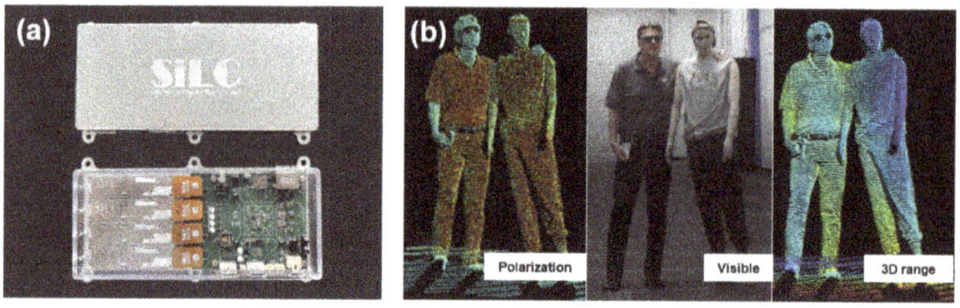

Figure 14. (**a**) The Eyeonic and (**b**) its output information [213].

Polarization metasurfaces + lidar: Even though one of the purposes behind metasurfaces is miniaturization, we still aspire to combine fascinating realms of the metasurfaces technology and P-lidar [219]. Metasurfaces possess the capability to manipulate and control the light state at sub-wavelength scales, presenting a unique opportunity to enhance lidar systems' performance [220]. For instance, in 2022, a team from South Korea were utilizing metasurfaces to create a solid-state lidar sensor, as illustrated in Figure 15a, which offers a 360° view of the surrounding environment [221]. Almost simultaneously, a team from France proposed an alternative high-frequency beam-scanning approach, harnessing the light-deflecting capabilities of large-area metasurfaces to expand the lidar FOV to $150 \times 150°$ and achieve simultaneous low- and high-resolution multi-zone imaging, as depicted in Figure 15b [222]. The convergence of polarization metasurfaces and lidar technologies holds immense promise in advancing the field of P-lidars [223,224]. By integrating polarization metasurfaces into lidar sensors, we can anticipate significant improvements in the precision and sensitivity of polarization measurements, enabling the extraction of richer environmental information. This synergy has the potential to revolutionize applications such as remote sensing, autonomous navigation, and atmospheric monitoring, where precise polarization information can unveil new insights and enhance systems' performance.

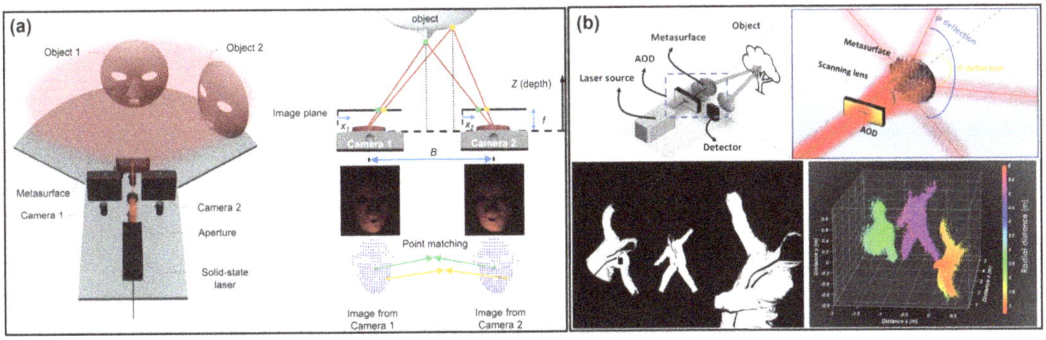

Figure 15. (**a**) Schematic of the optical setup in [221] and the stereo matching algorithm. (**b**) Schematic representation of the lidar system in [222] and its 3D ranging demonstration.

By addressing these research topics, we can push the boundaries of P-lidar technology, opening up new possibilities for its practical implementation and enhancing its impact in diverse fields.

Author Contributions: Conceptualization, X.L. (Xiaobo Li) and X.C.; formal analysis, X.L. (Xudong Liu); investigation, L.Z.; resources, X.L. (Xudong Liu) and L.Z.; data curation, X.L. (Xudong Liu) and X.L. (Xiaobo Li); writing—original draft preparation, X.L. (Xudong Liu), X.L. (Xiaobo Li) and X.C.; writing—review and editing, L.L., X.Z. and Q.Z.; visualization, X.L. (Xudong Liu); supervision, X.L. (Xiaobo Li); project administration, X.L. (Xiaobo Li); funding acquisition, X.L. (Xiaobo Li). All authors have read and agreed to the published version of the manuscript.

Funding: This work was supported by the National Natural Science Foundation of China (62205243) and the Key Laboratory of Intelligent Optoelectronics, the Ministry of Education (ZNGD20231001).

Acknowledgments: The authors would like to thank all the anonymous reviewers for their helpful comments and suggestions.

Conflicts of Interest: The authors declare no conflicts of interest.

Appendix A

In the paper, all abbreviations used are listed with their full definitions in Table A1.

Table A1. List of abbreviations.

Abbreviation	Definition
1C	Single-channel
2C	Dual-channel
4C	Four-channel
3D	Three-dimensional
2D	Two-dimensional
ACEPOL	Aerosol Characterization from Polarimeter and Lidar
ACHSRL	Aerosol and Cloud High-Spectral-Resolution Lidar
ALPS	Airborne Laser Polarization Sensor
APD	Avalanche photodiode
ARM	Atmospheric radiation measurement
CALIOP	Cloud-Aerosol Lidar with Orthogonal Polarization
CAPABL	Clouds Aerosol Polarization and Backscatter Lidar
CCD	Charge-coupled device
CMOS	Complementary metal–oxide semiconductor
DNN	Deep neural network
DoFP	Division-of-focal-plane
DoLP	Degree of linear polarization

Table A1. *Cont.*

Abbreviation	Definition
DoCP	Degree of circle polarization
EEL	Edge-emitting laser
EMCCD	Electron multiplying charge-coupled devices
FMCW	Frequency-modulated continuous wave
ER	Effective radius
ESA	European Space Agency
EUR	Europe
FOV	Field of view
FR	France
GR	Germany
HERA	Hybrid extinction retrieval algorithm
HSRL	High-spectral-resolution lidar
LASER	Light amplification by stimulated emissions of radiation
LED	Light-emitting diode
Lidar	Light detection and ranging
LOOP	Lidar for Ocean Optics Profiler
LSE	Light's stimulated emission
MEMS	Micro-electro mechanical system
MiniSAR	Miniaturized synthetic aperture radar
MIT	Massachusetts Institute of Technology
MULIS	Multichannel Lidar System
NASA	National Aeronautics and Space Administration
Nd:YAG	Neodymium-doped yttrium aluminum garnet
NOAA	National Oceanic and Atmospheric Administration
OPA	Optical phased array
PBS	Polarization beam splitter
PD	Photodiode
P-Lidar	Polarization lidar
PMT	Photo-multiplier tube
POLIS	Portable lidar system
PSA	Polarization state analyzer
PSG	Polarization state generator
QWP	Quarter-wave plate
RT	Radiative transfer
SCA	Scene classification algorithms
SiPM	Silicon photo-multiplier
SIBYL	Selective Iterative Boundary Locator
SLICER	Scanning Lidar Imager of Canopies by Echo Recovery
SNR	Signal-to-noise ratio
SP	Single-photon
SPAD	Single-photon APD
SVLE	Stokes vector lidar
SWIR	Short-wave infrared
TOF	Time of flight
UAV	Unmanned aerial vehicles
USA	The United States of America
UV	Ultraviolet
VC	Volume concentration
VCSEL	Vertical-cavity surface-emitting laser
Vis–NIR	Visible, near-infrared
VLDR	Volume linear depolarization ratio
WACAL	Water vapor, cloud, and aerosol lidar

References

1. Pérez, J.J.G.; Ossikovski, R. *Polarized Light and the Mueller Matrix Approach*; CRC Press: Boca Raton, FL, USA, 2017.
2. Goldstein, D.H. *Polarized Light*; CRC Press: Boca Raton, FL, USA, 2017.
3. Bass, M.; Van Stryland, E.W.; Williams, D.R.; Wolfe, W.L. *Handbook of Optics*; McGraw-Hill: New York, NY, USA, 1995; Volume 2.
4. Fowles, G.R. *Introduction to Modern Optics*; Courier Corporation: Washington, DC, USA, 1989.

5. Li, X.; Hu, H.; Zhao, L.; Wang, H.; Yu, Y.; Wu, L.; Liu, T. Polarimetric image recovery method combining histogram stretching for underwater imaging. *Sci. Rep.* **2018**, *8*, 12430. [CrossRef] [PubMed]
6. Li, X.; Han, Y.; Wang, H.; Liu, T.; Chen, S.C.; Hu, H. Polarimetric imaging through scattering media: A review. *Front. Phys.* **2022**, *10*, 815296. [CrossRef]
7. Ballesta-Garcia, M.; Peña-Gutiérrez, S.; Rodríguez-Aramendía, A.; García-Gómez, P.; Rodrigo, N.; Bobi, A.R.; Royo, S. Analysis of the performance of a polarized LiDAR imager in fog. *Opt. Express* **2022**, *30*, 41524–41540. [CrossRef] [PubMed]
8. Li, X.; Yan, L.; Qi, P.; Zhang, L.; Goudail, F.; Liu, T.; Zhai, J.; Hu, H. Polarimetric Imaging via Deep Learning: A Review. *Remote Sens.* **2023**, *15*, 1540. [CrossRef]
9. Chu, J.; Zhao, K.; Zhang, Q.; Wang, T. Construction and performance test of a novel polarization sensor for navigation. *Sens. Actuators Phys.* **2008**, *148*, 75–82. [CrossRef]
10. Wan, Z.; Zhao, K.; Chu, J. Robust azimuth measurement method based on polarimetric imaging for bionic polarization navigation. *IEEE Trans. Instrum. Meas.* **2019**, *69*, 5684–5692. [CrossRef]
11. Sun, Z.; Zhang, J.; Zhao, Y. Laboratory studies of polarized light reflection from sea ice and lake ice in visible and near infrared. *IEEE Geosci. Remote. Sens. Lett.* **2012**, *10*, 170–173. [CrossRef]
12. Sun, W.; Liu, Z.; Videen, G.; Fu, Q.; Muinonen, K.; Winker, D.M.; Lukashin, C.; Jin, Z.; Lin, B.; Huang, J. For the depolarization of linearly polarized light by smoke particles. *J. Quant. Spectrosc. Radiat. Transf.* **2013**, *122*, 233–237. [CrossRef]
13. Kong, Z.; Ma, T.; Zheng, K.; Cheng, Y.; Gong, Z.; Hua, D.; Mei, L. Development of an all-day portable polarization Lidar system based on the division-of-focal-plane scheme for atmospheric polarization measurements. *Opt. Express* **2021**, *29*, 38512–38526. [CrossRef]
14. Shibata, Y. Particle polarization Lidar for precipitation particle classification. *Appl. Opt.* **2022**, *61*, 1856–1862. [CrossRef]
15. Sassen, K. The polarization Lidar technique for cloud research: A review and current assessment. *Bull. Am. Meteorol. Soc.* **1991**, *72*, 1848–1866. [CrossRef]
16. Sassen, K. Polarization in Lidar. In *LIDAR: Range-Resolved Optical Remote Sensing of the Atmosphere*; Springer: Berlin/Heidelberg, Germany, 2005; pp. 19–42.
17. Einstein, A. Näherungsweise integration der feldgleichungen der gravitation. *Sitzungsberichte Königlich Preußischen Akad. Wiss.* **1916**, *22*, 688–696.
18. Herd, R.M.; Dover, J.S.; Arndt, K.A. Basic laser principles. *Dermatol. Clin.* **1997**, *15*, 355–372. [CrossRef]
19. Zinth, W.; Laubereau, A.; Kaiser, W. The long journey to the laser and its rapid development after 1960. *Eur. Phys. J. H* **2011**, *36*, 153–181. [CrossRef]
20. Collis, R. Lidar. *Applied Optics* **1970**, *9*, 1782–1788. [CrossRef]
21. Dickey, J.O.; Bender, P.; Faller, J.; Newhall, X.; Ricklefs, R.; Ries, J.; Shelus, P.; Veillet, C.; Whipple, A.; Wiant, J.; et al. Lunar laser ranging: A continuing legacy of the Apollo program. *Science* **1994**, *265*, 482–490. [CrossRef]
22. Bender, P.; Currie, D.; Poultney, S.; Alley, C.; Dicke, R.; Wilkinson, D.; Eckhardt, D.; Faller, J.; Kaula, W.; Mulholland, J.; et al. The Lunar Laser Ranging Experiment: Accurate ranges have given a large improvement in the lunar orbit and new selenophysical information. *Science* **1973**, *182*, 229–238. [CrossRef]
23. Wandinger, U. Introduction to Lidar. In *Lidar: Range-Resolved Optical Remote Sensing of the Atmosphere*; Springer: Berlin/Heidelberg, Germany, 2005; pp. 1–18.
24. Dubayah, R.O.; Drake, J.B. Lidar remote sensing for forestry. *J. For.* **2000**, *98*, 44–46.
25. Dong, P.; Chen, Q. *LiDAR Remote Sensing and Applications*; CRC Press: Boca Raton, FL, USA, 2017.
26. Han, Y.; Li, Z.; Wu, L.; Mai, S.; Xing, X.; Fu, H. High-Speed Two-Dimensional Spectral-Scanning Coherent LiDAR System Based on Tunable VCSEL. *J. Light. Technol.* **2022**, *41*, 412–419. [CrossRef]
27. Sun, X.; Zhang, L.; Zhang, Q.; Zhang, W. Si photonics for practical LiDAR solutions. *Appl. Sci.* **2019**, *9*, 4225. [CrossRef]
28. Shiina, T. LED mini Lidar for atmospheric application. *Sensors* **2019**, *19*, 569. [CrossRef] [PubMed]
29. Sarbolandi, H.; Plack, M.; Kolb, A. Pulse based time-of-flight range sensing. *Sensors* **2018**, *18*, 1679. [CrossRef] [PubMed]
30. Stove, A.G. Linear FMCW radar techniques. In *IEE Proceedings F (Radar and Signal Processing)*; IET: Washington, DC, USA, 1992; Volume 139, pp. 343–350.
31. McManamon, P.F.; Banks, P.; Beck, J.; Fried, D.G.; Huntington, A.S.; Watson, E.A. Comparison of flash Lidar detector options. *Opt. Eng.* **2017**, *56*, 031223. [CrossRef]
32. Raj, T.; Hanim Hashim, F.; Baseri Huddin, A.; Ibrahim, M.F.; Hussain, A. A survey on LiDAR scanning mechanisms. *Electronics* **2020**, *9*, 741. [CrossRef]
33. Tossoun, B.; Stephens, R.; Wang, Y.; Addamane, S.; Balakrishnan, G.; Holmes, A.; Beling, A. High-speed InP-based pin photodiodes with InGaAs/GaAsSb type-II quantum wells. *IEEE Photonics Technol. Lett.* **2018**, *30*, 399–402. [CrossRef]
34. Villa, F.; Severini, F.; Madonini, F.; Zappa, F. SPADs and SiPMs arrays for long-range high-speed light detection and ranging (LiDAR). *Sensors* **2021**, *21*, 3839. [CrossRef]
35. Beer, M.; Haase, J.F.; Ruskowski, J.; Kokozinski, R. Background light rejection in SPAD-based LiDAR sensors by adaptive photon coincidence detection. *Sensors* **2018**, *18*, 4338. [CrossRef] [PubMed]
36. Behringer, M.; Johnson, K. Laser lightsources for LIDAR. In Proceedings of the 2021 27th International Semiconductor Laser Conference (ISLC), Potsdam, Germany, 10–14 October 2021; pp. 1–2.
37. Wang, D.; Watkins, C.; Xie, H. MEMS mirrors for LiDAR: A review. *Micromachines* **2020**, *11*, 456. [CrossRef]

38. Li, N.; Ho, C.P.; Xue, J.; Lim, L.W.; Chen, G.; Fu, Y.H.; Lee, L.Y.T. A progress review on solid-state LiDAR and nanophotonics-based LiDAR sensors. *Laser Photonics Rev.* **2022**, *16*, 2100511. [CrossRef]
39. Schotland, R.M.; Sassen, K.; Stone, R. Observations by Lidar of linear depolarization ratios for hydrometeors. *J. Appl. Meteorol. Climatol.* **1971**, *10*, 1011–1017. [CrossRef]
40. Platt, C.; Abshire, N.; McNice, G. Some microphysical properties of an ice cloud from Lidar observation of horizontally oriented crystals. *J. Appl. Meteorol. Climatol.* **1978**, *17*, 1220–1224. [CrossRef]
41. Kalshoven, J.E.; Dabney, P.W. Remote sensing of the Earth's surface with an airborne polarized laser. *IEEE Trans. Geosci. Remote. Sens.* **1993**, *31*, 438–446. [CrossRef]
42. Churnside, J.H. Can we see fish from an airplane? In *Airborne and In-Water Underwater Imaging*; SPIE: Bellingham, WA, USA, 1999; Volume 3761, pp. 45–48.
43. Kaul, B.V.; Samokhvalov, I.V.; Volkov, S.N. Investigating particle orientation in cirrus clouds by measuring backscattering phase matrices with Lidar. *Appl. Opt.* **2004**, *43*, 6620–6628. [CrossRef] [PubMed]
44. Del Guasta, M.; Vallar, E.; Riviere, O.; Castagnoli, F.; Venturi, V.; Morandi, M. Use of polarimetric Lidar for the study of oriented ice plates in clouds. *Appl. Opt.* **2006**, *45*, 4878–4887. [CrossRef] [PubMed]
45. Winker, D.; Hostetler, C.; Hunt, W. Caliop: The Calipso Lidar. In Proceedings of the 22nd International Laser Radar Conference (ILRC 2004), Matera, Italy, 12–16 July 2004; Volume 561, p. 941.
46. Churnside, J.H.; Sharov, A.F.; Richter, R.A. Aerial surveys of fish in estuaries: A case study in Chesapeake Bay. *ICES J. Mar. Sci.* **2011**, *68*, 239–244. [CrossRef]
47. Hayman, M.; Thayer, J.P. General description of polarization in Lidar using Stokes vectors and polar decomposition of Mueller matrices. *JOSA A* **2012**, *29*, 400–409. [CrossRef]
48. Hayman, M.; Spuler, S.; Morley, B.; VanAndel, J. Polarization Lidar operation for measuring backscatter phase matrices of oriented scatterers. *Opt. Express* **2012**, *20*, 29553–29567. [CrossRef] [PubMed]
49. Huang, Z.; Qi, S.; Zhou, T.; Dong, Q.; Ma, X.; Zhang, S.; Bi, J.; Shi, J. Investigation of aerosol absorption with dual-polarization Lidar observations. *Opt. Express* **2020**, *28*, 7028–7035. [CrossRef]
50. Qiu, J.; Xia, H.; Shangguan, M.; Dou, X.; Li, M.; Wang, C.; Shang, X.; Lin, S.; Liu, J. Micro-pulse polarization Lidar at 1.5 μm using a single superconducting nanowire single-photon detector. *Opt. Lett.* **2017**, *42*, 4454–4457. [CrossRef]
51. Stillwell, R.A.; Neely, R.R., III; Thayer, J.P.; Shupe, M.D.; Turner, D.D. Improved cloud-phase determination of low-level liquid and mixed-phase clouds by enhanced polarimetric Lidar. *Atmos. Meas. Tech.* **2018**, *11*, 835–859. [CrossRef]
52. Kokhanenko, G.P.; Balin, Y.S.; Klemasheva, M.G.; Nasonov, S.V.; Novoselov, M.M.; Penner, I.E.; Samoilova, S.V. Scanning polarization Lidar LOSA-M3: Opportunity for research of crystalline particle orientation in the ice clouds. *Atmos. Meas. Tech.* **2020**, *13*, 1113–1127. [CrossRef]
53. Qi, S.; Huang, Z.; Ma, X.; Huang, J.; Zhou, T.; Zhang, S.; Dong, Q.; Bi, J.; Shi, J. Classification of atmospheric aerosols and clouds by use of dual-polarization Lidar measurements. *Opt. Express* **2021**, *29*, 23461–23476. [CrossRef]
54. Kong, Z.; Ma, T.; Cheng, Y.; Fei, R.; Zhang, Z.; Li, Y.; Mei, L. A polarization-sensitive imaging Lidar for atmospheric remote sensing. *J. Quant. Spectrosc. Radiat. Transf.* **2021**, *271*, 107747. [CrossRef]
55. Kong, Z.; Yu, J.; Gong, Z.; Hua, D.; Mei, L. Visible, near-infrared dual-polarization Lidar based on polarization cameras: System design, evaluation and atmospheric measurements. *Opt. Express* **2022**, *30*, 28514–28533. [CrossRef]
56. Hulst, H.C.; van de Hulst, H.C. *Light Scattering by Small Particles*; Courier Corporation: Washington, DC, USA, 1981.
57. Chen, W.; Zheng, Q.; Xiang, H.; Chen, X.; Sakai, T. Forest canopy height estimation using polarimetric interferometric synthetic aperture radar (PolInSAR) technology based on full-polarized ALOS/PALSAR data. *Remote Sens.* **2021**, *13*, 174. [CrossRef]
58. Buurman, P.; Pape, T.; Muggler, C. Laser grain-size determination in soil genetic studies 1. Practical problems. *Soil Sci.* **1997**, *162*, 211–218. [CrossRef]
59. Sassen, K.; Zhu, J.; Webley, P.; Dean, K.; Cobb, P. Volcanic ash plume identification using polarization Lidar: Augustine eruption, Alaska. *Geophys. Res. Lett.* **2007**, *34*. [CrossRef]
60. David, G.; Thomas, B.; Nousiainen, T.; Miffre, A.; Rairoux, P. Retrieving simulated volcanic, desert dust and sea-salt particle properties from two/three-component particle mixtures using UV-VIS polarization Lidar and T matrix. *Atmos. Chem. Phys.* **2013**, *13*, 6757–6776. [CrossRef]
61. Pal, S.; Carswell, A. The polarization characteristics of Lidar scattering from snow and ice crystals in the atmosphere. *J. Appl. Meteorol. Climatol.* **1977**, *16*, 70–80. [CrossRef]
62. Gibbs, D.P.; Betty, C.L.; Bredow, J.W.; Fung, A.K. Polarized and cross-polarized angular reflectance characteristics of saline ice and snow. *Remote Sens. Rev.* **1993**, *7*, 179–195. [CrossRef]
63. Liu, Q.; Wu, S.; Liu, B.; Liu, J.; Zhang, K.; Dai, G.; Tang, J.; Chen, G. Shipborne variable-FOV, dual-wavelength, polarized ocean Lidar: Design and measurements in the Western Pacific. *Opt. Express* **2022**, *30*, 8927–8948. [CrossRef] [PubMed]
64. Behrenfeld, M.J.; Hu, Y.; O'Malley, R.T.; Boss, E.S.; Hostetler, C.A.; Siegel, D.A.; Sarmiento, J.L.; Schulien, J.; Hair, J.W.; Lu, X.; et al. Annual boom–bust cycles of polar phytoplankton biomass revealed by space-based Lidar. *Nat. Geosci.* **2017**, *10*, 118–122. [CrossRef]
65. Behrenfeld, M.J.; Gaube, P.; Della Penna, A.; O'malley, R.T.; Burt, W.J.; Hu, Y.; Bontempi, P.S.; Steinberg, D.K.; Boss, E.S.; Siegel, D.A.; et al. Global satellite-observed daily vertical migrations of ocean animals. *Nature* **2019**, *576*, 257–261. [CrossRef] [PubMed]

66. Hoge, F.E.; Wright, C.W.; Krabill, W.B.; Buntzen, R.R.; Gilbert, G.D.; Swift, R.N.; Yungel, J.K.; Berry, R.E. Airborne Lidar detection of subsurface oceanic scattering layers. *Appl. Opt.* **1988**, *27*, 3969–3977. [CrossRef] [PubMed]
67. Churnside, J.H. Polarization effects on oceanographic Lidar. *Opt. Express* **2008**, *16*, 1196–1207. [CrossRef] [PubMed]
68. Churnside, J.H.; Sullivan, J.M.; Twardowski, M.S. Lidar extinction-to-backscatter ratio of the ocean. *Opt. Express* **2014**, *22*, 18698–18706. [CrossRef] [PubMed]
69. Churnside, J.H.; Marchbanks, R.D. Inversion of oceanographic profiling Lidars by a perturbation to a linear regression. *Appl. Opt.* **2017**, *56*, 5228–5233. [CrossRef]
70. Chen, P.; Jamet, C.; Zhang, Z.; He, Y.; Mao, Z.; Pan, D.; Wang, T.; Liu, D.; Yuan, D. Vertical distribution of subsurface phytoplankton layer in South China Sea using airborne Lidar. *Remote Sens. Environ.* **2021**, *263*, 112567. [CrossRef]
71. Dassot, M.; Constant, T.; Fournier, M. The use of terrestrial LiDAR technology in forest science: Application fields, benefits and challenges. *Ann. For. Sci.* **2011**, *68*, 959–974. [CrossRef]
72. Kemeny, J.; Turner, K. *Ground-Based Lidar: Rock Slope Mapping and Assessment*; Technical report; Federal Highway Administration, Central Federal Lands Highway Division: Washington, DC, USA, 2008.
73. Williams, K.; Olsen, M.J.; Roe, G.V.; Glennie, C. Synthesis of transportation applications of mobile LiDAR. *Remote Sens.* **2013**, *5*, 4652–4692. [CrossRef]
74. Ecke, S.; Dempewolf, J.; Frey, J.; Schwaller, A.; Endres, E.; Klemmt, H.J.; Tiede, D.; Seifert, T. UAV-based forest health monitoring: A systematic review. *Remote Sens.* **2022**, *14*, 3205. [CrossRef]
75. Vasilkov, A.P.; Goldin, Y.A.; Gureev, B.A.; Hoge, F.E.; Swift, R.N.; Wright, C.W. Airborne polarized Lidar detection of scattering layers in the ocean. *Appl. Opt.* **2001**, *40*, 4353–4364. [CrossRef] [PubMed]
76. Goldin, Y.A.; Gureev, B.A.; Ventskut, Y.I. Shipboard polarized Lidar for seawater column sounding. In *Current Research on Remote Sensing, Laser Probing, and Imagery in Natural Waters*; SPIE: *Bellingham, WA, USA*, 2007; Volume 6615, pp. 152–159.
77. Okamoto, H.; Sato, K.; Borovoi, A.; Ishimoto, H.; Masuda, K.; Konoshonkin, A.; Kustova, N. Interpretation of Lidar ratio and depolarization ratio of ice clouds using spaceborne high-spectral-resolution polarization Lidar. *Opt. Express* **2019**, *27*, 36587–36600. [CrossRef] [PubMed]
78. Li, X.; Liu, T.; Huang, B.; Song, Z.; Hu, H. Optimal distribution of integration time for intensity measurements in Stokes polarimetry. *Opt. Express* **2015**, *23*, 27690–27699. [CrossRef] [PubMed]
79. Li, X.; Hu, H.; Wu, L.; Liu, T. Optimization of instrument matrix for Mueller matrix ellipsometry based on partial elements analysis of the Mueller matrix. *Opt. Express* **2017**, *25*, 18872–18884. [CrossRef] [PubMed]
80. Stokes, G.G. On the composition and resolution of streams of polarized light from different sources. *Trans. Camb. Philos. Soc.* **1851**, *9*, 399.
81. Li, X.; Hu, H.; Liu, T.; Huang, B.; Song, Z. Optimal distribution of integration time for intensity measurements in degree of linear polarization polarimetry. *Opt. Express* **2016**, *24*, 7191–7200. [CrossRef] [PubMed]
82. Song, Z.; Li, X.; Liu, T. Optimal distribution of integration time in degree of linear polarization polarimetry based on the expected variance. *Optik* **2017**, *136*, 123–128. [CrossRef]
83. Chen, W.N.; Chiang, C.W.; Nee, J.B. Lidar ratio and depolarization ratio for cirrus clouds. *Appl. Opt.* **2002**, *41*, 6470–6476. [CrossRef]
84. Noel, V.; Chepfer, H.; Ledanois, G.; Delaval, A.; Flamant, P.H. Classification of particle effective shape ratios in cirrus clouds based on the Lidar depolarization ratio. *Appl. Opt.* **2002**, *41*, 4245–4257. [CrossRef]
85. Lu, S.Y.; Chipman, R.A. Interpretation of Mueller matrices based on polar decomposition. *JOSA A* **1996**, *13*, 1106–1113. [CrossRef]
86. Morio, J.; Goudail, F. Influence of the order of diattenuator, retarder, and polarizer in polar decomposition of Mueller matrices. *Opt. Lett.* **2004**, *29*, 2234–2236. [CrossRef] [PubMed]
87. Gil, J.J.; San José, I.; Ossikovski, R. Serial–parallel decompositions of Mueller matrices. *JOSA A* **2013**, *30*, 32–50. [CrossRef] [PubMed]
88. Li, X.; Zhang, L.; Qi, P.; Zhu, Z.; Xu, J.; Liu, T.; Zhai, J.; Hu, H. Are indices of polarimetric purity excellent metrics for object identification in scattering media? *Remote Sens.* **2022**, *14*, 4148. [CrossRef]
89. Li, X.; Liu, W.; Goudail, F.; Chen, S.C. Optimal nonlinear Stokes—Mueller polarimetry for multi-photon processes. *Opt. Lett.* **2022**, *47*, 3287–3290. [CrossRef] [PubMed]
90. Li, X.; Goudail, F.; Chen, S.C. Self-calibration for Mueller polarimeters based on DoFP polarization imagers. *Opt. Lett.* **2022**, *47*, 1415–1418. [CrossRef] [PubMed]
91. Chandrasekar, V.; Keränen, R.; Lim, S.; Moisseev, D. Recent advances in classification of observations from dual polarization weather radars. *Atmos. Res.* **2013**, *119*, 97–111. [CrossRef]
92. Sassen, K.; Wang, Z.; Liu, D. Global distribution of cirrus clouds from CloudSat/Cloud-Aerosol Lidar and infrared pathfinder satellite observations (CALIPSO) measurements. *J. Geophys. Res. Atmos.* **2008**, *113*, D8. [CrossRef]
93. Ahmad, W.; Zhang, K.; Tong, Y.; Xiao, D.; Wu, L.; Liu, D. Water cloud detection with circular polarization Lidar: A semianalytic Monte Carlo simulation approach. *Sensors* **2022**, *22*, 1679. [CrossRef]
94. Evans, B.T.N. *On the Inversion of the Lidar Equation*; Department of National Defence, Research and Development Branch, Defence Research Establishment: Umea, Sweden, 1984.
95. Northend, C.; Honey, R.; Evans, W. Laser radar (Lidar) for meteorological observations. *Rev. Sci. Instrum.* **1966**, *37*, 393–400. [CrossRef]

96. Sasano, Y.; Shimizu, H.; Takeuchi, N.; Okuda, M. Geometrical form factor in the laser radar equation: An experimental determination. *Appl. Opt.* **1979**, *18*, 3908–3910. [CrossRef] [PubMed]
97. Biele, J.; Beyerle, G.; Baumgarten, G. Polarization Lidar: Corrections of instrumental effects. *Opt. Express* **2000**, *7*, 427–435. [CrossRef] [PubMed]
98. Freudenthaler, V. About the effects of polarising optics on Lidar signals and the Δ90 calibration. *Atmos. Meas. Tech.* **2016**, *9*, 4181–4255. [CrossRef]
99. Winker, D.M.; Vaughan, M.A.; Omar, A.; Hu, Y.; Powell, K.A.; Liu, Z.; Hunt, W.H.; Young, S.A. Overview of the CALIPSO mission and CALIOP data processing algorithms. *J. Atmos. Ocean. Technol.* **2009**, *26*, 2310–2323. [CrossRef]
100. Winker, D.M.; Hunt, W.H.; McGill, M.J. Initial performance assessment of CALIOP. *Geophys. Res. Lett.* **2007**, *34*, 19. [CrossRef]
101. Winker, D.; Pelon, J.; Coakley, J., Jr.; Ackerman, S.; Charlson, R.; Colarco, P.; Flamant, P.; Fu, Q.; Hoff, R.; Kittaka, C.; et al. The CALIPSO mission: A global 3D view of aerosols and clouds. *Bull. Am. Meteorol. Soc.* **2010**, *91*, 1211–1230. [CrossRef]
102. Sarabandi, K.; Ulaby, F.T.; Tassoudji, M.A. Calibration of polarimetric radar systems with good polarization isolation. *IEEE Trans. Geosci. Remote Sens.* **1990**, *28*, 70–75. [CrossRef]
103. Alvarez, J.; Vaughan, M.A.; Hostetler, C.A.; Hunt, W.; Winker, D.M. Calibration technique for polarization-sensitive Lidars. *J. Atmos. Ocean. Technol.* **2006**, *23*, 683–699. [CrossRef]
104. Platt, C. Lidar observation of a mixed-phase altostratus cloud. *J. Appl. Meteorol. Climatol.* **1977**, *16*, 339–345. [CrossRef]
105. Flynna, C.J.; Mendozaa, A.; Zhengb, Y.; Mathurb, S. Novel polarization-sensitive micropulse Lidar measurement technique. *Opt. Express* **2007**, *15*, 2785–2790. [CrossRef]
106. Pal, S.; Carswell, A. Polarization properties of Lidar backscattering from clouds. *Appl. Opt.* **1973**, *12*, 1530–1535. [CrossRef] [PubMed]
107. Houston, J.; Carswell, A. Four-component polarization measurement of Lidar atmospheric scattering. *Appl. Opt.* **1978**, *17*, 614–620. [CrossRef] [PubMed]
108. Xian, J.; Xu, W.; Long, C.; Song, Q.; Yang, S. Early forest-fire detection using scanning polarization Lidar. *Appl. Opt.* **2020**, *59*, 8638–8644. [CrossRef] [PubMed]
109. Noel, V.; Sassen, K. Study of planar ice crystal orientations in ice clouds from scanning polarization Lidar observations. *J. Appl. Meteorol. Climatol.* **2005**, *44*, 653–664. [CrossRef]
110. Sassen, K. Polarization in Lidar: A review. *Polariz. Sci. Remote Sens.* **2003**, *5158*, 151–160.
111. Li, C.; Xu, S.; Zhao, L.; Cheng, G. Research on MEMS biaxial electromagnetic micromirror based on radial magnetic field distribution. In Proceedings of the International Conference on Laser, Optics and Optoelectronic Technology (LOPET 2021), Xi'an, China, 28–30 May 2021; Volume 11885, pp. 206–212.
112. Roriz, R.; Cabral, J.; Gomes, T. Automotive LiDAR technology: A survey. *IEEE Trans. Intell. Transp. Syst.* **2021**, *23*, 6282–6297. [CrossRef]
113. Kokhanenko, G.P. Possibilities of using mirror scanners in polarizing Lidars. In Proceedings of the 28th International Symposium on Atmospheric and Ocean Optics: Atmospheric Physics, *Tomsk, Russia, 4–8 July* 2022; Volume 12341, pp. 328–333.
114. Li, X.; Li, Y.; Xie, X.; Xu, L. Laser polarization imaging models based on leaf moisture content (in Chinese). *Infrared Laser Eng.* **2017**, *46*, 1106004.
115. Behrendt, A.; Nakamura, T. Calculation of the calibration constant of polarization Lidar and its dependency on atmospheric temperature. *Opt. Express* **2002**, *10*, 805–817. [CrossRef]
116. Luo, J.; Liu, D.; Bi, L.; Zhang, K.; Tang, P.; Xu, P.; Su, L.; Yang, L. Rotating a half-wave plate by 45: An ideal calibration method for the gain ratio in polarization Lidars. *Opt. Commun.* **2018**, *407*, 361–366. [CrossRef]
117. Mattis, I.; Tesche, M.; Grein, M.; Freudenthaler, V.; Müller, D. Systematic error of Lidar profiles caused by a polarization-dependent receiver transmission: Quantification and error correction scheme. *Appl. Opt.* **2009**, *48*, 2742–2751. [CrossRef]
118. Hunt, W.H.; Winker, D.M.; Vaughan, M.A.; Powell, K.A.; Lucker, P.L.; Weimer, C. CALIPSO Lidar description and performance assessment. *J. Atmos. Ocean. Technol.* **2009**, *26*, 1214–1228. [CrossRef]
119. Sassen, K.; Benson, S. A midlatitude cirrus cloud climatology from the Facility for Atmospheric Remote Sensing. Part II: Microphysical properties derived from Lidar depolarization. *J. Atmos. Sci.* **2001**, *58*, 2103–2112. [CrossRef]
120. Tong, Y.; Tong, X.; Zhang, K.; Xiao, D.; Rong, Y.; Zhou, Y.; Liu, C.; Liu, D. Polarization Lidar gain ratio calibration method: A comparison. *Chin. Opt.* **2021**, *14*, 685–703.
121. Luo, J.; Liu, D.; Huang, Z.; Wang, B.; Bai, J.; Cheng, Z.; Zhang, Y.; Tang, P.; Yang, L.; Su, L. Polarization properties of receiving telescopes in atmospheric remote sensing polarization Lidars. *Appl. Opt.* **2017**, *56*, 6837–6845. [CrossRef] [PubMed]
122. Freudenthaler, V.; Esselborn, M.; Wiegner, M.; Heese, B.; Tesche, M.; Ansmann, A.; Müller, D.; Althausen, D.; Wirth, M.; Fix, A.; et al. Depolarization ratio profiling at several wavelengths in pure Saharan dust during SAMUM 2006. *Tellus Chem. Phys. Meteorol.* **2009**, *61*, 165–179. [CrossRef]
123. Fiocco, G.; Smullin, L. Detection of scattering layers in the upper atmosphere (60–140 km) by optical radar. *Nature* **1963**, *199*, 1275–1276. [CrossRef]
124. Polarization Lidar. Available online: https://www.tropos.de/en/research/projects-infrastructures-technology/technology-at-tropos/remote-sensing/polarization-Lidar (accessed on 3 September 2023).
125. Matrosov, S.Y.; Schmitt, C.G.; Maahn, M.; de Boer, G. Atmospheric ice particle shape estimates from polarimetric radar measurements and in situ observations. *J. Atmos. Ocean. Technol.* **2017**, *34*, 2569–2587. [CrossRef]

126. Wu, S.; Song, X.; Liu, B.; Dai, G.; Liu, J.; Zhang, K.; Qin, S.; Hua, D.; Gao, F.; Liu, L. Mobile multi-wavelength polarization Raman Lidar for water vapor, cloud and aerosol measurement. *Opt. Express* **2015**, *23*, 33870–33892. [CrossRef]
127. Tan, W.; Li, C.; Liu, Y.; Meng, X.; Wu, Z.; Kang, L.; Zhu, T. Potential of polarization Lidar to profile the urban aerosol phase state during haze episodes. *Environ. Sci. Technol. Lett.* **2019**, *7*, 54–59. [CrossRef]
128. Jimenez, C.; Ansmann, A.; Engelmann, R.; Donovan, D.; Malinka, A.; Jörg, S.; Patric, S.; Ulla, W. The dual-field-of-view polarization Lidar technique: A new concept in monitoring aerosol effects in liquid-water clouds—Theoretical framework. *Atmos. Chem. Phys.* **2020**, *20*, 15247–15263. [CrossRef]
129. Jimenez, C.; Ansmann, A.; Engelmann, R.; Donovan, D.; Malinka, A.; Seifert, P.; Wiesen, R.; Radenz, M.; Yin, Z.; Bühl, J.; et al. The dual-field-of-view polarization Lidar technique: A new concept in monitoring aerosol effects in liquid-water clouds—Case studies. *Atmos. Chem. Phys.* **2020**, *20*, 15265–15284. [CrossRef]
130. Zhang, S.; Huang, Z.; Alam, K.; Li, M.; Dong, Q.; Wang, Y.; Shen, X.; Bi, J.; Zhang, J.; Li, W.; et al. Derived Profiles of CCN and INP Number Concentrations in the Taklimakan Desert via Combined Polarization Lidar, Sun-Photometer, and Radiosonde Observations. *Remote Sens.* **2023**, *15*, 1216. [CrossRef]
131. Pal, S.; Carswell, A. Polarization properties of Lidar scattering from clouds at 347 nm and 694 nm. *Appl. Opt.* **1978**, *17*, 2321–2328. [CrossRef] [PubMed]
132. Sassen, K.; Zhao, H.; Dodd, G.C. Simulated polarization diversity Lidar returns from water and precipitating mixed phase clouds. *Appl. Opt.* **1992**, *31*, 2914–2923. [CrossRef]
133. Mamouri, R.E.; Ansmann, A. Potential of polarization Lidar to provide profiles of CCN-and INP-relevant aerosol parameters. *Atmos. Chem. Phys.* **2016**, *16*, 5905–5931. [CrossRef]
134. Ansmann, A.; Ohneiser, K.; Mamouri, R.E.; Knopf, D.A.; Veselovskii, I.; Baars, H.; Engelmann, R.; Foth, A.; Jimenez, C.; Seifert, P.; et al. Tropospheric and stratospheric wildfire smoke profiling with Lidar: Mass, surface area, CCN, and INP retrieval. *Atmos. Chem. Phys.* **2021**, *21*, 9779–9807. [CrossRef]
135. Murayama, T.; Müller, D.; Wada, K.; Shimizu, A.; Sekiguchi, M.; Tsukamoto, T. Characterization of Asian dust and Siberian smoke with multi-wavelength Raman Lidar over Tokyo, Japan in spring 2003. *Geophys. Res. Lett.* **2004**, *31*, 23. [CrossRef]
136. Sugimoto, N.; Lee, C.H. Characteristics of dust aerosols inferred from Lidar depolarization measurements at two wavelengths. *Appl. Opt.* **2006**, *45*, 7468–7474. [CrossRef]
137. Sugimoto, N.; Uno, I.; Nishikawa, M.; Shimizu, A.; Matsui, I.; Dong, X.; Chen, Y.; Quan, H. Record heavy Asian dust in Beijing in 2002: Observations and model analysis of recent events. *Geophys. Res. Lett.* **2003**, *30*, 12. [CrossRef]
138. Tesche, M.; Ansmann, A.; Müller, D.; Althausen, D.; Engelmann, R.; Freudenthaler, V.; Groß, S. Vertically resolved separation of dust and smoke over Cape Verde using multiwavelength Raman and polarization Lidars during Saharan Mineral Dust Experiment 2008. *J. Geophys. Res. Atmos.* **2009**, *114*, D13. [CrossRef]
139. Burton, S.; Hair, J.; Kahnert, M.; Ferrare, R.; Hostetler, C.; Cook, A.; Harper, D.; Berkoff, T.; Seaman, S.; Collins, J.; et al. Observations of the spectral dependence of linear particle depolarization ratio of aerosols using NASA Langley airborne High Spectral Resolution Lidar. *Atmos. Chem. Phys.* **2015**, *15*, 13453–13473. [CrossRef]
140. Haarig, M.; Althausen, D.; Ansmann, A.; Klepel, A.; Baars, H.; Engelmann, R.; Groß, S.; Freudenthaler, V. Measurement of the linear depolarization ratio of aged dust at three wavelengths (355, 532 and 1064 nm) simultaneously over Barbados. In *EPJ Web of Conferences*; EDP Sciences: Les Ulis, France, 2016; Volume 119, p. 18009.
141. Vaughan, M.; Garnier, A.; Josset, D.; Avery, M.; Lee, K.P.; Liu, Z.; Hunt, W.; Pelon, J.; Hu, Y.; Burton, S.; et al. CALIPSO Lidar calibration at 1064 nm: Version 4 algorithm. *Atmos. Meas. Tech.* **2019**, *12*, 51–82. [CrossRef]
142. Tsekeri, A.; Amiridis, V.; Louridas, A.; Georgoussis, G.; Freudenthaler, V.; Metallinos, S.; Doxastakis, G.; Gasteiger, J.; Siomos, N.; Paschou, P.; et al. Polarization Lidar for detecting dust orientation: System design and calibration. *Atmos. Meas. Tech.* **2021**, *14*, 7453–7474. [CrossRef]
143. Seckar, C.; Guy, L.; DiFronzo, A.; Weimer, C. Performance testing of an active boresight mechanism for use in the CALIPSO space bourne LIDAR mission. In *Optomechanics 2005*; SPIE: Bellingham, WA, USA, 2005; Volume 5877, pp. 319–330.
144. Atmospheric Aerosol Characterization. Available online: https://www.ll.mit.edu/r-d/projects/atmospheric-aerosol-characterization (accessed on 3 September 2023).
145. Knobelspiesse, K.; Barbosa, H.M.; Bradley, C.; Bruegge, C.; Cairns, B.; Chen, G.; Chowdhary, J.; Cook, A.; Di Noia, A.; van Diedenhoven, B.; et al. The Aerosol Characterization from Polarimeter and Lidar (ACEPOL) airborne field campaign. *Earth Syst. Sci. Data Discuss.* **2020**, *2020*, 1–38. [CrossRef]
146. Dulac, F.; Chazette, P. Airborne study of a multi-layer aerosol structure in the eastern Mediterranean observed with the airborne polarized Lidar ALEX during a STAARTE campaign (7 June 1997). *Atmos. Chem. Phys.* **2003**, *3*, 1817–1831. [CrossRef]
147. Bo, G.; Liu, D.; Wang, B.; Wu, D.; Zhong, Z. Two-wavelength polarization airborne Lidar for observation of aerosol and cloud. *Zhongguo Jiguang Chin. J. Lasers* **2012**, *39*, 1014002-6.
148. Fernald, F.G.; Herman, B.M.; Reagan, J.A. Determination of aerosol height distributions by Lidar. *J. Appl. Meteorol. Climatol.* **1972**, *11*, 482–489. [CrossRef]
149. Klett, J.D. Stable analytical inversion solution for processing Lidar returns. *Appl. Opt.* **1981**, *20*, 211–220. [CrossRef] [PubMed]
150. Davis, P. The analysis of Lidar signatures of cirrus clouds. *Appl. Opt.* **1969**, *8*, 2099–2102. [CrossRef]
151. Sasano, Y.; Nakane, H. Significance of the extinction/backscatter ratio and the boundary value term in the solution for the two-component Lidar equation. *Appl. Opt.* **1984**, *23*, 11–13. [CrossRef]

152. Hinkley, E.D. *Laser Monitoring of the Atmosphere*; Springer: Berlin, Germany, 1976.
153. Fernald, F.G. Analysis of atmospheric Lidar observations: Some comments. *Appl. Opt.* **1984**, *23*, 652–653. [CrossRef] [PubMed]
154. Dawson, K.; Ferrare, R.; Moore, R.; Clayton, M.; Thorsen, T.; Eloranta, E. Ambient aerosol hygroscopic growth from combined Raman Lidar and HSRL. *J. Geophys. Res. Atmos.* **2020**, *125*, e2019JD031708. [CrossRef]
155. Liu, Z.; Sugimoto, N.; Murayama, T. Extinction-to-backscatter ratio of Asian dust observed with high-spectral-resolution Lidar and Raman Lidar. *Appl. Opt.* **2002**, *41*, 2760–2767. [CrossRef] [PubMed]
156. Thorsen, T.J.; Fu, Q. Automated retrieval of cloud and aerosol properties from the ARM Raman Lidar. Part II: Extinction. *J. Atmos. Ocean. Technol.* **2015**, *32*, 1999–2023. [CrossRef]
157. Thorsen, T.J.; Fu, Q.; Newsom, R.K.; Turner, D.D.; Comstock, J.M. Automated retrieval of cloud and aerosol properties from the ARM Raman Lidar. Part I: Feature detection. *J. Atmos. Ocean. Technol.* **2015**, *32*, 1977–1998. [CrossRef]
158. Ferrare, R.A.; Thorsen, T.; Clayton, M.; Muller, D.; Chemyakin, E.; Burton, S.; Goldsmith, J.; Holz, R.; Kuehn, R.; Eloranta, E.; et al. *Vertically Resolved Retrievals of Aerosol Concentrations and Effective Radii from the DOE Combined HSRL and Raman Lidar Measurement Study (CHARMS) Merged High-Spectral-Resolution Lidar-Raman Lidar Data Set*; Technical report; DOE Office of Science Atmospheric Radiation Measurement (ARM) Program: Washington, DC, USA, 2017.
159. Sorrentino, A.; Sannino, A.; Spinelli, N.; Piana, M.; Boselli, A.; Tontodonato, V.; Castellano, P.; Wang, X. A Bayesian parametric approach to the retrieval of the atmospheric number size distribution from Lidar data. *Atmos. Meas. Tech.* **2022**, *15*, 149–164. [CrossRef]
160. Ke, J.; Sun, Y.; Dong, C.; Zhang, X.; Wang, Z.; Lyu, L.; Zhu, W.; Ansmann, A.; Su, L.; Bu, L.; et al. Development of China's first space-borne aerosol-cloud high-spectral-resolution Lidar: Retrieval algorithm and airborne demonstration. *PhotoniX* **2022**, *3*, 17. [CrossRef]
161. Harding, D. *SLICER Airborne Laser Altimeter Characterization of Canopy Structure and Sub-Canopy Topography for the BOREAS Northern and Southern Study Regions: Instrument and Data Product Description*; National Aeronautics and Space Administration, Goddard Space Flight Center: Pasadena, CA, USA, 2000.
162. Dubayah, R.; Prince, S.; JaJa, J.; Blair, J.; Bufton, J.L.; Knox, R.; Luthcke, S.B.; Clarke, D.B.; Weishampel, J. The vegetation canopy Lidar mission. In *Land Satellite Information in the Next Decade II: Sources and Applications*; 1997. Available online: https://www.umiacs.umd.edu/publications/vegetation-canopy-lidar-mission (accessed on 3 September 2023).
163. Kalshoven, J.E.; Tierney, M.R.; Daughtry, C.S.; McMurtrey, J.E. Remote sensing of crop parameters with a polarized, frequency-doubled Nd: YAG laser. *Appl. Opt.* **1995**, *34*, 2745–2749. [CrossRef]
164. Tan, S.; Narayanan, R.M. A multiwavelength airborne polarimetric Lidar for vegetation remote sensing: Instrumentation and preliminary test results. *IEEE Int. Geosci. Remote. Sens. Symp.* **2002**, *5*, 2675–2677.
165. Tan, S.; Narayanan, R.M. Design and performance of a multiwavelength airborne polarimetric Lidar for vegetation remote sensing. *Appl. Opt.* **2004**, *43*, 2360–2368. [CrossRef]
166. Tan, S.; Narayanan, R.M.; Helder, D.L. Polarimetric reflectance and depolarization ratio from several tree species using a multiwavelength polarimetric Lidar. In *Polarization Science and Remote Sensing II*; SPIE: Bellingham, WA, USA, 2005; Volume 5888, pp. 180–188.
167. Tan, S.; Johnson, S.; Gu, Z. Laser depolarization ratio measurement of corn leaves from the biochar and non-biochar applied plots. *Opt. Express* **2018**, *26*, 14295–14306. [CrossRef] [PubMed]
168. Andreucci, F.; Arbolino, M. A study on forest fire automatic detection systems: I.—Smoke plume model. *Il Nuovo C. C* **1993**, *16*, 35–50. [CrossRef]
169. Vaughan, G.; Draude, A.P.; Ricketts, H.M.; Schultz, D.M.; Adam, M.; Sugier, J.; Wareing, D.P. Transport of Canadian forest fire smoke over the UK as observed by Lidar. *Atmos. Chem. Phys.* **2018**, *18*, 11375–11388. [CrossRef]
170. Taboada, J.; Tamburino, L.A. Laser Imaging and Ranging System Using Two Cameras. U.S. Patent 5,157,451, 20 October 1992.
171. Chen, Z.; Liu, B.; Liu, E.; Peng, Z. Adaptive polarization-modulated method for high-resolution 3D imaging. *IEEE Photonics Technology Letters* **2015**, *28*, 295–298. [CrossRef]
172. Chen, Z.; Liu, B.; Wang, S.; Liu, E. Polarization-modulated three-dimensional imaging using a large-aperture electro-optic modulator. *Appl. Opt.* **2018**, *57*, 7750–7757. [CrossRef] [PubMed]
173. Li, X.; Hu, H.; Goudail, F.; Liu, T. Fundamental precision limits of full Stokes polarimeters based on DoFP polarization cameras for an arbitrary number of acquisitions. *Opt. Express* **2019**, *27*, 31261–31272. [CrossRef] [PubMed]
174. Li, X.; Le Teurnier, B.; Boffety, M.; Liu, T.; Hu, H.; Goudail, F. Theory of autocalibration feasibility and precision in full Stokes polarization imagers. *Opt. Express* **2020**, *28*, 15268–15283. [CrossRef] [PubMed]
175. Jo, S.; Kong, H.J.; Bang, H.; Kim, J.W.; Kim, J.; Choi, S. High resolution three-dimensional flash LIDAR system using a polarization modulating Pockels cell and a micro-polarizer CCD camera. *Opt. Express* **2016**, *24*, A1580–A1585. [CrossRef]
176. Nunes-Pereira, E.; Peixoto, H.; Teixeira, J.; Santos, J. Polarization-coded material classification in automotive LIDAR aiming at safer autonomous driving implementations. *Appl. Opt.* **2020**, *59*, 2530–2540. [CrossRef]
177. Ronen, A.; Agassi, E.; Yaron, O. Sensing with polarized Lidar in degraded visibility conditions due to fog and low clouds. *Sensors* **2021**, *21*, 2510. [CrossRef]
178. Kattawar, G.W.; Plass, G.N.; Guinn, J.A., Jr. Monte Carlo calculations of the polarization of radiation in the earth's atmosphere-ocean system. *J. Phys. Oceanogr.* **1973**, *3*, 353–372. [CrossRef]

179. Chowdhary, J.; Cairns, B.; Travis, L.D. Contribution of water-leaving radiances to multiangle, multispectral polarimetric observations over the open ocean: Bio-optical model results for case 1 waters. *Appl. Opt.* **2006**, *45*, 5542–5567. [CrossRef] [PubMed]
180. Chowdhary, J.; Cairns, B.; Waquet, F.; Knobelspiesse, K.; Ottaviani, M.; Redemann, J.; Travis, L.; Mishchenko, M. Sensitivity of multiangle, multispectral polarimetric remote sensing over open oceans to water-leaving radiance: Analyses of RSP data acquired during the MILAGRO campaign. *Remote Sens. Environ.* **2012**, *118*, 284–308. [CrossRef]
181. Chami, M. Importance of the polarization in the retrieval of oceanic constituents from the remote sensing reflectance. *J. Geophys. Res. Ocean* **2007**, *112*, C5. [CrossRef]
182. Tonizzo, A.; Zhou, J.; Gilerson, A.; Twardowski, M.S.; Gray, D.J.; Arnone, R.A.; Gross, B.M.; Moshary, F.; Ahmed, S.A. Polarized light in coastal waters: Hyperspectral and multiangular analysis. *Opt. Express* **2009**, *17*, 5666–5683. [CrossRef] [PubMed]
183. Voss, K.J.; Souaidia, N. POLRADS: Polarization radiance distribution measurement system. *Opt. Express* **2010**, *18*, 19672–19680. [CrossRef] [PubMed]
184. Churnside, J.H. Review of profiling oceanographic Lidar. *Opt. Eng.* **2014**, *53*, 051405–051405. [CrossRef]
185. Yang, Y.; Pan, H.; Zheng, D.; Zhao, H.; Zhou, Y.; Liu, D. Characteristics and Formation Conditions of Thin Phytoplankton Layers in the Northern Gulf of Mexico Revealed by Airborne Lidar. *Remote Sens.* **2022**, *14*, 4179. [CrossRef]
186. Woods, S.; Piskozub, J.; Freda, W.; Jonasz, M.; Bogucki, D. Laboratory measurements of light beam depolarization on turbulent convective flow. *Appl. Opt.* **2010**, *49*, 3545–3551. [CrossRef]
187. Bogucki, D.J.; Domaradzki, J.A.; von Allmen, P. Polarimetric Lidar measurements of aquatic turbulence-laboratory experiment. *Opt. Express* **2018**, *26*, 6806–6816. [CrossRef]
188. Churnside, J.H.; Wilson, J.J.; Tatarskii, V.V. Lidar profiles of fish schools. *Appl. Opt.* **1997**, *36*, 6011–6020. [CrossRef] [PubMed]
189. Churnside, J.H.; Marchbanks, R.D.; Donaghay, P.L.; Sullivan, J.M.; Graham, W.M.; Wells, R.D. Hollow aggregations of moon jellyfish (*Aurelia* spp.). *J. Plankton Res.* **2016**, *38*, 122–130. [CrossRef]
190. Collister, B.L.; Zimmerman, R.C.; Sukenik, C.I.; Hill, V.J.; Balch, W.M. Remote sensing of optical characteristics and particle distributions of the upper ocean using shipboard Lidar. *Remote Sens. Environ.* **2018**, *215*, 85–96. [CrossRef]
191. Murphree, D.L.; Taylor, C.D.; Mcclendon, R.W. Mathematical modeling for the detection of fish by an airborne laser. *Aiaa J.* **1974**, *12*, 1686–1692. [CrossRef]
192. Fredriksson, K.; Galle, B.; Nystrom, K. Underwater laser-radar experiments for bathymetry and fish school detection: Report GJPR-162, Göteborg Inst. *Phys. Göteborg.* **1978**, *162*, 1–28.
193. Shamanaev, V. Detection of schools of marine fish using polarization laser sensing. *Atmos. Ocean. Opt.* **2018**, *31*, 358–364. [CrossRef]
194. Wang, X.; Zhao, K.; Zhou, Y.; Zhang, F.; Xu, P.; Liu, Q.; Liu, C.; Liu, D. Characteristics of jellyfish in the Yellow Sea detected by polarized oceanic Lidar *Infrared Laser Eng.* **2021**, *50*, 20211038-1. (In Chinese)
195. Fingas, M.; Brown, C.E. A review of oil spill remote sensing. *Sensors* **2017**, *18*, 91. [CrossRef]
196. Hengstermann, T.; Reuter, R. Lidar fluorosensing of mineral oil spills on the sea surface. *Appl. Opt.* **1990**, *29*, 3218–3227. [CrossRef]
197. Jha, M.N.; Levy, J.; Gao, Y. Advances in remote sensing for oil spill disaster management: State-of-the-art sensors technology for oil spill surveillance. *Sensors* **2008**, *8*, 236–255. [CrossRef]
198. Wang, Z.; Zhang, Y.; Liu, D.; Wang, X.; Yuan, J.; Pan, C.; Zhao, Y.; Han, X.; Zhou, Y.; Liu, Q.; et al. Research on the development of detection satellite technology in the novel multi-beam land and ocean Lidar. *Infrared Laser Eng.* **2021**, *50*, 20211041. (In Chinese)
199. Zhao, C.; Li, X.; Ma, Y. Multi-channel ocean fluorescence Lidar system for oil spill monitoring. *Infrared Laser Eng.* **2011**, *40*, 1263–1269. (In Chinese)
200. Jamet, C.; Ibrahim, A.; Ahmad, Z.; Angelini, F.; Babin, M.; Behrenfeld, M.J.; Boss, E.; Cairns, B.; Churnside, J.; Chowdhary, J.; et al. Going beyond standard ocean color observations: Lidar and polarimetry. *Front. Mar. Sci.* **2019**, *6*, 251. [CrossRef]
201. Zhang, W.; Li, X.; Xu, S.; Li, X.; Yang, Y.; Xu, D.; Liu, T.; Hu, H. Underwater Image Restoration via Adaptive Color Correction and Contrast Enhancement Fusion. *Remote Sens.* **2023**, *15*, 4699. [CrossRef]
202. Hair, J.; Hostetler, C.; Hu, Y.; Behrenfeld, M.; Butler, C.; Harper, D.; Hare, R.; Berkoff, T.; Cook, A.; Collins, J.; et al. Combined atmospheric and ocean profiling from an airborne high spectral resolution Lidar. In *EPJ Web of Conferences*; EDP Sciences: Les Ulis, France, 2016; Volume 119, p. 22001.
203. Zhao, H.; Shi, H.; Xu, P.; Qian, Z.; Xu, X.; Li, Z.; Zhai, J.; Li, X.; Xue, B. Direct Measurement of Underwater Sound Velocity via Dual-Comb System and Matched Filtering Algorithm. *IEEE Trans. Instrum. Meas.* **2023**, *72*, 3293552.
204. Li, X.; Li, H.; Lin, Y.; Guo, J.; Yang, J.; Yue, H.; Li, K.; Li, C.; Cheng, Z.; Hu, H.; et al. Learning-based denoising for polarimetric images. *Opt. Express* **2020**, *28*, 16309–16321. [CrossRef] [PubMed]
205. Hu, H.; Huang, Y.; Li, X.; Jiang, L.; Che, L.; Liu, T.; Zhai, J. UCRNet: Underwater color image restoration via a polarization-guided convolutional neural network. *Front. Mar. Sci.* **2022**, *9*, 1031549. [CrossRef]
206. Qi, P.; Li, X.; Han, Y.; Zhang, L.; Xu, J.; Cheng, Z.; Liu, T.; Zhai, J.; Hu, H. U2R-pGAN: Unpaired underwater-image recovery with polarimetric generative adversarial network. *Opt. Lasers Eng.* **2022**, *157*, 107112. [CrossRef]
207. Liu, X.; Li, X.; Chen, S.C. Enhanced polarization demosaicking network via a precise angle of polarization loss calculation method. *Opt. Lett.* **2022**, *47*, 1065–1068. [CrossRef]

208. Li, Y.; Ma, L.; Zhong, Z.; Liu, F.; Chapman, M.A.; Cao, D.; Li, J. Deep learning for Lidar point clouds in autonomous driving: A review. *IEEE Trans. Neural Netw. Learn. Syst.* **2020**, *32*, 3412–3432. [CrossRef]
209. Hu, H.; Yang, S.; Li, X.; Cheng, Z.; Liu, T.; Zhai, J. Polarized image super-resolution via a deep convolutional neural network. *Opt. Express* **2023**, *31*, 8535–8547. [CrossRef] [PubMed]
210. Huang, Z.; Dong, Q.; Chen, B.; Wang, T.; Bi, J.; Zhou, T.; Alam, K.; Shi, J.; Zhang, S. Method for retrieving range-resolved aerosol microphysical properties from polarization Lidar measurements. *Opt. Express* **2023**, *31*, 7599–7616. [CrossRef]
211. Di Noia, A.; Hasekamp, O.; Van Harten, G.; Rietjens, J.; Smit, J.; Snik, F.; Henzing, J.; De Boer, J.; Keller, C.; Volten, H. Use of neural networks in ground-based aerosol retrievals from multi-angle spectropolarimetric observations. *Atmos. Meas. Tech.* **2015**, *8*, 281–299. [CrossRef]
212. Di Noia, A.; Hasekamp, O.P.; Wu, L.; van Diedenhoven, B.; Cairns, B.; Yorks, J.E. Combined neural network/Phillips—Tikhonov approach to aerosol retrievals over land from the NASA Research Scanning Polarimeter. *Atmos. Meas. Tech.* **2017**, *10*, 4235–4252. [CrossRef]
213. Eyeonic Vision System The Industry's Most Compact, Powerful Coherent Machine Vision Solution. Available online: https://www.silc.com/product/ (accessed on 3 September 2023).
214. Li, X.; Hu, H.; Zhao, L.; Wang, H.; Han, Q.; Cheng, Z.; Liu, T. Pseudo-polarimetric method for dense haze removal. *IEEE Photonics J.* **2019**, *11*, 1–11. [CrossRef]
215. Hu, H.; Qi, P.; Li, X.; Cheng, Z.; Liu, T. Underwater imaging enhancement based on a polarization filter and histogram attenuation prior. *J. Phys. Appl. Phys.* **2021**, *54*, 175102. [CrossRef]
216. Hu, H.; Lin, Y.; Li, X.; Qi, P.; Liu, T. IPLNet: A neural network for intensity-polarization imaging in low light. *Opt. Lett.* **2020**, *45*, 6162–6165. [CrossRef] [PubMed]
217. Li, X.; Xu, J.; Zhang, L.; Hu, H.; Chen, S.C. Underwater image restoration via Stokes decomposition. *Opt. Lett.* **2022**, *47*, 2854–2857. [CrossRef]
218. Xin, Z.; Changrui, Q.; Jianfeng, S.; Jiang, P.; Qi, W. Research on triggering properties enhancement of polarization detection geiger-mode APD LIDAR. *J. Quant. Spectrosc. Radiat. Transf.* **2020**, *254*, 107182.
219. Lio, G.E.; Ferraro, A. LIDAR and beam steering tailored by neuromorphic metasurfaces dipped in a tunable surrounding medium. *Photonics* **2021**, *8*, 65. [CrossRef]
220. Kim, I.; Martins, R.J.; Jang, J.; Badloe, T.; Khadir, S.; Jung, H.Y.; Kim, H.; Kim, J.; Genevet, P.; Rho, J. Nanophotonics for light detection and ranging technology. *Nat. Nanotechnol.* **2021**, *16*, 508–524. [CrossRef]
221. Kim, G.; Kim, Y.; Yun, J.; Moon, S.W.; Kim, S.; Kim, J.; Park, J.; Badloe, T.; Kim, I.; Rho, J. Metasurface-driven full-space structured light for three-dimensional imaging. *Nat. Commun.* **2022**, *13*, 5920. [CrossRef]
222. Juliano Martins, R.; Marinov, E.; Youssef, M.A.B.; Kyrou, C.; Joubert, M.; Colmagro, C.; Gâté, V.; Turbil, C.; Coulon, P.M.; Turover, D.; et al. Metasurface-enhanced light detection and ranging technology. *Nat. Commun.* **2022**, *13*, 5724. [CrossRef]
223. Liang, Y.; Lin, H.; Koshelev, K.; Zhang, F.; Yang, Y.; Wu, J.; Kivshar, Y.; Jia, B. Full-stokes polarization perfect absorption with diatomic metasurfaces. *Nano Lett.* **2021**, *21*, 1090–1095. [CrossRef]
224. Rubin, N.A.; D'Aversa, G.; Chevalier, P.; Shi, Z.; Chen, W.T.; Capasso, F. Matrix Fourier optics enables a compact full-Stokes polarization camera. *Science* **2019**, *365*, eaax1839. [CrossRef]

photonics

Article
Accurate Passive 3D Polarization Face Reconstruction under Complex Conditions Assisted with Deep Learning

Pingli Han [1,2], Xuan Li [1,3], Fei Liu [1,2], Yudong Cai [1], Kui Yang [1], Mingyu Yan [1], Shaojie Sun [1], Yanyan Liu [4] and Xiaopeng Shao [1,2,*]

1 School of Optoelectronic Engineering, Xidian University, Xi'an 710071, China
2 Xi'an Key Laboratory of Computational Imaging, Xi'an 710071, China
3 Hangzhou Institute of Technology, Xidian University, Hangzhou 311231, China
4 Science and Technology on Electro-Optical Information Security Control Laboratory, Tianjin 300308, China
* Correspondence: xpshao@xidian.edu.cn

Abstract: Accurate passive 3D face reconstruction is of great importance with various potential applications. Three-dimensional polarization face reconstruction is a promising approach, but one bothered by serious deformations caused by an ambiguous surface normal. In this study, we propose a learning-based method for passive 3D polarization face reconstruction. It first calculates the surface normal of each microfacet at a pixel level based on the polarization of diffusely reflected light on the face, where no auxiliary equipment, including artificial illumination, is required. Then, the CNN-based 3DMM (convolutional neural network; 3D morphable model) generates a rough depth map of the face with the directly captured polarization image. The map works as an extra constraint to correct the ambiguous surface normal obtained from polarization. An accurate surface normal finally allows for an accurate 3D face reconstruction. Experiments in both indoor and outdoor conditions demonstrate that accurate 3D faces can be well-reconstructed. Moreover, with no auxiliary equipment required, the method ensures a total passive 3D face reconstruction.

Keywords: three-dimensional face reconstruction; polarization; ambiguous normal correction; CNN; rough depth map

Citation: Han, P.; Li, X.; Liu, F.; Cai, Y.; Yang, K.; Yan, M.; Sun, S.; Liu, Y.; Shao, X. Accurate Passive 3D Polarization Face Reconstruction under Complex Conditions Assisted with Deep Learning. *Photonics* **2022**, *9*, 924. https://doi.org/10.3390/photonics9120924

Received: 31 August 2022
Accepted: 21 November 2022
Published: 30 November 2022

1. Introduction

With the rapid development of visual technology, 3D face reconstruction has extended many applications, including security authentication [1], face media manipulation [2,3] and face animation [4,5]. Generally speaking, 3D face reconstruction can be categorized into three groups: software-based, hardware-based and image-based [6]. Software-based methods usually employ 3D processing software to establish a face model. Examples include 3DS MAX, Maya, Unity 3D, etc. Hardware-based methods usually adopt a specialized projector or laser to obtain the face shape in a noncontact way. For example, LiDAR (light detection and ranging) derives 3D shapes by detecting the time difference of reflected lasers. It has the advantages of high precision and a long detection distance, but costs a lot and has a hidden danger to the eyes. Three-dimensional imaging with structured light can also enable high-precision shape recovery. However, multicolored object reconstruction is a big challenge due to dependence on striped light. Image-based methods use one or more pictures to build a 3D face shape based on computer vision models. Examples include methods based on Lambert, Phong, Cook–Torrance, etc. [7]. Image-based methods using only one or more pictures have the advantages of a low cost and simple operation, but also cannot guarantee high accuracy due to limited input data. Compared to active 3D reconstruction methods, passive shape recovery has more potential applications. For example, binocular stereo vision mimics how our eyes perceive 3D information using two cameras. However, the reconstruction accuracy highly depends on the base line, leading to an accuracy decline along with increasing detection distance.

Polarization provides another possibility for high-precision passive 3D face reconstruction. In 1979, K. Koshikawa initiated a 3D shape recovery utilizing polarization. The quantitative characterization relationship between the polarization characteristic of reflected light and the zenith of an object's surface normal was established via the illumination of active lighting sources [8]. Since then, 3D polarization imaging has been widely studied. In 1988, L. B. Wolff established a model that used the reflected light of an object's surface to calculate 3D information for the first time [9]. This provided an essential theoretical foundation for 3D polarization imaging technology. In 1999, O. Drbohlav et al. used the polarization characteristic of diffuse reflected light to reconstruct objects' 3D shapes, together with the SFS (shape from shading) method [10]. In 2006, G. A. Atkinson et al. concluded that using the polarization characteristic of diffuse reflected light could solely determining the zenith [11]. Furthermore, K. P. Gurton accomplished 3D polarization facial reconstruction in the long-wave infrared band using the polarization characteristic of diffuse reflected light [12]. Subsequently, A. Kadambi et al. proved the feasibility of passive 3D polarization imaging technology based on experimental data. The 3D polarization imaging system of natural illumination was verified in accordance with the rough depth map obtained using ToF [13]. Y. Ba et al. utilized deep learning to directly estimate the surface normal and further recover 3D shapes [14]. P. Gotardo et al. employed a subset of cross-polarized and parallel-polarized cameras for the illumination in combination with photogrammetry systems. Very favorable results were obtained [15,16]. Although much progress has been accomplished in 3D polarization imaging, the imaging method resorting to auxiliary equipment failed to substantially achieve a major breakthrough on accurate passive 3D face reconstruction. Major limitations exist, especially for complex ambient light under natural illumination.

In this study, an accurate passive 3D polarization face reconstruction method is proposed, assisted with deep learning. By first capturing polarization images, the normal of each microfacet on a pixel level can then be solved based upon the polarization characteristic of diffusely reflected light from the target surface. Instead of directly reconstructing a 3D face, a CNN-based 3DMM (3D morphable model) is adopted here in response to the ambiguous normal. It provides conditional constraints to determine an accurate unique normal and avoid further distortions. Then, an accurate 3D face can be reconstructed. It attained a breakthrough in passively reconstructing 3D faces, even at a long detection distance under natural conditions. Detailed models are introduced in Section 2 and experiments and results are shown in Section 3.

2. Models

Figure 1 illustrates the overall schematic of the proposed passive 3D polarization face reconstruction method. By introducing a CNN-based 3DMM into the imaging process, the azimuth ambiguity, the major problem of 3D polarization imaging, can be amended without auxiliary equipment. We detail the models as follows:

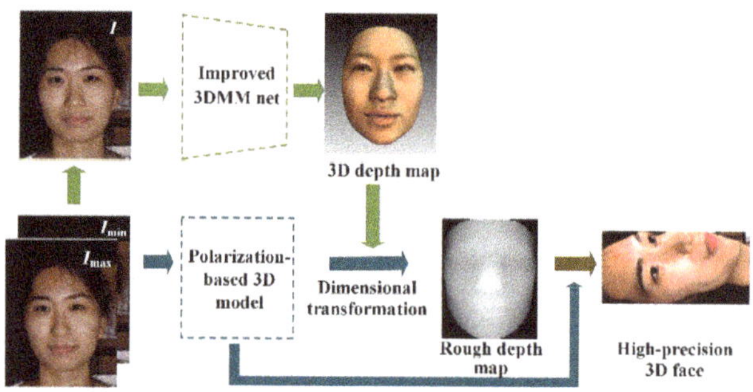

Figure 1. Overall schematic of the 3D face reconstruction method.

2.1. Normal Estimation from Polarization

The normal vector n of a microfacet on an object's surface is constrained by two parameters: the zenith θ and the azimuth φ [17]. Fresnel's formula provides a relation between the normal vector and the polarization information of the reflected light as shown through Equation (1)

$$p = \frac{\left(n - \frac{1}{n}\right)^2 \sin^2\theta}{2 + 2n^2 - \left(n + \frac{1}{n}\right)^2 \sin^2\theta + 4\cos\theta\sqrt{n^2 - \sin^2\theta}} \tag{1}$$

where p is the degree of polarization (DoP), θ is the zenith of the normal of a microfacet and n is the refractive index of the target, typically, $n = 1.5$ [18]. Then, the zenith θ can be uniquely determined using Equation (1) [19].

Polarization imaging detection usually requires four polarization subplots with a linear polarizer mounted in front the detector [20]. According to Malus' law, the detected intensity varies as expressed through Equation (2) [21]

$$I = \frac{I_{\max} + I_{\min}}{2} + \frac{I_{\max} - I_{\min}}{2}\cos\left(2\theta_{pol} - 2\phi\right) \tag{2}$$

where $I_{\max}$ and $I_{\min}$ are the observed maximum and minimum pixel brightness during polarizer rotation and ϕ is the phase of the detected light. For the detected diffuse reflection of every microfacet, the azimuth of the surface normal satisfies $\varphi = \phi$ at $I_{\max}$. According to (2), when $\theta_{pol} = \phi$, $I = I_{\max}$. We could determine azimuth φ when $I_{\max}$ was generated. However, as Figure 2a shows, $I_{\max}$ would appear at ϕ or $\phi + \pi$, but only one of them would be correct, as Figure 2b illustrates. This is the problem of an ambiguous azimuth, which makes the surface normal uncertain and further deforms the reconstructed 3D face. Figure 3 illustrates the deformation caused by the uncertain normal in face reconstruction from polarization. In the following discussion, we attempted to fix this issue by introducing a CNN into the process and allowing for an accurate 3D face reconstruction.

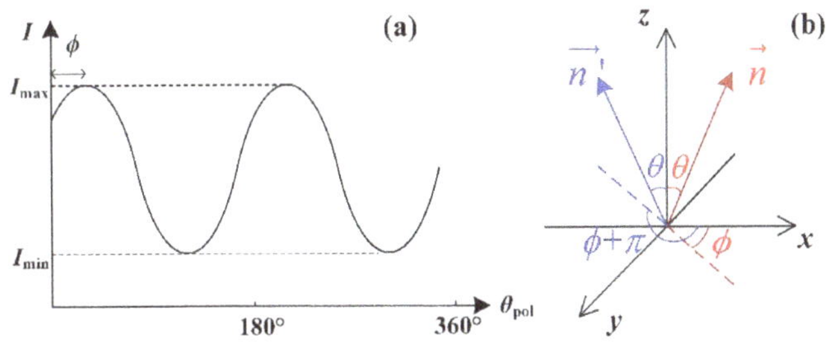

Figure 2. (**a**) Sinusoidal nature of reflected partially polarized light through a linear polarizer as a function of polarizer orientation; (**b**) illustration of ambiguous normal.

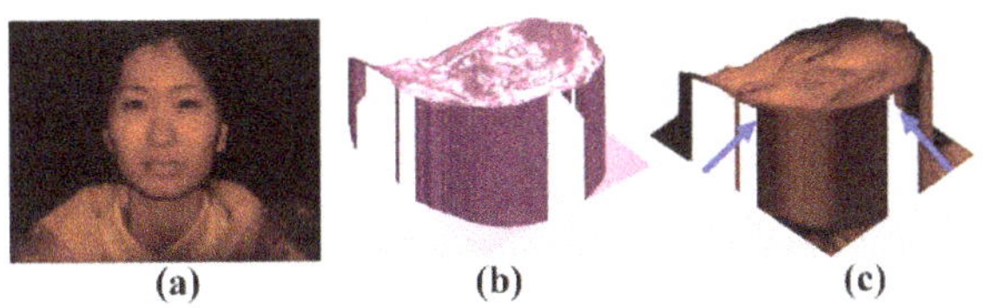

Figure 3. Example of deformation caused by ambiguous azimuth, (**a**) Directly captured human face, (**b**) Depth map and (**c**) recovered 3D face deformation caused by ambiguous azimuth.

2.2. Normal Constraint from CNN-Based 3DMM

The major thought for correcting an ambiguous normal is to determine an extra constraint, and, here, we used a CNN-based 3DMM to calculate a rough depth map. The 3DMM is a typical method used to recover a 3D face, including the CNN-based 3DMM and optimization-based 3DMM. In comparison to the optimization-based 3DMM, a CNN-based method enables the avoidance of an ill-posed problem, and is more robust in reconstruction accuracy. It incorporates the feature information extraction of the input face using the convolutional nerve and matching the input facial feature information using the 3DMM method. Therefore, it meets the requirements better than the optimization-based 3DMM in the study. It can avoid potential errors caused by facial expressions and the lack of 3D facial sample data [2]. By introducing factors, including intensity, expression, texture details and illumination into the training network, the rough depth map could be obtained through only one single image.

A training process, as shown in Figure 4, was set up. A 2D image I with no label was input. Then, facial feature points were located and key facial areas were segmented according to facial information in the image. With the feature points inputting into the CNN-based ResNet-50 network model, an initial set of parameters was generated and, further, an initial depth map was obtained. By adjusting the weight coefficients to a minimum loss, the final depth map expressed through Equation (3) was achieved [22]

$$\begin{aligned} \mathbf{S} &= \mathbf{S}(\alpha, \beta) = \mathbf{S}^{avg} + \mathbf{B}_{id}\alpha + \mathbf{B}_{\exp}\beta \\ \mathbf{T} &= \mathbf{T}(\delta) = \mathbf{T}^{avg} + \mathbf{B}_t\delta \end{aligned} \tag{3}$$

where $\mathbf{S}^{avg}$ and $\mathbf{T}^{avg}$ represent the average facial form and the average facial texture information, respectively; $\mathbf{B}_{id}$, $\mathbf{B}_{\exp}$ and $\mathbf{B}_t$ are the principal component analysis (PCA) bases of identity, expression and texture, respectively; and α, β and δ ($\alpha \in R^{80}$, $\beta \in R^{64}$ and $\delta \in R^{80}$) are parameters for estimating the 3D target in-depth map [23]. Furthermore, the influences of illumination and facial expression on the result estimation should also be considered and expressed with illumination parameter $\gamma \in R^9$ and the facial expression parameter $\mathbf{p}$.

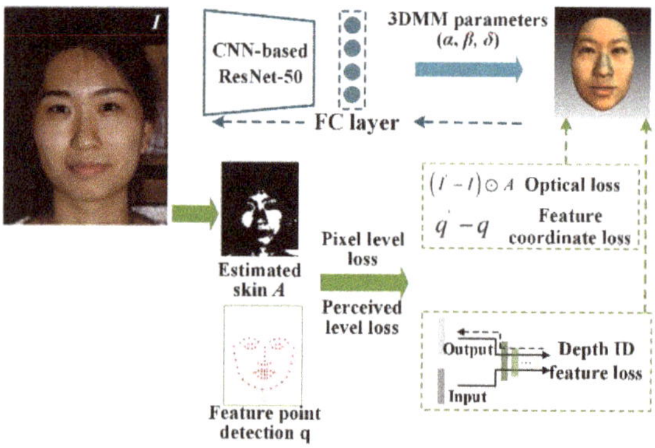

Figure 4. Flow chart of network training.

The unknown parameters could be expressed as a vector $\mathbf{x} = (\alpha, \beta, \delta, \gamma, p) \in \mathrm{R}^{239}$ [24]. With the given training image I, the parameter $\mathbf{x}$ was estimated without labels using ResNet-50 to obtain the initial 3D facial estimation I'. Next, the mixed-level loss (including the pixel level and perception level) in the initial estimation result I' was evaluated and propagated reversely. Therefore, the iterative corrections of network model parameters could be realized. Furthermore, a multilayer cascaded loss function was constructed to provide reference to the parameter correction against the application scenarios of the face target. This could be expressed as [2]:

$$L(\mathbf{x}) = \omega_{pto} L_{pto}(\mathbf{x}) + \omega_{fea} L_{fea}(\mathbf{x}) + \omega_{per} L_{per}(\mathbf{x}) + \omega_{coef} L_{coef}(\mathbf{x}) + \omega_{alb} L_{alb}(\mathbf{x}) \quad (4)$$

where $L_{pto}(\mathbf{x})$, $L_{fea}(\mathbf{x})$, $L_{per}(\mathbf{x})$, $L_{coef}(\mathbf{x})$ and $L_{alb}(\mathbf{x})$ indicate the photometric loss, the landmark loss, the perceptual-level loss, the regularization terms and flat constraint penalty function and ω is the corresponding coefficients determined during the training process.

Since changes in appearances, such as glasses, hair, beards or even heavy makeup, are challenging for facial target recognition, the skin color probability P_i was introduced to each pixel in the image in the loss function [22]. Through this, it could prevent the increase in errors caused by the above interferences during the network training [2,3]. Moreover, facial feature classification could be avoided using facial segmentation, thanks to the Bayesian classifier obtained from training. Meanwhile, the reconstruction area concerning the network could focus on the face by overcoming interferences, such as the occlusion of glasses and hair. The photometric loss function of the facial target could be defined as [25,26]

$$L_{pto}(\mathbf{x}) = \frac{\sum\limits_{i \in \mathcal{M}} A_i \cdot \| I_i - I_i'(\mathbf{x}) \|_2}{\sum\limits_{i \in \mathcal{M}} A_i} \quad (5)$$

where i is the position index of the image pixel, M is the reprojection area of the face target and the L2 norms were solved at the corresponding positions for the input image I and the output image I'. In addition, when $P_i > 0.5$, $A_i = 1$; otherwise, $A_i = P_i$ [27]. The loss of pixel intensity information could be estimated in the 2D image.

When processing the loss function of the target feature point, a 3D face alignment should have first been performed. Then, 68 facial feature points $\{\mathbf{q}_n\}$ in the facial training image were detected using the A. Bulat method [28]. Moreover, a 2D projection should have been performed on the output data to obtain 68 facial feature points $\{\mathbf{q}'_n\}$. This was because $\{\mathbf{q}_n\}$ were the 3D data. The loss function between the input and output information could be expressed as [29]:

$$L_{fea}(\mathbf{x}) = \frac{1}{N} \sum_{n=1}^{N} \omega_n \| \mathbf{q}_n - \mathbf{q}_n'(\mathbf{x}) \|^2 \quad (6)$$

where ω_n represents the weight of the target feature point. In general, the weight values of the mouth and nose feature points in the image were set to 20 as a matter of experience, and other feature areas were set to 1 [23]. The introduction of such a loss would raise the smoothness and continuity in various facial areas of the 3DMM.

The local minimum often appeared in the output results of the CNN. This was due to various external factors of changes and disturbances of facial targets, leading to the nonconvergence of the optimization functions [2,3]. In response to this problem, the loss of target perception was added to the model to enhance the network accuracy in facial reconstruction [30]. The cosine distance between the feature points for calculating the loss function shown in Equation (7) was constructed using the extracted depth features of the target image

$$L_{per}(\mathbf{x}) = 1 - \frac{< f(I), f(I'(\mathbf{x})) >}{\| f(I) \| \cdot \| f(I'(\mathbf{x})) \|} \quad (7)$$

where, $f(\cdot)$ stands for the depth feature coding and $<\cdot,\cdot>$ represents the vector inner product. The overall profile was more consistent with the real situation of the target in terms of intuitive perception. This was achieved upon the introduction of target perception loss into the CNN model.

The target depth map estimated with the network consisted of various discrete points with depth information. Thus, a regularization constraint was added to parameters α, β

and δ in Equation (4). This was to prevent the degradation of facial shapes and texture details during the estimation of the depth information of discrete points from the unlabeled dataset. The loss function could be expressed as:

$$L_{coef}(\mathbf{x}) = \omega_\alpha ||\boldsymbol{\alpha}||^2 + \omega_\beta ||\boldsymbol{\beta}||^2 + \omega_\gamma ||\delta||^2 \tag{8}$$

The flat constraint penalty function is usually used to correct color distortions. In this study, real-color images captured by using the polarization image were used. Therefore, L_{alb} in Equation (4) was ignored.

With the help of the four loss functions and the 3DMM method, given a single target image as the input, the rough depth map could be effectively obtained. Then, high-precision 3D target reconstruction could be implemented by correcting the surface normal calculated based on polarization.

The rough depth map could provide a constraint for the ambiguous normal solved using polarization. The constraint was provided by means of effectively showing the accurate gradient variation trend in microfacets on a target's surface. Thus, the polarization normal could be corrected using Equation (9):

$$\hat{\Lambda} = \underset{\Lambda}{\operatorname{argmin}} ||G^{\text{net}} - \Lambda\left(G^{\text{polar}}\right)||_2^2, \ \Lambda = \{0, 1\} \tag{9}$$

where G^{net} is the normal data obtained from the CNN-based 3DMM G^{polar} is the normal data from polarization, $\hat{\Lambda}$ is a set of binary operands and Λ is a binary operator. $\hat{\Lambda} = 1$ was assumed to indicate that the estimated azimuth from the polarization was accurate. Conversely, if $\hat{\Lambda} = 0$, the estimated azimuth from the polarization should have been corrected by reversing 180°. The corrected azimuth could be expressed as

$$\varphi^{\text{cor}} = \varphi^{\text{polar}} + (1 - \Lambda) \cdot \pi \tag{10}$$

where φ^{polar} is the azimuth directly recovered from polarization and φ^{cor} stands for the final accurate azimuth, i.e., the normal of a target's microfacet could be solely solved.

2.3. *Reconstructing 3D Profile*

Theoretically, an accurate 3D face could be recovered based on a corrected (θ, ϕ) using integration. However, in some conditions, there are discrete nonintegrable points that can fail the final reconstruction. Here, we reconstructed a 3D face with the help of the Frankot–Chellappa 3D surface restoration functions [31] in Equation (12)

$$\begin{aligned} p_g &= \tan\theta \cos\varphi \\ q_g &= \tan\theta \sin\varphi \end{aligned} \tag{11}$$

$$z = F^{-1}\left\{ -j\frac{\frac{2\pi u}{N} F\{p_g\} + \frac{2\pi v}{M} F\{q_g\}}{\left(\frac{2\pi u}{N}\right)^2 + \left(\frac{2\pi v}{M}\right)^2} \right\} = F^{-1}\left\{ -\frac{j}{2\pi} \frac{\frac{u}{N} F\{p_g\} + \frac{v}{M} F\{q_g\}}{\left(\frac{u}{N}\right)^2 + \left(\frac{v}{M}\right)^2} \right\} \tag{12}$$

where $F\{\cdot\}$ and $F^{-1}\{\cdot\}$ stand for the discrete Fourier transform and inverse transform, respectively, u and v are the frequency in the Fourier domain and M and N are the dimensions of the input images.

3. Experiments and Results Analysis

Relying solely on the CNN-based 3DMM also allowed for 3D face reconstruction. However, two major problems could be encountered, 2D photo misrecognition and inaccurate recovery for Asian people, since training data were based on occidental people. Figure 5 shows an illustration where four targets were used, including a photo of a Chinese man, a Chinese woman wearing no glasses, a Chinese man wearing near-infrared glasses and a plaster statue.

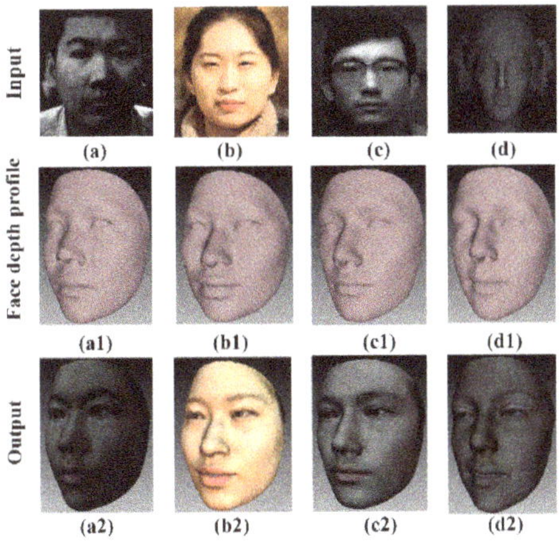

Figure 5. Reconstructed 3D faces using CNN-based 3DMM; (**a**) a photo of a Chinese man; (**b**) a Chinese woman wearing no glasses; (**c**) a Chinese man wearing near-infrared glasses; (**d**) a plaster statue; (**a1–d1**) reconstructed face depth profile of (**a–d**); (**a2–d2**) finally, reconstructed 3D face of (**a–d**).

The depth information recovered with the CNN-based 3DMM is shown in the second row in Figure 5, where interferences in the information, such as hair and glasses, were overcome. However, what should be noted is that the 3D information was also recovered from a male photo resulting from the training set input in the CNN-based 3DMM training process. In addition, in the final recovered results in the third row in Figure 5, a significant difference could be noticed between the recovered 3D face and the real face. This was affected by the training set of the CNN-based 3DMM, which was based on occidentals, causing the output result approaching the "average face" as much as possible.

Though the CNN-based 3DMM method was disadvantageous in the two referred aspects, it could assist 3D polarization imaging to recover accurate 3D faces as discussed before. The real surface normal could be directly obtained through 3D polarization imaging, but a constraint from the CNN-based 3DMM method would be required to handle the ambiguous azimuth. Through this, the 3D face could be reconstructed in an accurate manner.

In Figure 6, we illustrated in detail why the normal correction was required and how to correct it with the assistance of the CNN-based 3DMM. Figure 6a shows four intensity images at 0°, 45°, 90° and 135°, directly obtained with the detector. Figure 6b1,b2 are the azimuth φ and the zenith θ from the polarization. Figure 6c1,c2 are φ and θ from the CNN-based 3DMM. In general, the data from the CNN-based 3DMM presented relatively mild variations, indicating possible lost details. We chose two rows of pixels from each angle image in Figure 6b1,b2,c1,c2, and drew up the intensity distribution over the pixel position. Figure 6d1,d2 are azimuth data from the results of the polarization and CNN-based 3DMM, respectively. The curve in Figure 6d1 showed that both statistical curves had no regular changes at zero, which was consistent with the analysis on the azimuth uncertainty problem in 3D polarization imaging. However, the variations indicated that abundant facial details were detected. According to the graph in Figure 6d2, both the blue and red curves were at the pixel value of approximately 90 in the middle of the image, while the azimuth changed from a positive value to a negative value. The curve variation showed that a similar trend on both sides could be witnessed, which was consistent with the variation trend of the real target. The data variation in Figure 6d1,d2 indicate that it was feasible to employ the CNN-based 3DMM to the azimuth from polarization by providing an overall reference and retaining detail information at the same time.

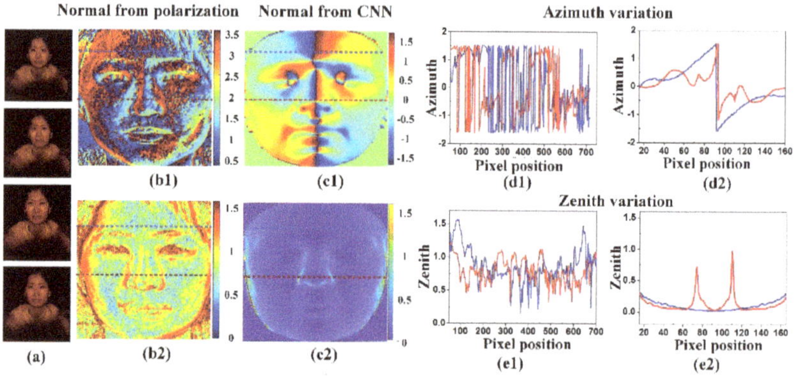

Figure 6. Surface normal data of the Chinese woman wearing no glasses; (**a**) four polarization intensity images; (**b1**,**b2**) azimuth and zenith derived from polarization; (**c1**,**c2**) azimuth and zenith derived from CNN-based 3DMM; (**d1**,**d2**) azimuth from polarization and CNN-based 3DMM variation along the red and blue lines; (**e1**,**e2**) zenith from polarization and CNN-based 3DMM variation along the red and blue lines.

Figure 6e1,e2 display the zenith data from the results of the polarization and CNN-based 3DMM, respectively. According to (1), the zenith could be solely determined based on polarization. Data in Figure 6e1 from the polarization also present the general shape, and data in Figure 6e2 from the CNN-based 3DMM present relatively smooth variations. Moreover, in Figure 6b2,e1, high-frequency details of the face were remarkably perceived. To be specific, the zenith changed rapidly in the areas with abundant high-frequency detail information, such as the nose, eyes, mouth and hair; i.e., high-frequency detail information was obtained in a more abundant way using the polarization information.

Based on the normal data from the CNN-based 3DMM and polarization, the 3D face reconstruction was implemented according to Equation (12). The targets in Figure 3a–c were the directly reconstructed results from polarization, where 3D information could not be estimated and recovered. For instance, under the influence of the ambiguous azimuth, the relative height information that was higher than the nose area was presented in the facial edge, and the sunken part of the facial area did not correspond with the facial variation trend in reality. In Figure 7a,b, a relative complete 3D face profile consistent with face variations was obtained using the CNN-based 3DMM. Assisted with the data from the CNN-based 3DMM, the final corrected and reconstructed 3D face is shown in Figure 7d,e. The profile information was consistent with the variation trend of the real face. Specifically, the information that was not acquired with the detector at the lower part of the nose was also estimated. Thus, contributing to not only the effective solving of the reconstructed distortion caused due to the uncertainty of the azimuth, but also to the 3D face reconstruction.

Furthermore, relative height value statistics were performed on the face reconstruction results from the CNN-based 3DMM and corrected polarization in the horizontal and vertical directions, as shown in Figure 7c,f. The 3D face profile could be reconstructed using both ways. The variation trends in the horizontal and vertical directions were also consistent with those observed in the real situation. However, we should notice that the faces in the dataset of the CNN were mainly occidental, significantly different from the oriental, with broad foreheads and high noses. Therefore, it could be noticed from the results obtained with the CNN-based 3DMM estimation that remarkable variations could be found in terms of the nose and forehead, which were greatly different from those of the real targets. This issue could also be avoided with the polarization employed for reconstructing a realistic 3D face.

In order to verify the robustness of the proposed method in diversified application scenarios, a 3D face reconstruction was performed under indoor lighting of a male face, outdoor natural illumination of a male face and indoor lighting of a plaster statue, as shown in Figure 8. In all the experiments, no modulated illumination was required. For example, when imaging an outdoor human face, natural illumination is complex, with uncontrolled direction, intensity and polarization. The features visualized in the 3D renderings fit well with their original appearance, and some fine features could be easily identified. Using passive 3D face reconstruction allowed for the developed method to show promise for various potential applications.

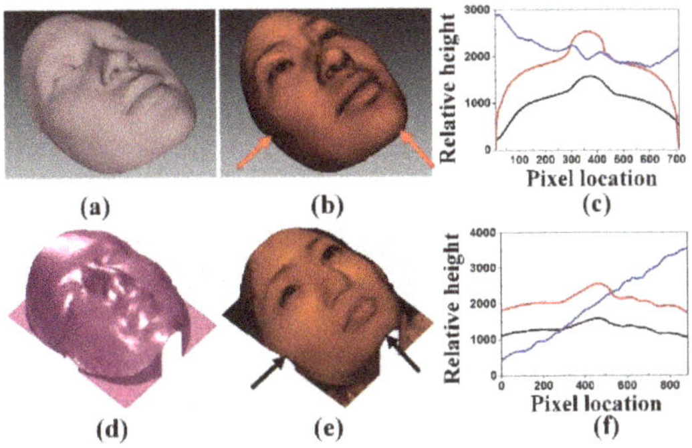

Figure 7. Reconstruction of the Chinese woman wearing no glasses; (**a**) depth profile and (**b**) 3D face recovered from CNN-based 3DMM; (**d**) depth profile and (**e**) 3D face recovered from the proposed method; (**c**) face profile comparison in the horizontal and (**f**) vertical directions from Figure 3c and (**b**,**e**). Blue, red and black lines indicate profiles from Figure 3c, (**b**) and (**e**), respectively.

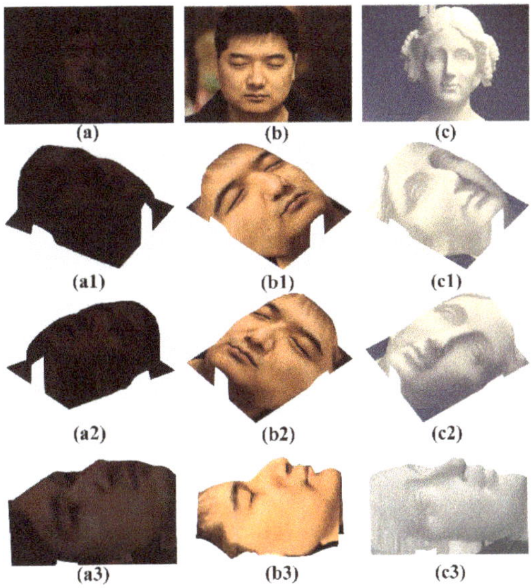

Figure 8. Three-dimensional face reconstruction using the developed method. (**a**) Indoor lighting of a male face, (**b**) outdoor natural illumination of a male face, (**c**) indoor lighting of a plaster statue, (**a1–a3**) Three different views of recovered 3D face of (**a**), (**b1–b3**) Three different views of recovered 3D face of (**b**), (**c1–c3**) Three different views of recovered 3D face of (**c**).

To quantify the reconstruction accuracy of the proposed method, we employed laser scanning technology to conduct the comparison. The 3D data of the human face and plaster statue obtained with an AltairScan Elite scanner, which can achieve an accuracy of 0.01mm, were taken as the ground truth. The scenes and scanned results are shown in Figure 9a–c. The 3D reconstruction results obtained using the proposed method were matched with the scanned results, as shown in Figure 9d,e. The 3D polarization datapoint sets nearly all coincided with the 3D scanned results, showing a high precision in reconstruction. The RMS of Figure 9d,e was approximately 2.707 mm and 2.074 mm, respectively.

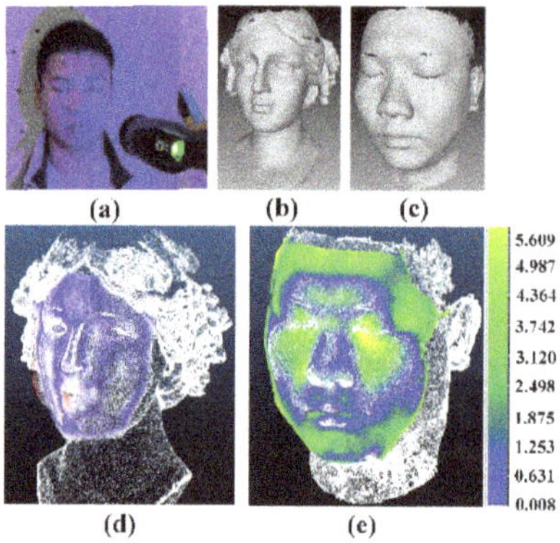

Figure 9. Three-dimensional reconstruction accuracy quantification of the proposed method using laser scanning. (**a**) Laser scanner and measuring illustration, (**b**,**c**) Recovered 3D face by laser scanner; (**d**,**e**) point cloud comparison between laser scanner and the developed method.

4. Conclusions

In this study, we showed that an accurate passive 3D face can be reconstructed from polarization assisted with the CNN-based 3DMM. We acquired a rough depth map from the CNN-based 3DMM method. In addition, it was used as an extra constraint to correct the ambiguous surface normal from polarization and to ensure an accurate 3D face reconstruction. Experiments and results showed that an accurate 3D face could be passively reconstructed with the proposed method. Moreover, the problem of the "average occidental face" generated with the CNN-based 3DMM was also avoided by taking the map as a constraint to the surface normal obtained with polarization. The passive 3D face reconstruction method could potential be used in various applications, including security authentication, face media manipulation, etc.

Author Contributions: Conceptualization, P.H. and X.L.; Data curation, Y.C.; Formal analysis, P.H.; Funding acquisition, Y.L. and X.S.; Investigation, P.H.; Methodology, F.L.; Project administration, X.S.; Software, Y.C.; Validation, K.Y., M.Y. and S.S.; Writing—review and editing, P.H. and X.L. All authors have read and agreed to the published version of the manuscript.

Funding: This research was funded by the National Natural Science Foundation of China (62075175, 62005203); the Opening Funding of Key Laboratory of Space Precision Measurement Technology, the Xi'an Institute of Optics and Precision Mechanics, the Chinese Academy of Sciences (90109220005); and the Opening Funding of Science and Technology on Electro-Optical Information Security Control Laboratory (61421070203).

Informed Consent Statement: Informed consent was obtained from all subjects involved in the study. The human faces displayed in this manuscript are all of the authors of the manuscript, including Fei Liu, Kui Yang, Shaojie Sun and Mingyu Yan. The authors confirmed that informed consent was obtained.

Conflicts of Interest: The authors declare no conflict of interest.

References

1. Uzair, M.; Mahmood, A.; Shafait, F.; Nansen, C.; Mian, A. Is spectral reflectance of the face a reliable biometric? *Opt. Experss* **2015**, *23*, 15160–15173. [CrossRef] [PubMed]
2. Blanz, V.; Vetter, T. A morphable model for the synthesis of 3D faces. In Proceedings of the 26th Annual conference on Computer Graphics and Interactive Techniques, Los Angeles, CA, USA, 8–13 August 1999; pp. 187–194.
3. Thies, J.; Zollhofer, M.; Stamminger, M.; Theobalt, C.; Nießner, M. Face2face: Real-time face capture and reenactment of rgb videos. In Proceedings of the IEEE Conference on computer Vision and Pattern Recognition, Las Vegas, NV, USA, 27–30 June 2016; pp. 2387–2395.
4. Cao, C.; Weng, Y.; Lin, S.; Zhou, K. 3D shape regression for real-time facial animation. *ACM Trans. Graph. (TOG)* **2013**, *32*, 1–10. [CrossRef]
5. Hu, L.; Saito, S.; Wei, L.; Nagano, K.; Seo, J.; Fursund, J.; Sadeghi, I.; Sun, C.; Chen, Y.-C.; Li, H. Avatar digitization from a single image for real-time rendering. *ACM Trans. Graph. (TOG)* **2017**, *36*, 1–14. [CrossRef]
6. Xie, W.; Kuang, Z.; Wang, M. SCIFI: 3D face reconstruction via smartphone screen lighting. *Opt. Express* **2021**, *29*, 43938–43952. [CrossRef]
7. Ma, L.; Lyu, Y.; Pei, X.; Hu, Y.M.; Sun, F.M. Scaled SFS method for Lambertian surface 3D measurement under point source lighting. *Opt. Express* **2018**, *26*, 14251–14258. [CrossRef] [PubMed]
8. Koshikawa, K. A polarimetric approach to shape understanding of glossy objects. *Adv. Robot.* **1979**, *2*, 190.
9. Wolff, L.B.; Boult, T.E. Constraining object features using a polarization reflectance model. *Phys. Based Vis. Princ. Pract. Radiom* **1993**, *1*, 167. [CrossRef]
10. Mahmoud, A.H.; El-Melegy, M.T.; Farag, A.A. Direct method for shape recovery from polarization and shading. In Proceedings of the 2012 19th IEEE International Conference on Image Processing, Orlando, FL, USA, 30 September–3 October 2012; pp. 1769–1772.
11. Atkinson, G.A.; Hancock, E.R. Recovery of surface orientation from diffuse polarization. *IEEE Trans. Image Process.* **2006**, *15*, 1653–1664. [CrossRef] [PubMed]
12. Gurton, K.P.; Yuffa, A.J.; Videen, G.W. Enhanced facial recognition for thermal imagery using polarimetric imaging. *Opt. Lett.* **2014**, *39*, 3857–3859. [CrossRef]
13. Kadambi, A.; Taamazyan, V.; Shi, B.; Raskar, R. Polarized 3D: Synthesis of polarization and depth cues for enhanced 3D sensing. In *SIGGRAPH 2015: Studio*; Association for Computing Machinery: New York, NY, USA, 2015.
14. Ba, Y.; Gilbert, A.; Wang, F.; Yang, J.; Chen, R.; Wang, Y.; Yan, L.; Shi, B.; Kadambi, A. Deep shape from polarization. In Proceedings of the European Conference on Computer Vision, Glasgow, UK, 23–28 August 2020; pp. 554–571.
15. Riviere, J.; Gotardo, P.; Bradley, D.; Ghosh, A.; Beeler, T. Single-shot high-quality facial geometry and skin appearance capture. *ACM Trans. Graph.* **2020**, *39*, 81:1–81:12. [CrossRef]
16. Gotardo, P.; Riviere, J.; Bradley, D.; Ghosh, A.; Beeler, T. Practical dynamic facial appearance modeling and acquisition. *ACM Trans. Graph.* **2018**, *37*, 232. [CrossRef]
17. Miyazaki, D.; Shigetomi, T.; Baba, M.; Furukawa, R.; Hiura, S.; Asada, N. Surface normal estimation of black specular objects from multiview polarization images. *Opt. Eng.* **2016**, *56*, 041303. [CrossRef]
18. Kadambi, A.; Taamazyan, V.; Shi, B.; Raskar, R. Depth sensing using geometrically constrained polarization normals. *Int. J. Comput. Vis.* **2017**, *125*, 34–51. [CrossRef]
19. Smith, W.A.; Ramamoorthi, R.; Tozza, S. Linear depth estimation from an uncalibrated, monocular polarisation image. In Proceedings of the European Conference on Computer Vision, Amsterdam, The Netherlands, 11–14 October 2016; pp. 109–125.
20. Schechner, Y.Y.; Narasimhan, S.G.; Nayar, S.K. Polarization-based vision through haze. *Appl. Opt.* **2003**, *42*, 511–525. [CrossRef] [PubMed]
21. Zhu, H.; Li, D.; Song, L.; Wang, X.; Zhan, K.; Wang, N.; Yun, M. Precise analysis of formation and suppression of intensity transmittance fluctuations of glan-taylor prisms. *Laser Optoelectron. Prog.* **2013**, *50*, 052302.
22. Deng, Y.; Yang, J.; Xu, S.; Chen, D.; Jia, Y.; Tong, X. Accurate 3d face reconstruction with weakly-supervised learning: From single image to image set. In Proceedings of the IEEE/CVF Conference on Computer Vision and Pattern Recognition Workshops, Long Beach, CA, USA, 16–17 June 2019.
23. Guo, Y.; Cai, J.; Jiang, B.; Zheng, J. CNN-based real-time dense face reconstruction with inverse-rendered photo-realistic face images. *IEEE Trans. Pattern Anal. Mach. Intell.* **2018**, *41*, 1294–1307. [CrossRef] [PubMed]
24. He, K.; Zhang, X.; Ren, S.; Sun, J. Deep residual learning for image recognition. In Proceedings of the IEEE Conference on Computer Vision and Pattern Recognition, Las Vegas, NV, USA, 27–30 June 2016; pp. 770–778.
25. Tewari, A.; Zollhöfer, M.; Garrido, P.; Bernard, F.; Kim, H.; Pérez, P.; Theobalt, C. Self-supervised multi-level face model learning for monocular reconstruction at over 250 hz. In Proceedings of the IEEE Conference on Computer Vision and Pattern Recognition, Salt Lake City, UT, USA, 18–23 June 2018; pp. 2549–2559.
26. Tewari, A.; Zollhofer, M.; Kim, H.; Garrido, P.; Bernard, F.; Perez, P.; Theobalt, C. Mofa: Model-based deep convolutional face autoencoder for unsupervised monocular reconstruction. In Proceedings of the IEEE International Conference on Computer Vision Workshops, Venice, Italy, 22–29 October 2017; pp. 1274–1283.
27. Jones, M.J.; Rehg, J.M. Statistical color models with application to skin detection. *Int. J. Comput. Vis.* **2002**, *46*, 81–96. [CrossRef]

28. Jackson, A.S.; Bulat, A.; Argyriou, V.; Tzimiropoulos, G. Large pose 3D face reconstruction from a single image via direct volumetric CNN regression. In Proceedings of the IEEE International Conference on Computer Vision, Venice, Italy, 22–29 October 2017; pp. 1031–1039.
29. Bulat, A.; Tzimiropoulos, G. How far are we from solving the 2D & 3D face alignment problem? (And a dataset of 230,000 3D facial landmarks). In Proceedings of the IEEE International Conference on Computer Vision, Venice, Italy, 22–29 October 2017; pp. 1021–1030.
30. Genova, K.; Cole, F.; Maschinot, A.; Sarna, A.; Vlasic, D.; Freeman, W.T. Unsupervised training for 3d morphable model regression. In Proceedings of the IEEE Conference on Computer Vision and Pattern Recognition, Salt Lake City, UT, USA, 18–23 June 2018; pp. 8377–8386.
31. Frankot, R.T.; Chellappa, R. A method for enforcing integrability in shape from shading algorithms. *IEEE Trans. Pattern Anal. Mach. Intell.* **1988**, *10*, 439–451. [CrossRef]

photonics

Review

Working Mechanism and Progress of Electromagnetic Metamaterial Perfect Absorber

Xiajun Liu [1], Feng Xia [1], Mei Wang [1], Jian Liang [2] and Maojin Yun [1,*]

[1] College of Physics, Center for Marine Observation and Communications, Qingdao University, Qingdao 266071, China
[2] School of Physics and Information Technology, Shaanxi Normal University, Xi'an 710119, China
* Correspondence: mjyun@qdu.edu.cn

Abstract: Electromagnetic metamaterials are artificial subwavelength composites with periodic structures, which can interact strongly with the incident light to achieve effective control of the light field. Metamaterial absorbers can achieve nearly 100% perfect absorption of incident light at a specific frequency, so they are widely used in sensors, optical switches, communication, and other fields. Based on the development history of metamaterials, this paper discusses the research background and significance of metamaterial perfect absorbers. Some perfect absorption mechanisms, such as impedance matching and coherent perfect absorption, are discussed. According to the functional division, the narrowband, dual frequency, multi-frequency, broadband, and tunable metamaterial perfect absorbers are briefly described.

Keywords: electromagnetic metamaterials; perfect absorption; terahertz; impedance matching; coherent absorption; narrow band absorbers; dual-frequency absorbers; multi-frequency absorbers; broadband absorbers; tunable absorbers

Citation: Liu, X.; Xia, F.; Wang, M.; Liang, J.; Yun, M. Working Mechanism and Progress of Electromagnetic Metamaterial Perfect Absorber. *Photonics* **2023**, *10*, 205. https://doi.org/10.3390/photonics10020205

Received: 2 January 2023
Revised: 9 February 2023
Accepted: 9 February 2023
Published: 14 February 2023

1. Introduction

1.1. Development History of Metamaterials

Electromagnetic metamaterial, an artificial material with sub-wavelength size and periodic array structure that has optical properties that traditional natural optical materials do not have in nature, has been used to realize more extensive and rich optical functions [1,2]. Such optical properties depend not only on chemical composition but also on the geometric parameters of the structure [3]. The development of metamaterials can be traced back to the 1960s, when Veselago first proposed a material with a negative dielectric constant and negative magnetic permeability [4]; it is also called left-handed material because it conforms to the left-handed rule [5,6]. In 1996, Pendry et al.designed the periodically arranged thin metal wire and split-ring resonator (SRR) structure [1] and proved that this structure can realize negative dielectric constant and negative permeability, respectively. The concept of metamaterial emerged at this point. In the last twenty years, the development of nanotechnology has provided opportunities for the precise machining and control of materials far smaller than the wavelength, and has further promoted the theoretical research and functional expansion of metamaterials [7,8]. The theory of transformation optics [9], the plasmon-induced transparency (PIT) [10], the generalized reflection and refraction law, including the linear phase change of interface space [11], and the concept of Huygens metasurface [12] have been put forward one after another. At the same time, scientists are skillfully combining metamaterials with various optical phenomena to realize the effective regulation of the interaction between light and matter [13], such as perfect absorption [14], regulation of the polarization state of the light field [15], Fano resonance [16], wavefront modulation [3], electromagnetic wave phase regulation, and amplitude regulation [17], etc.

During the development of metamaterials in the past 60 years, there have been many difficulties and challenges hindering the practical application of metamaterials, such as high loss and strong dispersion related to resonant response, and the use of metal structures, high dependence on wavelength, and difficulties in manufacturing micro–nano three-dimensional structures. The metasurface, a two-dimensional (2D) metamaterial with sub-wavelength thickness, can be prepared by traditional micro–nano processing methods, such as lithography and nanoimprinting [18]. The metasurface has provided answers to the difficulties present in three-dimensional materials. The unit structure of the metasurface is composed of meta-atoms, and through the types and arrangement of meta-atoms, the interaction between light and matter can be efficiently regulated. Compared with traditional optical components, the metasurface can better meet the development requirements of miniaturization, integration, and multifunction. According to Huygens' principle, the hypersurface can subtly reshape the electromagnetic wave, resulting in new physical defects, such as abnormal refraction and reflection of light, surface wave coupling, etc. In addition, the metasurface can break the optical diffraction limit and realize super-resolution. At present, scientists use metasurface exploration to discover more extraordinary electromagnetic characteristics and expand into more application fields, such as filtering, holographic imaging, communication, energy, sensing, and so on.

1.2. Metamaterial Perfect Absorber and Its Research Significance

Perfect absorption (PA) refers to the phenomenon where the absorption is nearly 100% for a certain frequency, or for a certain frequency at a specific incident angle [19]. Metamaterials can supply many unconventional optical properties, and the electromagnetic properties are simple to control. Based on the impedance matching between metamaterials and free space, Landy et al. realized the perfect absorption of metamaterials in microwave band for the first time in 2008 [14]. Since then, the metamaterial perfect absorber (MPA) has gradually expanded to terahertz [20–22], mid-infrared [23,24], infrared [25,26], and visible light [27].

The typical metamaterial absorber is composed of a metal resonant layer, dielectric layer, and base layer from top to bottom [28,29]. Scientists have fused the absorber with the metasurface structure to achieve excellent absorption performance in a sub-wavelength size, and further realized multi-functional control by optimizing the parameters, such as the type and arrangement of metal particles in the structured metal layer. The development of the metasurface provides a unique opportunity for the planarization and miniaturization of the perfect absorber. The ultra-thin perfect absorber based on the metasurface is of great importance for unified application.

The bandwidth of the absorber is very important for many scientific and technological applications. Narrow-band absorbers are widely used in sensing fields, such as temperature sensing, refractive index sensing, absorption filtering, and optical flare processing. Broadband absorbers have pioneering applications in photovoltaic cells, photodetectors, and other applications. In Section 3, according to the function of MPA, narrow-band absorbers, dual-frequency and multi-frequency absorbers, broadband absorbers, and tunable absorbers are systematically introduced.

2. Perfect Absorption Mechanism

2.1. Impedance Matching

Impedance matching is a necessary state for perfect absorption [30]. Broadly, the metamaterial absorber is usually configured with three layers, including the metamaterial layer, the dielectric compartment layer, and the grounding plane. The traditional structure takes the form of metal–dielectric–metal (MDM), that is, the top metal resonance structure, the middle dielectric layer, and the bottom metal reflection layer. The relationship between absorption, transmission, and reflection is as follows [14]:

$$A(\omega) = 1 - T(\omega) - R(\omega), T(\omega) = |t(\omega)|^2, R(\omega) = |r(\omega)|^2 \quad (1)$$

$T(\omega) = R(\omega) = 0$ is required for the absorption to arrive at 100%.

The bottom layer of MDM is metal, and its thickness is much greater than the skin depth of light in this metal, which can block the transfer of light waves, $T(\omega) = 0$.

Suppose the impedance of free space and the equivalent impedance of absorber are Z_0 and Z_1, respectively. Then the expression of reflection coefficient R is:

$$R = \frac{Z_1 - Z_0}{Z_1 + Z_0} \quad (2)$$

The impedance of the free space and the equivalent impedance of the MPA can be expressed as:

$$Z_0 = \sqrt{\frac{\varepsilon_0}{\mu_0}}, Z_1 = \sqrt{\frac{\varepsilon_1}{\mu_1}} \quad (3)$$

The impedance matching condition requires that $R = 0$ when the surface impedance of metamaterial matches the impedance in free space, and $Z_0 = 1$ when the surrounding environment is air. It can be seen from the above formula that $Z_1 = 1$ can match the surface impedance of metamaterial with the air impedance. Therefore, it is necessary to adjust the efficient dielectric continuous ε_1 and the permeability μ_1 by designing the metamaterial structure and making them identical [31,32].

2.2. *Four Theoretical Models*

Impedance matching introduced in the previous section is a necessary condition for absorption. In the research process of metamaterials, many theoretical models have been developed, among which four are popular, namely, effective medium theory, transmission line modelling, coupled mode theory, and interference theory. The following briefly introduces various models and analyzes their advantages and disadvantages [33].

2.2.1. Effective Medium Model

Metamaterials can produce an electromagnetic response through their special structures. Often the working wavelength is much larger than the periodic size of metamaterials, so for the convenience of analysis, metamaterials can be approximated as homogeneous materials. Because of the sub-wavelength structure, the average electromagnetic response of metamaterials can be described by frequency-dependent effective dielectric constant $\varepsilon_{eff}(\omega)$ and magnetic permeability $\mu_{eff}(\omega)$. Ignoring anisotropy, effective dielectric constant and permeability can be extracted from S parameters. The effective impedance of metamaterials can be expressed as:

$$Z = \sqrt{\frac{\mu_{eff}}{\varepsilon_{eff}}} Z_0 \quad (4)$$

When the material has high electromagnetic wave loss, the imaginary parts of $\varepsilon_{eff}(\omega)$ and $\mu_{eff}(\omega)$ are very large. When impedance matching and high loss are realized at the same time, metamaterials can achieve perfect absorption. The effective medium model provides an intuitive explanation for achieving perfect absorption. However, because the absorber is regarded as a uniform model of reflection and transmission coefficients, the effective medium model cannot decouple the contributions of each part of the metamaterial absorber.

2.2.2. Transmission Line Modelling

At first, the transmission line modelling was applied to the interpretation of Salisbury screen, Jaumann absorption, and circuit analog absorption. Due to the expansibility of electromagnetic theory, the metamaterial absorber can also use transmission line modelling to explain its working principle after appropriate modification. In transmission line modelling, the metamaterial layer can be regarded as a coupled resonance mode connected in parallel.

For example, the metamaterial layer is simulated by LC oscillation circuit and dipole resonance mode, as shown in Figure 1. Through mutual inductance coupling, the impedance of LC and dipole resonance mode (Z_L, Z_d) can be expressed as:

$$Z_{L,d} = R_{L,d} + \frac{2\varepsilon_i}{\omega C_{L,d}(\varepsilon_r + 1)^2} + i\omega L_{L,d} + \frac{2}{i\omega C_{l,d}(\varepsilon_r + 1)} \tag{5}$$

where ε_r and ε_i are the real and imaginary parts of the dielectric constant of the dielectric layer, and R_{Ld}, L_{Ld}, and C_{Ld} are the impedance, inductance, and capacitance of the structure, respectively. To simplify the calculation and analysis, the coupling between the two modes is ignored, so the impedance of metamaterial can be calculated as:

$$Z = \frac{Z_L \cdot Z_d}{Z_L + Z_d} \tag{6}$$

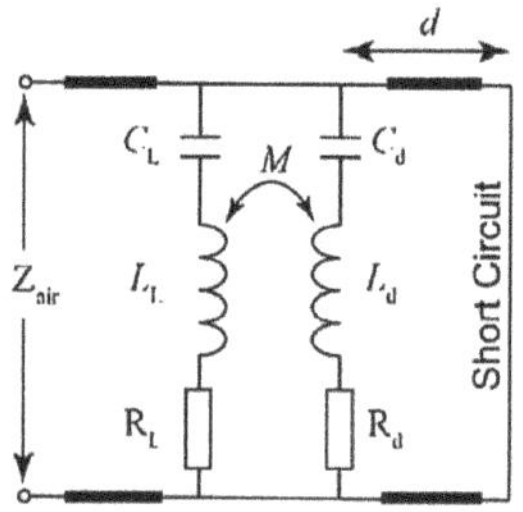

Figure 1. Schematic diagram of transmission line modelling of metamaterial absorber [33].

When impedance matching is satisfied, perfect absorption occurs. The transmission line model can analyze the contribution of each part of the metamaterial to absorption by analyzing the circuit, but the near-field coupling between the metamaterial resonant layer and substrate is ignored in the analysis, which leads to the blue shift of the fitted absorption peak. To solve this problem, in the use of transmission line modelling, this near-field coupling effect can be included in capacitance or inductance to compensate for frequency mismatch [34].

2.2.3. Coupled Mode Theory

Coupled mode theory is a theory that studies the coupling of a harmonic oscillator with one or more ports or other harmonic oscillators. This theory describes a system with lumped parameters, and each parameter corresponds to the specific physical meaning of the whole system. Coupled mode theory is widely used in the research of optical resonators, waveguides, photonic crystals, and metamaterials because of its low requirements for algebra and simple physical concepts.

In coupled mode theory, the metamaterial absorber can be regarded as a single-port single-mode resonator, which can be described by the following formula:

$$\frac{da}{dt} = i\omega_0 a - \frac{1}{\tau_0}a - \frac{1}{\tau_e}a + \sqrt{\frac{2}{\tau_e}}s_+ \tag{7}$$

$$s_- = -s_+ + \sqrt{\frac{2}{\tau_e}}a \tag{8}$$

where a is the amplitude of resonance, ω_0 is the angular frequency of resonance, $1/\tau_0$ is the internal loss of Joule heat and other materials, $1/\tau_e$ is the energy loss, such as radiation loss from resonance, and s_+ and s_- are the amplitudes of the incident wave and reflected wave, respectively. The first equation represents the energy change in the resonant cavity, and the second equation represents the total reflected wave. Therefore, the reflection coefficient can be expressed as:

$$r = \frac{s_-}{s_+} = \frac{\frac{1}{\tau_e} - \frac{1}{\tau_0} - i(\omega - \omega_0)}{\frac{1}{\tau_e} + \frac{1}{\tau_0} + i(\omega - \omega_0)} \tag{9}$$

It can be seen that the expression of the metamaterial absorber is relatively simple, able to be described only by three parameters: ω_0, τ_0, and τ_e. However, these parameters are difficult for complex metamaterial analysis, and more parameters need to be considered to analyze the contribution of other resonance modes.

2.2.4. Interference Theory

In the traditional interpretation, the superposition of multiple reflections between the metamaterial layer and the substrate constitutes the overall reflection of the metamaterial absorber, as shown in Figure 2, and perfect absorption can be achieved through the destructive interference of reflection amplitude. The interference theory model provides a relatively simple mathematical model to understand the metamaterial absorber from the optical point of view. This provides a platform for studying the properties of metamaterial absorbers and exploring the internal relationship between material properties and absorber properties [35].

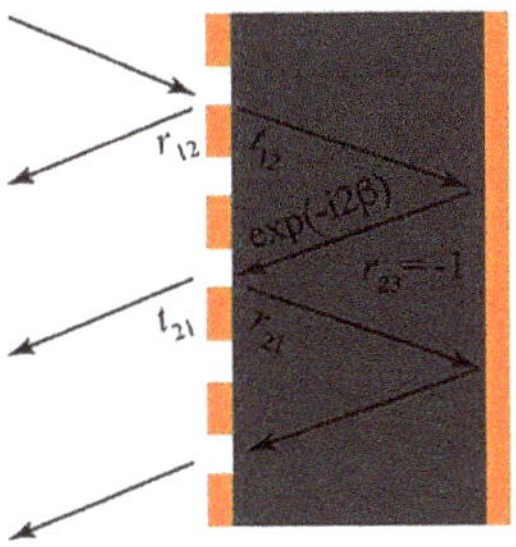

Figure 2. Multi−reflection diagram of the metamaterial layer and base reflection layer [33].

With the refractive index n of the dielectric layer, the angular wave number k in a vacuum, and the thickness d of the dielectric layer, the one-way phase delay can be expressed:

$$\beta = nkd \tag{10}$$

Assuming that the reflection coefficient of the substrate is -1, the overall reflection coefficient can be expressed as:

$$r = r_{12} - \frac{t_{12}t_{21}}{r_{21} + e^{-2i\beta}} \tag{11}$$

The reflection and transmission coefficients of metallic interfaces (r_{12}, t_{12}, r_{21} and t_{21}) can be obtained from the impedance of the metamaterial layer. The reflection coefficient is optimized to 0 to achieve perfect absorption. By analyzing the above formula, it can be found that when other parameters are fixed, the critical spacer thickness is mainly determined by the loss of the metamaterial absorber, and the frequency is mainly determined by the resonance frequency of the metamaterial layer.

Although the interference model is intuitive and concise, it usually ignores the near-field coupling between the metamaterial layer and the base reflective layer. Generally, the near-field coupling can reduce the resonant frequency by introducing additional reactance. However, because the near-field coupling decays quickly away from the material layer, we can ignore the near-field coupling when the reflective substrate is far away from the metamaterial layer.

2.3. Coherent Perfect Absorption

The traditional metal conductive film's absorption in the microwave series is very low because of the impedance mismatch between metal and free space. Although reducing the film thickness can improve the matching performance and absorption effectiveness, absorption cannot surpass the highest absorption rate of 50% with ultra-thin film. A coherent perfect absorber (CPA) based on coherent principle provides a new idea to solve such problems [36,37].

When several coherent waves overlap, interference occurs, which can redistribute the energy in space. CPA is a common phenomenon caused by the interaction of interference and scattering. It can completely absorb electromagnetic radiation by controlling the interference of multiple incident waves. Typical coherent perfect absorption is a two-port linear system, in which two coherent beams with equal intensity and opposite directions are vertically incident on the absorbing material. In a two-port linear system, the absorption of light can be dynamically adapted from 0% to 100% only by changing the relative phases of two disturbance beams, thus realizing the transition of the metasurface from a transparent state to an opaque state [38]. In addition to the dual-port system, the typical CPA has an FP dielectric cavity, where all the incident energy can be captured and dissipated. Up to now, many CPA structures have been proposed, such as grating, thin film, metal–insulator–metal structures, etc.

In a two-port linear system, the scattering matrix can be used to theoretically analyze the coherent perfect absorption. When the coherent light in the facing direction is perpendicularly incident on the absorbing material, the relation between the incident wave and the outgoing wave is:

$$\begin{bmatrix} O_1 \\ O_2 \end{bmatrix} = S \begin{bmatrix} I_1 \\ I_2 \end{bmatrix}, \; S = \begin{bmatrix} r_{11} & t_{12} \\ t_{21} & r_{22} \end{bmatrix} \tag{12}$$

I_i and O_i represent the amplitude of the incident wave and the amplitude of the outgoing wave in the ith direction, respectively. S is the scattering matrix, and r_{ii} and t_{ii} are the reflection coefficient and transmission coefficient, respectively, which are determined by the component materials and geometric structure of CPA. The relationship between I_1 and I_2 is:

$$I_2 = \alpha I_1 e^{i\varphi + ikz} \tag{13}$$

where α, φ and z are the relation amplitude, phase dissimilarity, and phase reference points between I_1 and I_2, respectively. When the incoherent absorption limit (R_{11} = R_{22} = 0.5, T_{12} = T_{21} = 0.5) is met, if z = 0, the absorption of CPA can be obtained as follows:

$$A = 1 - \frac{|O_1|^2 + |O_2|^2}{|I_1|^2 + |I_2|^2} = 1 - \frac{1 + \alpha^2 - 2\alpha cos(\varphi)}{2(1 + \alpha^2)} \tag{14}$$

Therefore, it can be tuned by changing α and φ.

3. Research Progress of MPA

3.1. Narrow Band Absorber

Before the development of the metamaterial absorber, the traditional absorber needed to spot a resistance plate with a thickness of at least one-quarter wavelength in front of the metal plate to defeat the restriction of the traditional absorber working at one-quarter wavelength. This resistor plate becomes very thick and heavy in the long wavelength frequency range (such as microwave frequency), which means that it does not meet the requirements of light and miniaturized devices. In 2008, Landy et al. designed the first metamaterial absorber, successfully overcoming the above difficulties. Within a single unit of the absorber, the electrical coupling is provided by an electric ring resonator (ERR) [14]. The element consists of two standard split ring resonators connected by an induction ring parallel to the split line, as shown in Figure 3. Through the impedance matching principle, the perfect narrowband absorption is realized in the microwave band. The absorber is favored by the industry due to its light structure and high absorption rate.

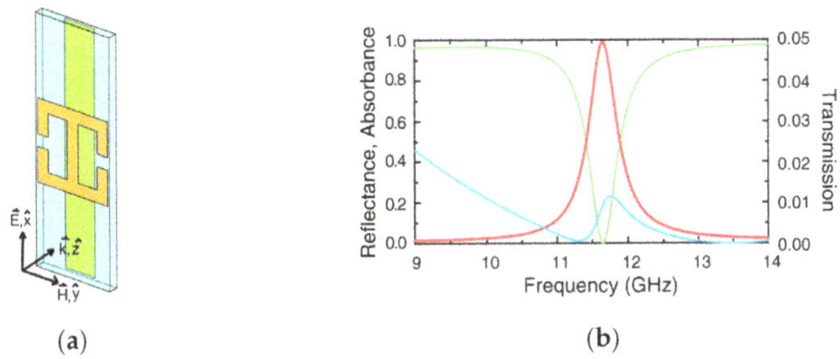

Figure 3. (**a**) The structure of the MPA unit was designed by Landy et al. [14]; (**b**) absorption, reflection, and projection curves [14].

According to Maxwell's equations, the adjustment of the resonance frequency can be realized by scaling the size of the metamaterial structure. In the same year, Tao et al. designed the first THz metamaterial absorber with an absorption rate of 70%. After continuous optimization and improvement, Tao et al. designed a kind of MPA composed of an ERR and split wires [39]. Its structure is similar to that designed by Landy et al., as shown in Figure 4, and it is a typical MDM structure of MPA today. In addition, the absorption can be up to 97% for the incident wave at 1.6 THz.

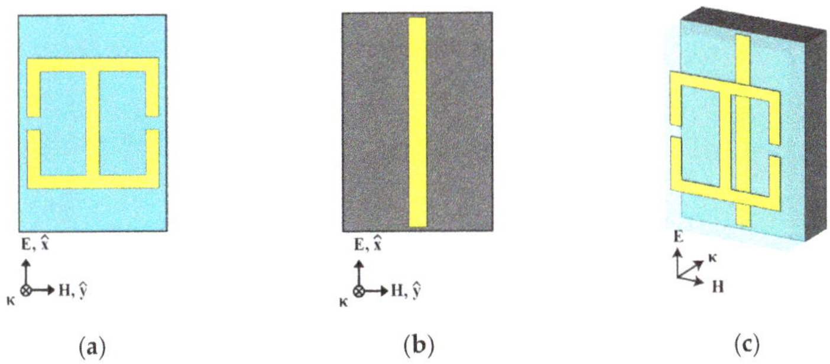

Figure 4. Schematic diagram of the perfect absorber for terahertz band designed by Tao et al. [39] (**a**) Electric resonator on the top of a polyimide spacer; (**b**) cut wire on GaAs wafer; (**c**) single unit cell showing the direction of propagation of incident EM wave.

In 2010, Ye et al. proposed an absorber with a composite structure of metallic crosses, which is a near omnidirectional terahertz absorber with high absorption for transverse electric wave (TE) and transverse magnetic wave (TM) [40]. By exciting the magnetic pole in the metal–dielectric layer, the incident light is perfectly absorbed in the thin layer structure, which is about 25 times smaller than the resonance wavelength. Next, Ye et al. designed the absorber with an overlapping structure on this basis to achieve broadband absorption, which is mentioned in Section 3.3 of this paper.

At first, most MPA can only perfectly absorb a certain polarized electromagnetic wave, which means it is very sensitive to the polarization angle of the incident light. This leads to its limited application in many fields, such as solar cells and biosensors. Therefore, polarization-insensitive MPA has become the development requirement for MPA in some fields [41]. In 2017, Astorino et al. designed a polarization-insensitive ultra-thin narrowband absorber [42], which is composed of a top metallic ERR and a bottom ground plane, both made of lossy gold, separated by a thick dielectric layer of benzocyclobutane, as shown in Figure 5. Its absorption structure has four rotational symmetries. It can be found that the high symmetry of the absorber structure can help realize polarization insensitivity.

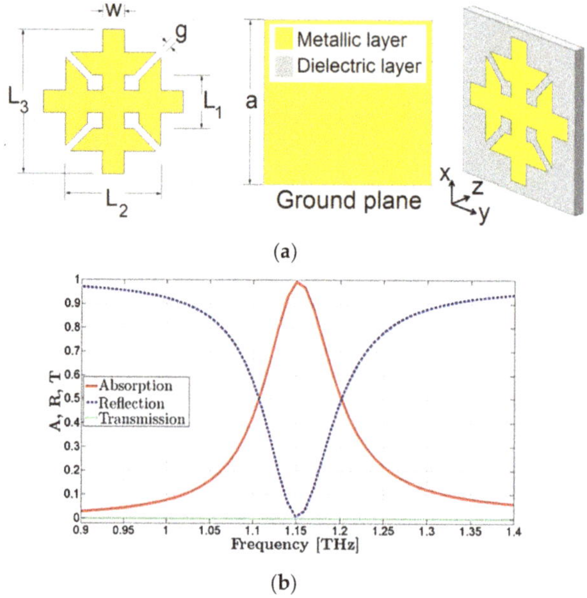

Figure 5. (**a**) Structure of polarization-insensitive ultra-thin narrowband absorber; (**b**) absorption, reflection, and transmission curves [42].

In recent years, with the development of MPA in the sensing field, the requirements for sensor multi-scene applications are gradually increasing. Flexible materials are widely used in sensing fields, such as biological detection, pressure sensing, flexible artificial skin, etc., due to their excellent characteristics, such as flexibility, low dielectric constant, stable performance, and easy integration into wearable devices.

In 2020, Cheng et al. proposed a planar array Fano asymmetric split ring resonator fabricated on flexible Polyimide (PI) substrate [43] for protein sensing. Its sensitivity was 240 GHz/RIU, which provided the idea for the combination of flexible materials and MPA. In 2021, Wang et al. improved the structure. As shown in Figure 6 [44], the unit structure of the absorber is gold, dielectric (polyimide), and gold, from top to bottom. The upper surface comprises two identical split rings, which are rotated and spliced 180 degrees to form an asymmetric split ring resonator. When there is no substance to be measured, the highest absorption peak is 4.83 THz. By changing the refractive index of the object to be measured, its sensitivity can reach 1018 GHz/RIU [45].

Although the metamaterial absorber with MDM structure can realize the correspondence of most electromagnetic spectrums with novel performance, it also faces some shortcomings due to the existence of metal structure, such as high ohmic loss, high thermal conductivity, and low melting point [45]. These shortcomings limit its application scope. Therefore, the researchers designed a metamaterial absorber with all-dielectric structure, which put forward a new idea for the regulation of electromagnetic field. An all-dielectric metamaterial absorber made of materials with good temperature stability can effectively avoid Joule heating, and has potential applications in energy collection, imaging, and sensing. In 2017, Liu et al. presented a terahertz all-dielectric metasurface absorber based on hybrid dielectric waveguide resonances [46]. By adjusting the geometric structure, the resonance of the electric and magnetic dipole overlapped, thus achieving perfect absorption, as shown in Figure 7. The results show that the absorbance is 97.5% at 1.011 THz. Compared with metal-based methods, all-dielectric absorbers can use near-infrared materials with high melting points and low loss, so they will be used in imaging, thermophotovoltaic, and other sensing applications.

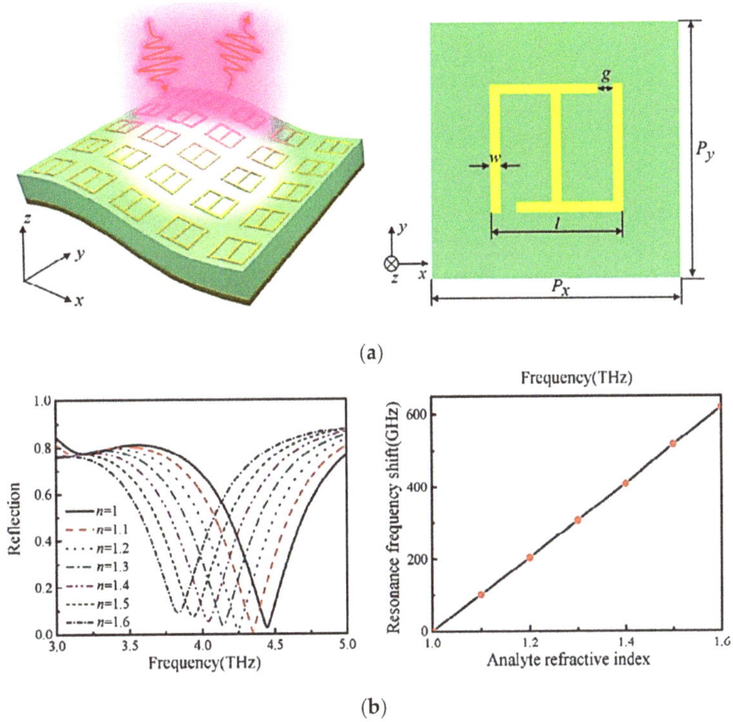

Figure 6. MPA biosensor. (**a**) Sensor structure; (**b**) reflection curve with the change of refractive index of the detected object [44].

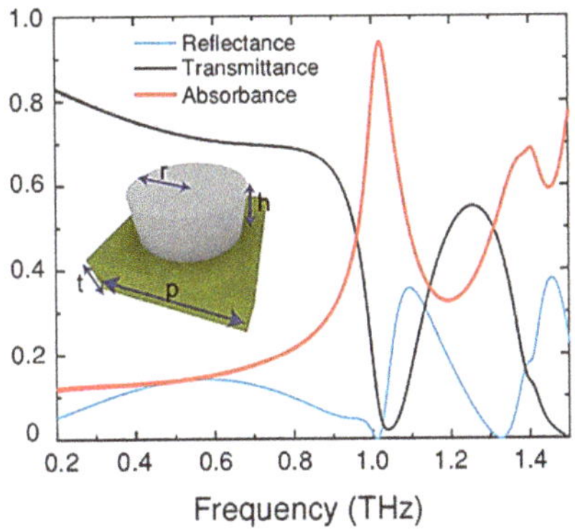

Figure 7. Simulated spectral reflectance (blue), transmittance (black), and absorbance (red) [46].

In the same year, Fan et al. realized a metasurface consisting of sub-wavelength cylindrical resonators that achieve diffraction-limited imaging at THz frequencies without cooling [45]. The ingenuity of this design lies in the use of conversion imaging method. As shown in Figure 8, the unit structure is cylindrical dielectric particles, and the medium used here is silicon. As a universal converter of radiation, the all-dielectric metasurface absorber absorbs incident terahertz waves, converts them into heat, and then is detected by infrared cameras. The results show that at the frequency of 603 GHz, the absorbance of all-dielectric metasurface is as high as 96%, and the thermal response rate is as high as 2.16×10^4 K/W. This result can be extended to other spectra, which provides a new way of measuring thermal imaging.

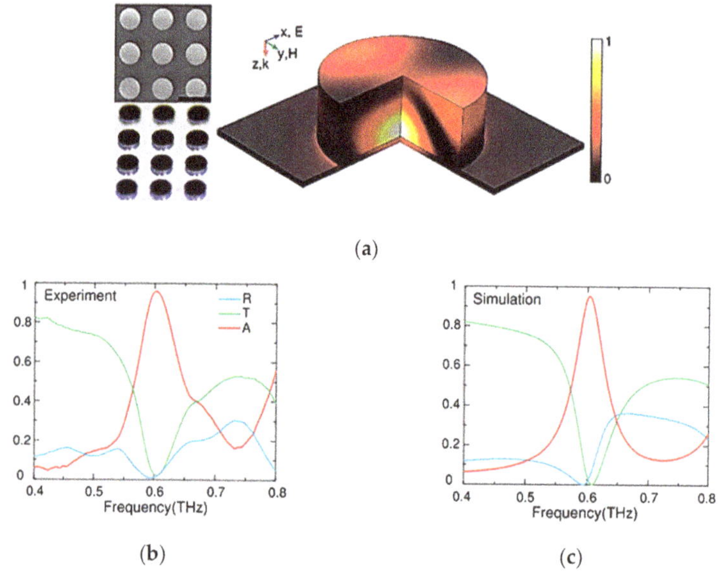

Figure 8. (**a**) Scanning electron microscopic image of fabricated cylinder array; (**b**) measured reflectance R (blue), transmittance T (green), and absorbance A (red) of the dielectric absorber; (**c**) numerically simulated R, T, and A [45].

3.2. *Dual-Frequency and Multi-Frequency Absorbers*

In the design of the MPA structure, usually, there are two ways to achieve dual-frequency absorption, multi-frequency absorption, and even broadband absorption. In the first method, sub-lattice units with various sizes and structures are assembled in one lattice unit, so that different resonance effects occur in each sub-lattice unit or among sub-lattice units. In the second method, MPA is designed as a multi-layer superposition structure, and the structure and size of resonators in each layer are different. The principle is similar to that of the first method. When different resonance absorption peaks are far away, we can get dual-frequency or multi-frequency MPA; when the distance between different resonance peaks is close, or the average absorption between resonance peaks is high, we can get broadband MPA.

In 2009, Wen et al. designed a dual-frequency THz metamaterial absorber. Its ERR unit consists of two symmetrical single resonant metamaterials, one embedded in the other [47]. The results show that MPA has two obvious strong absorption peaks near 0.45 THz and 0.92 THz, both of which are related to the LC resonance of metamaterials, as shown in Figure 9. Compared with single-frequency absorbers, such a design has a wider application range and a broader development prospect. In 2011, Ma et al. also designed a terahertz dual-band metamaterial absorber with metal nested square ring arrays [48]. Its structure is simpler than the former. Two obvious absorption peaks are found at 2.7 THz and 5.2 THz, which are in good agreement with the simulation results, as shown in Figure 10. In addition, the design is highly symmetrical, so it is insensitive to polarization.

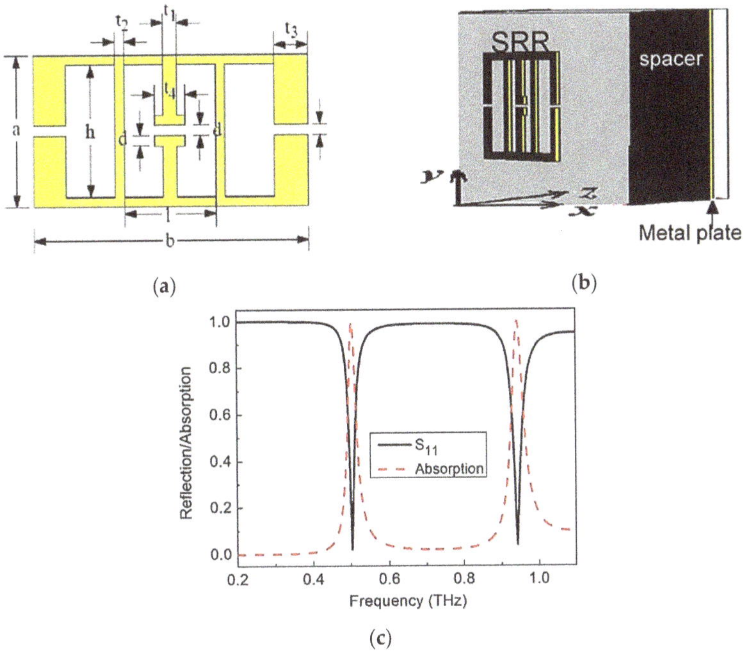

Figure 9. Dual frequency terahertz metamaterial absorber. (a) Top view; (b) the main view; (c) reflection and absorption curves [47].

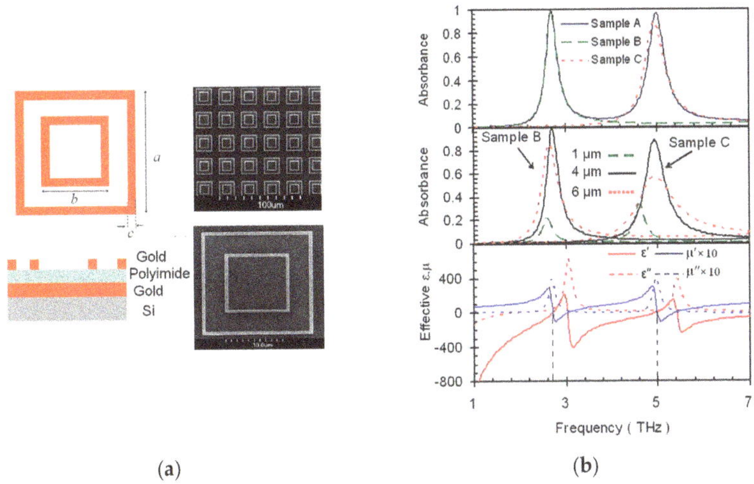

Figure 10. (a) Structure and SEM of the terahertz dual−band metamaterial absorber; (b) absorption curves and calculated effective dielectric constant and permeability under different structural parameters [48].

In 2015, Shan et al. proposed an ultra-thin flexible dual-band terahertz wave absorber based on metamaterials [49]. The metamaterial structure has two periodic split ring resonators with asymmetric gaps. The results show that the absorber has two resonance absorption peaks at 0.41 THz and 0.75 THz, respectively, and the absorption rates are 92.2% and 97.4%, respectively, as shown in Figure 11. Because the structure is symmetrical, the absorber also has polarization insensitivity. The middle dielectric layer of the absorber is made of polyimide, a flexible material with a thickness of 25 μm, which makes the absorber highly flexible and gives it a non-planar application potential, for instance, for microbolometers and stealth aircraft.

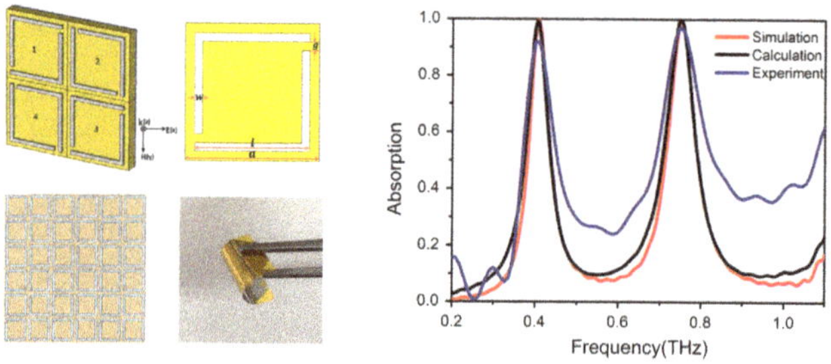

(a) (b)

Figure 11. (**a**) Structure and (**b**) absorption curve of the ultra-thin flexible dual-band terahertz absorber [49].

In 2022, Yun et al. designed a dual narrow-band perfect metamaterial absorber suitable for the optical communication band [50], as shown in Figure 12, in which the silicon nanodisc array was placed on a thin gold film separated by a dielectric layer. The results showed that perfect absorption peaks appeared at 1310 nm and 1550 nm, respectively. The absorber had a wide application prospect in the fields of optical communication, frequency-selective optical detection, and inter-satellite laser communication.

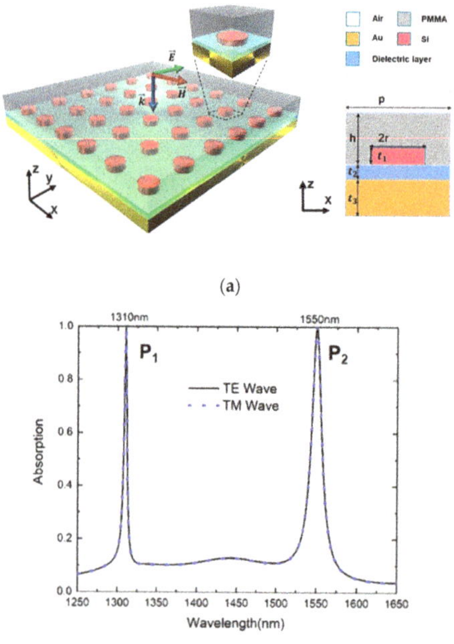

(a)

(b)

Figure 12. A dual-narrowband perfect metamaterial absorber suitable for optical communication band [50]. (**a**) The structure; (**b**) the curve of absorption.

In 2019, Wang et al. designed two all-dielectric terahertz plasmonic metamaterial absorbers (PMA) [51]. The cell structures of both PMAs are composed of square arrays of all-dielectric rings and cylindrical disks, and the same heavily doped silicon is used. However, their geometric parameters are different. As shown in Figure 13a, the gap between the two rings is narrow. At this time, PMA can achieve dual-frequency absorption, and there are two different absorption peaks at 0.96 THz and 1.92 THz, with absorption rates of 99.7% and 99.9%, respectively. By changing the inner radius of the ring and the radius of the cylinder, multiple resonance modes can be overlapped, and the Q value can be reduced, so that broadband operation can be obtained, as shown in Figure 13b. Because the quality factor Q of dual-band resonance is larger than that of broadband resonance, it has better sensing performance.

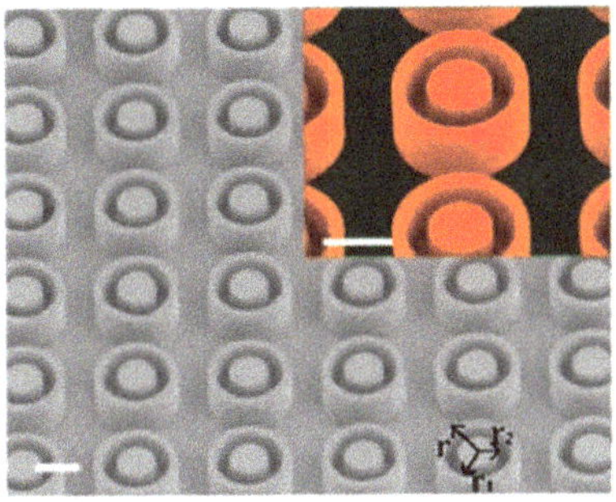

(**a**)

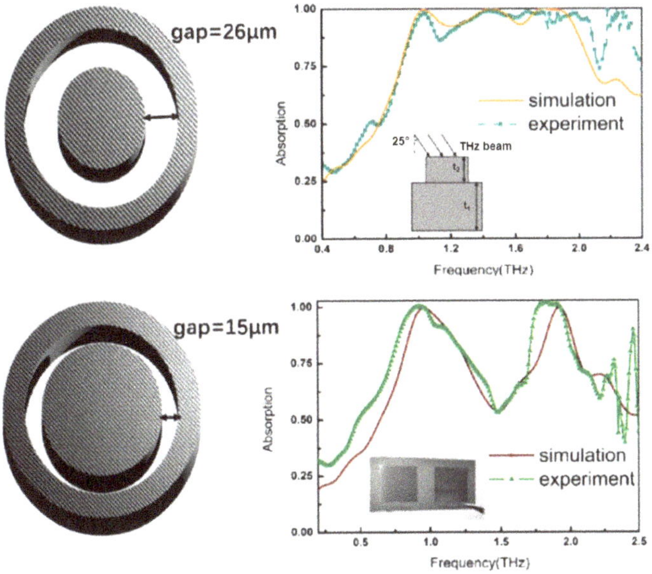

(**b**)

Figure 13. (**a**) SEM image of the designed PMAs; (**b**) illustrations of unit cells of SRRs and simulated (yellow curve) and measured (green curve) absorption characteristics of the broadband PMAs [51].

With the development of MPA, perfect multi-band absorption has become a reality. In 2012, Shen et al. designed a terahertz three-band MPA consisting of a square metal piece and two concentric metal square rings with different sizes, as shown in Figure 14 [52]. The results show that there are three different absorption peaks at 0.5 THz, 1.03 THz, and 1.71 THz, and the absorption rates are 96.4%, 96.3%, and 96.7%, respectively. It is simple in structure, easy to fabricate, insensitive to polarization, and has high absorption. In 2016, Wang et al. designed a five-band MPA. The resonant structure of the absorber is two split rings [53], as shown in Figure 15. The working mechanism of five-band near-perfect absorption is due to the joint action of LC, dipole, and surface resonance. The device can be used in the fields of biosensing, material detection, and thermal imaging.

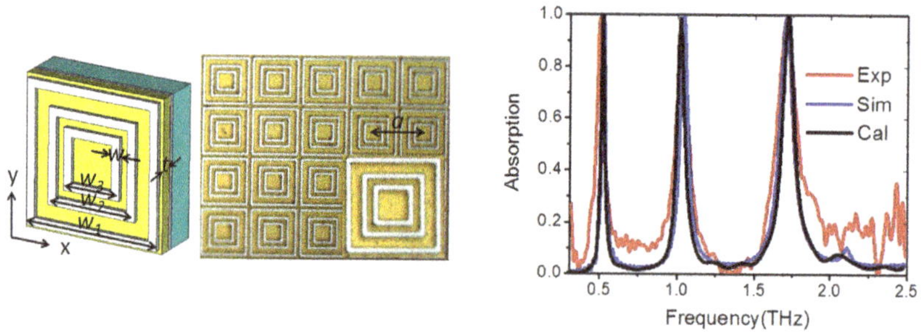

(a) (b)

Figure 14. Terahertz tri-band MPA. (**a**) The structure; (**b**) the curve of absorption [52].

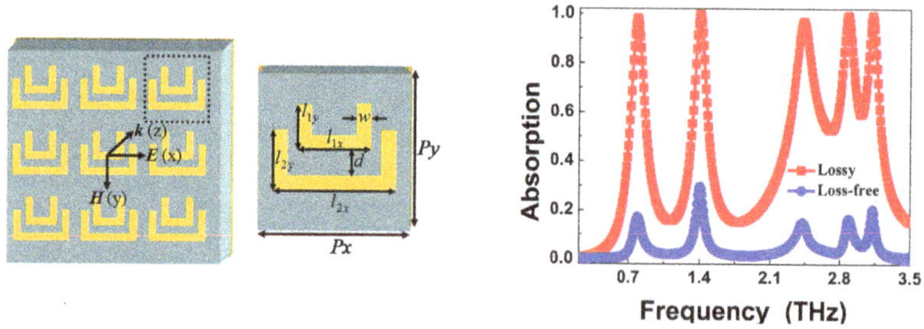

(a) (b)

Figure 15. Five-band MPA. (**a**) Structure; (**b**) absorption curve [53].

3.3. Broadband Absorber

In many practical applications, such as solar energy collection and photoelectric detectors, researchers hope to obtain absorbers with high average absorption. However, the absorption peaks of multi-frequency absorbers are often discrete, and the average absorption is not high enough. Thus, the perfect absorber with continuous broadband has become the inevitable goal and an important focus for development.

As mentioned above, one of the methods to realize broadband absorption is by assembling sub-lattice units with various sizes in one lattice unit. In 2012, Huang et al. designed a terahertz broadband absorber with three different sizes of "I"-shaped metal structures [54]. As shown in Figure 16, the realization of broadband absorption depends on the superposition of two similar formats. The highest absorbance measured at 0.905 and 0.956 THz is 99.9%, and the minimum absorbance between these two frequencies is still as high as 93%. The absorber can absorb TE or TM polarized terahertz radiation according to the incident direction.

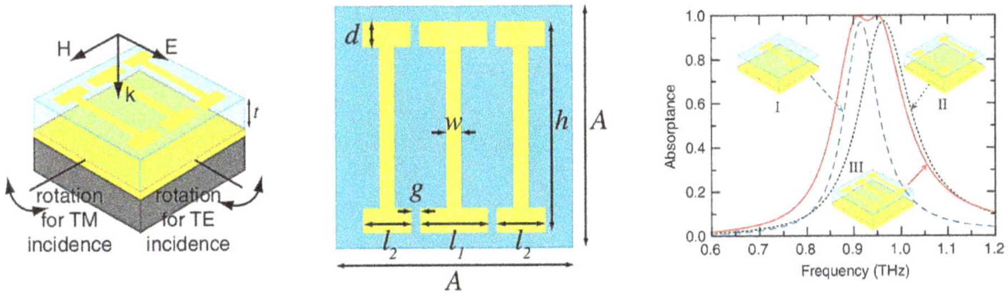

(a) (b) (c)

Figure 16. The whole unit cell of the Terahertz broadband absorber [54]. (**a**) Schematic of the whole unit cell; and (**b**) top view of the metamaterial absorber with dimensions. (**c**) Numerical simulation results of absorption spectra at normal incidence for three different configurations of the I-shaped resonators.

In 2021, Yun et al. designed an ultra-wideband absorber based on multiple resonances [55]. The absorber is composed of amorphous silicon and metallic titanium, and its whole structure comprises grating, as shown in Figure 17. When the visible light is incident in the X polarization direction, the ultra-wideband absorption is realized in the wavelength range of 382 nm to 1100 nm, and three near-perfect absorption bands are obtained. To realize the polarization insensitivity of the absorber and keep excellent absorption performance, Yun et al. improved the above structure to make the unit periodic structure centrosymmetric. The modified absorber can achieve absorption in the range of 0~90° polarization angle.

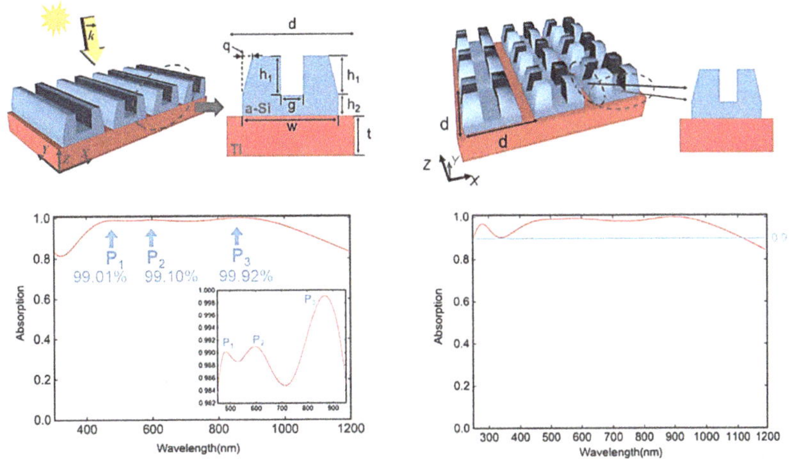

(a) (b)

Figure 17. (**a**) Structure and absorption curve of the ultra-wideband absorber; (**b**) polarization insensitive structure, and absorption curve after optimization [55].

As mentioned in the previous section, to realize broadband absorption, the absorber is designed as a multi-layer superposition structure and the structure and size of resonators in each layer are different. In 2010, Ye et al. proposed an absorber that could expand the absorption bandwidth by simply stacking several structural layers with different geometric sizes [40], as shown in Figure 18. Three close resonances are observed at 4.55 THz,

4.96 THz, and 5.37 THz. The bandwidth of this strong absorption can be effectively enhanced due to the hybridization of magnetic polarization elements in different layers.

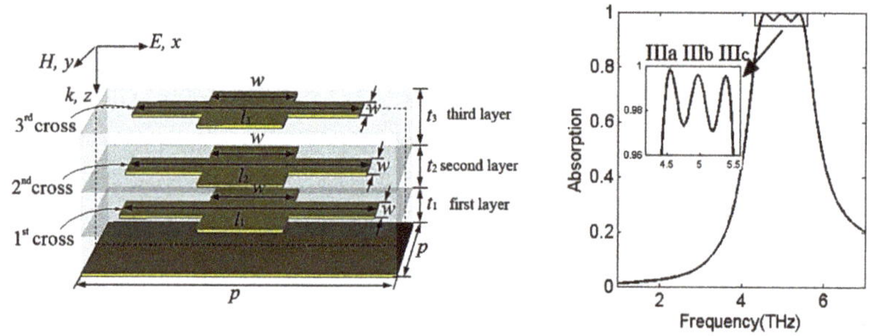

(a) (b)

Figure 18. (**a**) Structure of multilayer broadband absorber; (**b**) the absorption curve [40].

In 2022, Li et al. designed an ultra-wideband absorber with stacked round hole discs [56], as shown in Figure 19. Ultra-wideband absorption can be realized in the whole spectrum (0.25 μm–4 μm). Due to the plasma resonance, the absorber's absorption rate is more than 90% in the ultra-wide spectral range of 3450 nm, the average absorption rate is 97.5%, and the average absorption rate in the visible light range is more than 99%. Therefore, the absorber shows good performance in the field of solar energy collection and conversion.

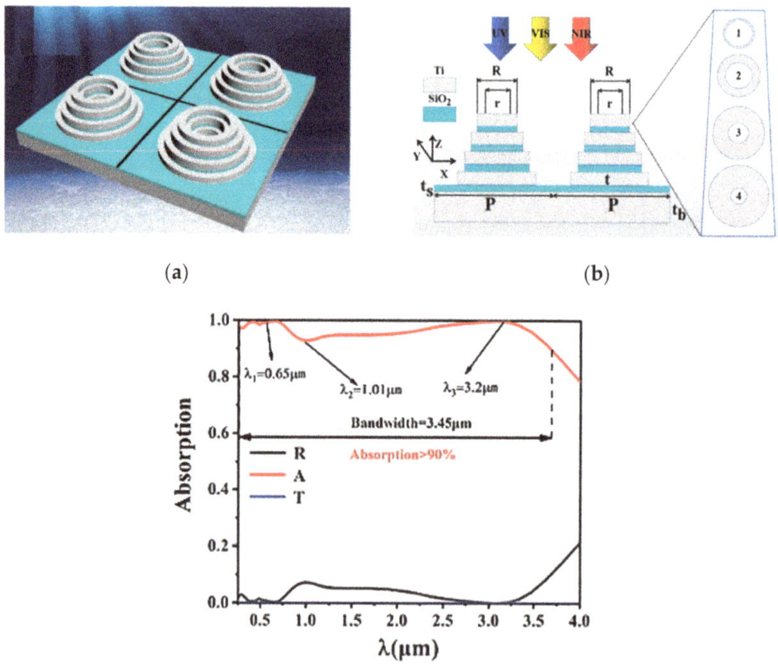

(a) (b)

(c)

Figure 19. (**a**) Full spectrum absorber structure diagram; (**b**) full spectrum absorber plan; (**c**) full spectrum absorption spectrum, reflection spectrum, and transmission spectrum [56].

In addition to the broadband absorber based on metal structure, the designed all-dielectric metasurface absorber can also achieve broadband absorption. In 2022, Huang et al. introduced a thin-film silicon metasurface absorber with periodic elliptical holes [57], as shown in Figure 20. The absorption performance theory of the absorber is consistent with the experimental results. The absorption of ≥90% starts at 1.1 THz and keeps close to unity until 1.6 THz under normal incidence, and the corresponding bandwidth is 500 GHz. The absorber provides a new idea for modulators, switches, and detectors in the terahertz band.

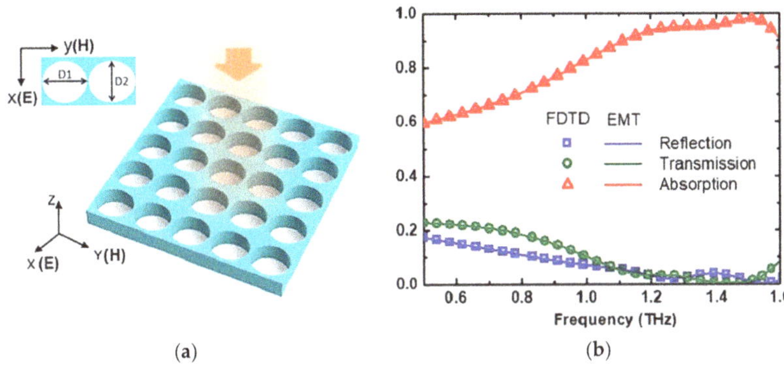

Figure 20. (**a**) Schematic of the broadband THz silicon membrane metasurface absorber (SMMA); (**b**) simulated reflection (blue), transmission (green), and absorption (red) of SMMA [57].

3.4. *Tunable Absorber*

Many metamaterial absorbers have been widely studied. However, there are typical defects in the working wavelength and absorption efficiency of metamaterial absorbers, which are determined by the original structure. Once the absorber structure is determined, the optical response will be fixed, and it is difficult to tune flexibly. This defect seriously limits its application as an optical switch or modulator. Therefore, by flexibly adding photosensitive materials, temperature-sensitive materials, liquid crystal materials, or phase-change materials, the absorber can be tuned, effectively broadening the application field. Several tunable absorbers based on different materials will be introduced below.

3.4.1. Tunable Absorber Based on Photosensitive Silicon

A photosensitive semiconductor has an excellent photoelectric conversion effect. When the energy of the pump light source increases, the carrier density in the photosensitive semiconductor will increase synchronously, thus optical control can be realized. Photosensitive silicon is a typical representative. In 2015, Xu et al. showed a metamaterial absorber that can be tuned at terahertz frequency by integrating photosensitive silicon into a metamaterial unit [58], as shown in Figure 21. By changing the pump beam, the conductivity of silicon changes, and thus its optical response can be modified.

In 2019, Zhao et al. demonstrated tunable ultra-wideband terahertz wave absorption by using a single-layer H-type all-silicon array [59], as shown in Figure 22. The absorption is optically tunable. When the pump flux increases from 0 to 4000 J/cm^2, the absorption frequency shifts, which in turn changes the absorption width and intensity. The dynamic response of light excitation depends on the penetration depth of pump light in silicon.

In 2020, Wang et al. proposed a single narrow-band THz absorber based on cylindrically shaped periodical p-type doped silicon [60], as shown in Figure 23. The absorber is made of a square polyimide substrate and a cylindrical P-doped silicon array. At 0.57 THz, the absorption rate is close to 99.75%, and the absorption characteristics are good. In addition, by changing the pump luminous flux from 0 to 3000 μJ/cm^2, its absorption rate can be flexibly adjusted from above 99% to below 35%. The above two designs provide new ideas for dynamic functional terahertz devices.

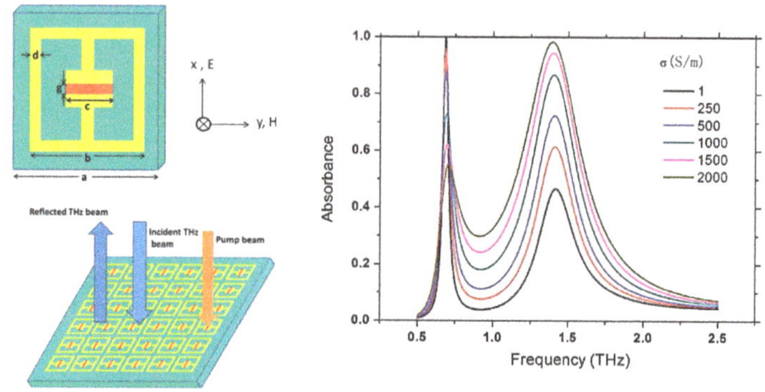

(a) (b)

Figure 21. Tunable metamaterial absorber based on photosensitive silicon (**a**) structure and (**b**) absorption curve with silicon conductivity [58].

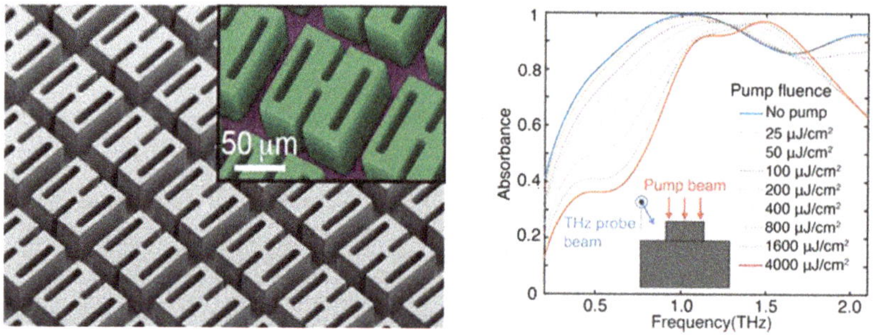

(a) (b)

Figure 22. (**a**) Scanning electron microscope image of the all-silicon; (**b**) OPTP spectrum of the all-dielectric MPA for different pump fluences [59].

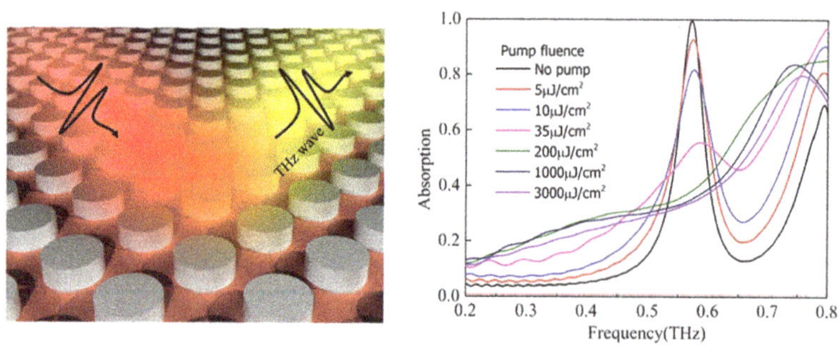

(a) (b)

Figure 23. (**a**) Schematic structure of the sample; (**b**) the absorption of the proposed absorber for different pump fluences [60].

3.4.2. Tunable Absorber Based on Graphene

Graphene is a two-dimensional material. The conductivity of graphene can be controlled by changing the Fermi energy level (EF). EF can be adjusted in a certain range by chemical doping or electrostatic gating [61].

In 2018, Yun et al. demonstrated a dual-band independent tunable absorber composed of stacked graphene nanodiscs, graphene layers with nanohole structure, and metal reflective layers separated by insulator layers [62]. The two absorption peaks A and B in the spectrum are respectively contributed by graphene nanodiscs and graphene layers with nanoholes. The resonance wavelengths corresponding to the two absorption peaks are λ_A = 11.92 μm and λ_B = 14.55 μm. Moreover, the TE and TM polarization keep the absorption above 90% in a wide range of incident angles, so the proposed structure is a wide-angle polarization-independent absorber. By changing the Fermi level of graphene nanodiscs and graphene layers with nanoholes, the two absorption peaks of the structure can be tuned independently, as shown in Figure 24. Because of the corresponding wave band and absorption characteristics of the absorber, it can be used in applications related to chemical sensors, detectors, and multi-band infrared absorbers.

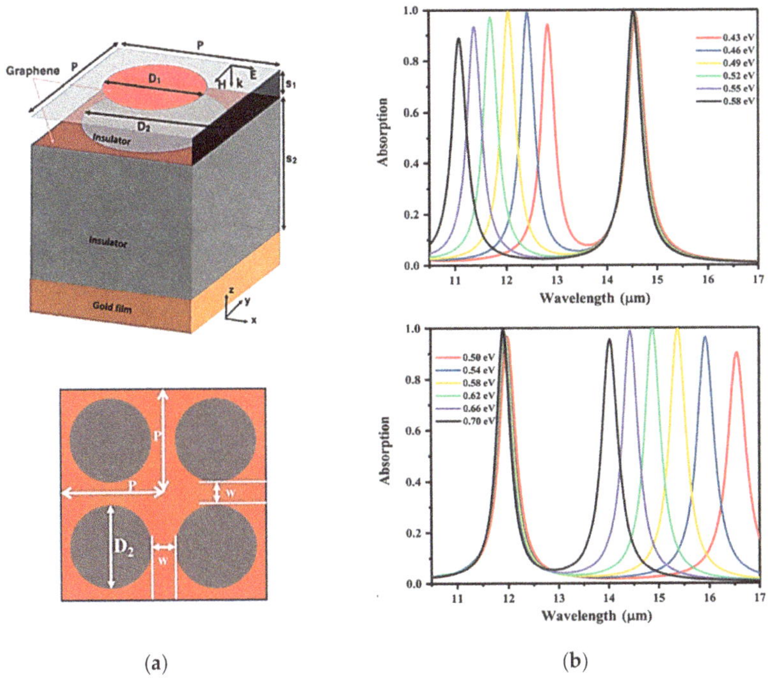

Figure 24. Dual-band independent tunable absorber (**a**) structure, and (**b**) absorption curves of graphene nanodiscs and graphene layers at different Fermi energy levels [62].

In 2019, Yun et al. designed a tunable infrared tri-band perfect absorber based on bilayer graphene, which can adjust the absorption peak by changing the Fermi energy level or the geometrical parameters of the graphene layer [63]. It can be seen from Figure 25 that the intermediate absorption peak depends on the upper graphene layer. With the increase in EF in the upper graphene, the intermediate absorption peak experiences a blue shift. Meanwhile, the left and right absorption peaks depend on the lower layer of L-graphene. When the EF of the lower graphene increases, both the long-wave and short-wave absorption peaks shift blue. This means that there is no coupling effect between the upper and lower layers of graphene during the change of EF. Therefore, the absorber can independently tune the absorption spectrum in a wide spectral range. The absorber can be used in the field of photoelectric detectors, thermal emitters, and photovoltaics.

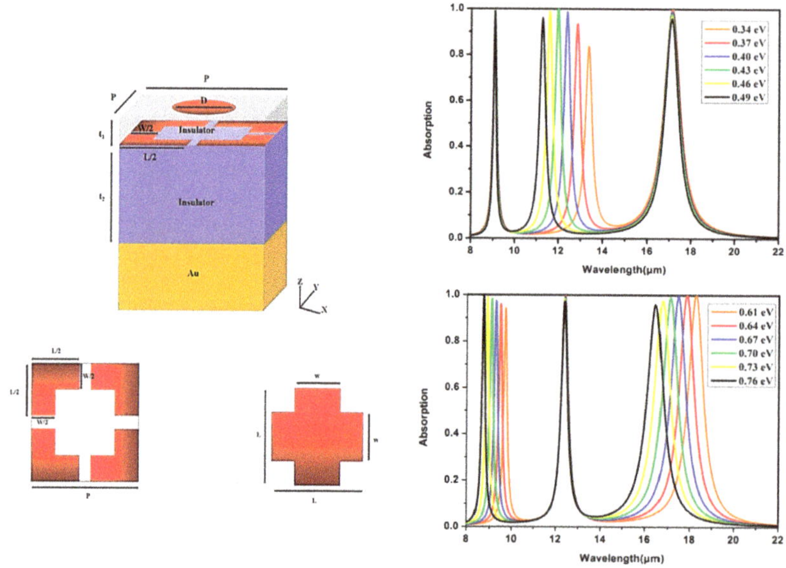

(a) (b)

Figure 25. (**a**) Structure of graphene-based infrared band absorber and (**b**) corresponding absorption spectra of graphene nanodisc and L-type single-layer graphene array at different Fermi levels [63].

In 2021, Yun et al. proposed a polarization-insensitive broadband metamaterial absorber based on local surface plasma (LSP) and propagating surface plasma (PSP) [64], as shown in Figure 26. The proposed absorber is designed by the combination of a gold mirror, dielectric layer, and graphene nanostructures. Under the normal irradiation of polarized wave, the absorber has a high absorption, which can be up to 99% from 1.23 to 1.68 THz. The absorption can be nearly tuned from 1% to 99% by adjusting the EF of the graphene. Given the high modulation depth of the absorber, it can be used in thermal emission devices, photovoltaic devices, intelligent absorbers, and active optical switches.

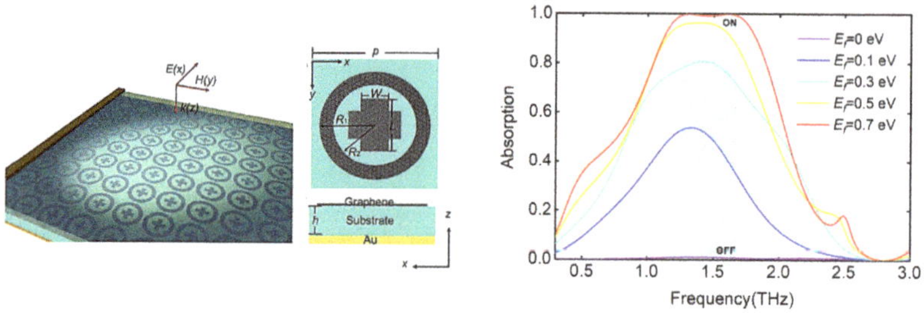

(a) (b)

Figure 26. (**a**) The structure of the polarization-insensitive broadband metamaterial absorber based on graphene and (**b**) the broadband absorption curve with EF [64].

3.4.3. Tunable Absorber Based on Other Materials

At present, phase change materials are also used in the field of tunable absorbers. When the external environment changes, the phase change material can change from one phase to another. VO_2 is a typical phase change material that changes with temperature, and the change is reversible. With 341 K as the phase transition point, VO_2 changes from the insulation state to the semiconductor state and then to the metal state when the ambient temperature rises gradually. In 2022, Yun et al. proposed a tunable CPA based on VO_2 metamaterial in the terahertz frequency range [65], as shown in Figure 27. The CPA realizes intelligent reconfigurable switching modulation from ultra-wideband absorption mode to dual-band absorption mode through VO_2 thermal control. When the absorber temperature is controlled at T = 328 K, the conductivity of VO_2 can reach 11,850 S/m and then realize the ultra-wideband absorber mode. In this mode, CPA shows an absorption efficiency of more than 90% in the ultra-wideband from 0.1 THz to 10.8 THz. When the conductivity of VO_2 reaches 2×10^5 S/m (T = 340 K), CPA switches to dual frequency absorption mode. As described in Section 2.2, the absorption of CPA can be controlled by the phase difference of the incident light. By modulating the phase of the incident light, the absorber can adjust the absorption efficiency to achieve intelligent control from full absorption to high transmittance transmission. In a word, this achievement ingeniously combines a variety of control modes, which can be used to enhance stealth equipment, all-optical switches, and coherent photodetectors.

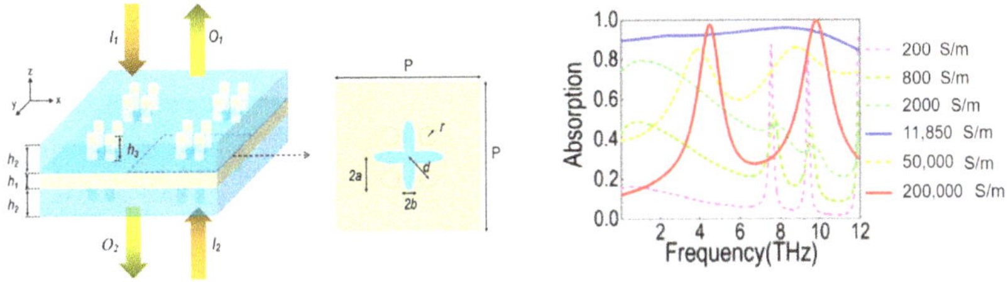

(a) (b)

Figure 27. (**a**) The structure of a tunable coherent perfect absorber based on VO_2 metamaterial and (**b**) the absorption curve changing with VO_2 conductivity [65].

In addition to VO_2, the active material strontium titanate (STO) can also be temperature-regulated, to realize the frequency tunable characteristic. In 2019, Huang et al. designed a thermally tunable metamaterial absorber based on THz strontium titanate [66], as shown in Figure 28. The elliptical pattern on the top of the absorber and the reflective layer are both made of gold material, and the dielectric layer is filled with STO active material. When the temperature dropped from 400 K to 200 K, the center frequency showed a uniform redshift, the tuning range reached 0.77 THz, and the peak absorption remained above 99%. In addition, the absorption bandwidth remains stable, only partially reduced. The structure is simple and ultra-thin, which makes manufacturing easy. In addition, the structure can be extended to other frequencies and has application prospects in imaging, detection, and tunable sensors.

Liquid crystal is a material with variable properties. The voltage-dependent birefringence of nematic liquid crystals has been developed and utilized. By combining liquid crystal with metamaterial, an electronically tunable metamaterial absorber at terahertz frequency can be realized. It provides a new idea for the development of a tunable metamaterial absorber. In 2013, Shrekenhamer and others proposed a liquid crystal tunable metamaterial absorber [67]. The absorber is MDM structure, and the liquid crystal 4′-n-pentyl-4-cyanodiphenyl (5CB) is deposited on the metamaterial array, as shown in Figure 29. When a potential is applied between ERR and the ground plane, LC will be oriented along the electric field line. By changing the applied bias voltage, the absorber achieves 30% amplitude tuning absorption.

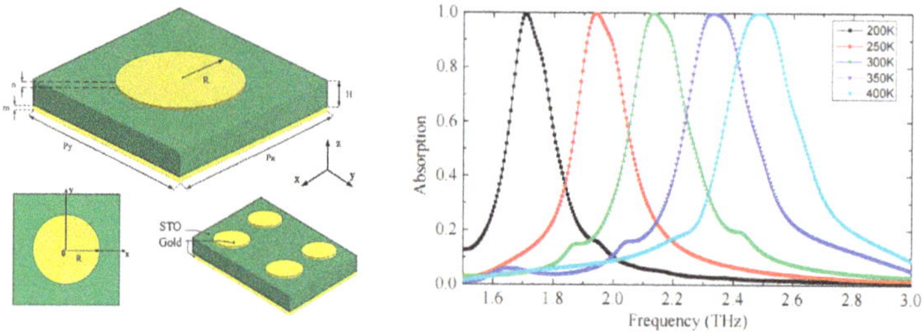

(a) (b)

Figure 28. (a) The proposed tunable terahertz absorber with a classical sandwiched structure consisted of a metallic top layer and ground plane, spaced by STO material film; (b) absorption spectra of the absorber with different temperatures of 200 K, 250 K, 300 K, 350 K, and 400 K under normal incidence [66].

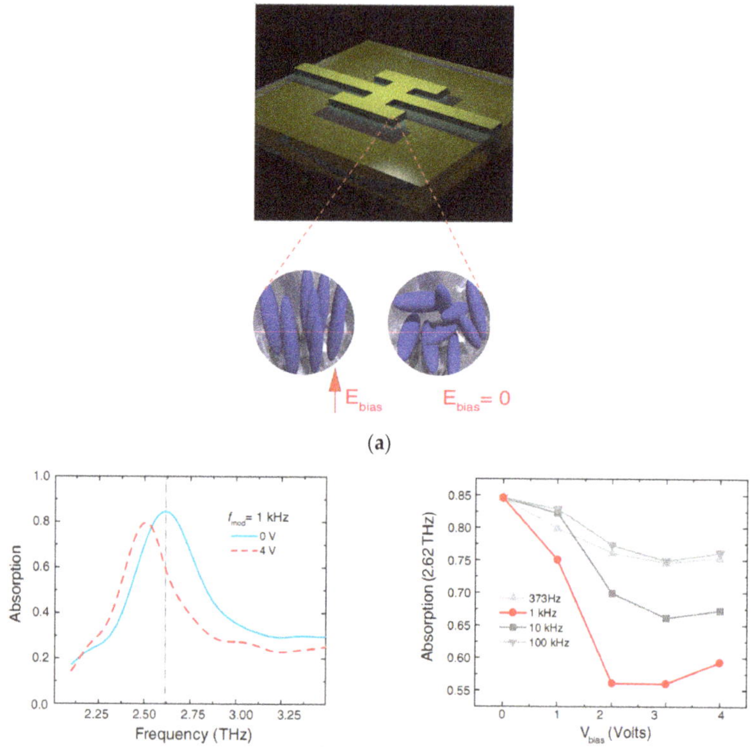

(b) (c)

Figure 29. (a) Rendering of a single unit cell of the liquid crystal metamaterial absorber; (b) frequency dependent absorption for 0 V (blue solid curve) and 4 V (red dashed curve) at f_{mod} = 1 kHz; (c) the absorption value at 2.62 THz as a function of bias voltage (V_{bias}) for various modulation frequencies [67].

In 2021, Zhang et al. proposed a tunable transparent metamaterial absorber with high optical transparency and broadband microwave absorption performance [68]. The absorber consists of three layers. The resonant layer and the reflective layer are made of indium tin oxide (ITO) films, and the dielectric layer is made of distilled water combined with polymethyl methacrylate, as shown in Figure 30. The absorber can achieve ultra-wideband absorption in 5.8~16.2 GHz, and the absorption rate is greater than 90%. The absorption bandwidth and absorption intensity of the absorber can be adjusted by changing the thickness of the water substrate. Increasing the thickness of the water substrate can improve the absorption capacity of high frequency waves and increase the absorption bandwidth. This achievement can be applied to aviation, medical treatment, research facilities and other fields. However, the structure of the absorber is complex, and the requirements for the manufacturing process are high.

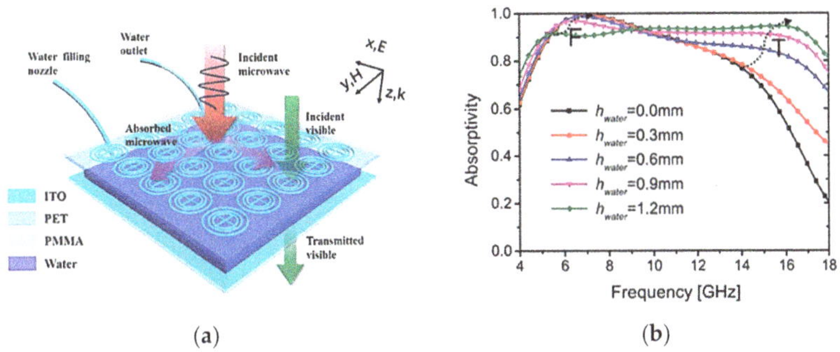

Figure 30. (**a**) Schematic diagram of the tunable, transparent metamaterial absorber with a water-based substrate; (**b**) simulated absorption spectra for normal incident waves with different h_{water}s from 0 mm to 1.2 mm [68].

4. Conclusions

As MPA develops, it also faces higher development requirements, such as simpler structure, richer functions, and more accurate tuning. In conclusion, metamaterials can overcome the defects of traditional materials. The MPA can be designed in a personalized fashion, and its structure is smaller and lighter, so it is convenient to realize more novel electromagnetic functions, and it has broad application prospects in the field of absorbing materials. It can be seen from this paper that compared with traditional absorbers, MPAs can meet the absorption requirements from narrowband to broadband. Combined with photosensitive silicon, graphene, VO_2, and other phase change materials, it can realize optical control, electrical control, and temperature control, and is widely used in sensing, optical switch, communication, stealth equipment, photoelectric detection, and other fields. In theory, MPA can meet the development requirements of miniaturization and integration. However, one of the implementation methods of MPAs is to stack several resonant layers, and its production process inevitably involves multi-layer thin film deposition, which is too complex, so MPAs still face technical challenges in the practical integration of metasurface devices.

In the future, with the maturity of science and technology, MPA will also develop rapidly, and the relevant theory and manufacturing process will also continue to progress. It is expected that MPA will have better performance and be widely used.

Author Contributions: Conceptualization, X.L.; methodology, F.X.; software, X.L.; validation, F.X.; formal analysis, M.W.; resources, J.L.; data curation, X.L.; writing—original draft preparation, X.L.; writing—review and editing, M.Y.; visualization, X.L.; supervision, M.Y. All authors have read and agreed to the published version of the manuscript.

Funding: This research received no external funding.

Institutional Review Board Statement: Not applicable.

Informed Consent Statement: Not applicable.

Data Availability Statement: Data sharing not applicable.

Conflicts of Interest: The authors declare no conflict of interest.

References

1. Pendry, J.B.; Holden, A.J.; Stewart, W.J.; Youngs, I. Extremely Low Frequency Plasmons in Metallic Mesostructures. *Phys. Rev. Lett.* **1996**, *76*, 4773–4776. [CrossRef] [PubMed]
2. Kildishev, A.V.; Boltasseva, A.; Shalaev, V.M. Planar Photonics with Metasurfaces. *Science* **2013**, *339*, 6. [CrossRef] [PubMed]
3. He, J.W.; Dong, T.; Chi, B.H.; Zhang, Y. Metasurfaces for Terahertz Wavefront Modulation: A Review. *J. Infrared Millim. Terahertz Waves* **2020**, *41*, 607–631. [CrossRef]
4. Veselago, V.G. The Electrodynamics of Substance with Simultaneously Negative Values of E and M. *Physics-Uspekhi* **1968**, *10*, 509. [CrossRef]
5. Wei, Z.; Xiaopeng, Z. Advances in Research on Left-Handed Metamaterials. *Mater. Rev.* **2006**, *20*, 26.
6. Liu, C.Q.; Zhang, S.; Wang, S.J.; Cai, Q.N.; Wang, P.; Tian, C.S.; Zhou, L.; Wu, Y.Z.; Tao, Z.S. Active Spintronic-Metasurface Terahertz Emitters with Tunable Chirality. *Adv. Photonics* **2021**, *3*, 19. [CrossRef]
7. Li, X.J.; Yin, J.; Liu, J.J.; Shu, F.Z.; Lang, T.T.; Jing, X.F.; Hong, Z. Resonant Transparency of a Planar Anapole Metamaterial at Terahertz Frequencies. *Photonics Res.* **2021**, *9*, 125–130. [CrossRef]
8. Silalahi, H.M.; Chen, Y.P.; Shih, Y.H.; Chen, Y.S.; Lin, X.Y.; Liu, J.H.; Huang, C.Y. Floating Terahertz Metamaterials with Extremely Large Refractive Index Sensitivities. *Photonics Res.* **2021**, *9*, 1970–1978. [CrossRef]
9. Pendry, J.B.; Schurig, D.; Smith, D.R. Controlling Electromagnetic Fields. *Science* **2006**, *312*, 1780–1782. [CrossRef] [PubMed]
10. Gao, E.D.; Liu, Z.M.; Li, H.J.; Xu, H.; Zhang, Z.B.; Lu, X.; Xiong, C.X.; Liu, C.; Zhang, B.H.; Zhou, F.Q. Dynamically Tunable Dual Plasmon-Induced Transparency and Absorption Based on a Single-Layer Patterned Graphene Metamaterial. *Opt. Express* **2019**, *27*, 13884–13894. [CrossRef]
11. Yu, N.F.; Genevet, P.; Kats, M.A.; Aieta, F.; Tetienne, J.P.; Capasso, F.; Gaburro, Z. Light Propagation with Phase Discontinuities: Generalized Laws of Reflection and Refraction. *Science* **2011**, *334*, 333–337. [CrossRef]
12. Pfeiffer, C.; Grbic, A. Metamaterial Huygens' Surfaces: Tailoring Wave Fronts with Reflectionless Sheets. *Phys. Rev. Lett.* **2013**, *110*, 197401. [CrossRef] [PubMed]
13. Liu, N.; Mesch, M.; Weiss, T.; Hentschel, M.; Giessen, H. Infrared Perfect Absorber and Its Application as Plasmonic Sensor. *Nano Lett.* **2010**, *10*, 2342–2348. [CrossRef]
14. Landy, N.I.; Sajuyigbe, S.; Mock, J.J.; Smith, D.R.; Padilla, W.J. Perfect Metamaterial Absorber. *Phys. Rev. Lett.* **2008**, *100*, 4. [CrossRef] [PubMed]
15. Xu, J.; Li, R.Q.; Qin, J.; Wang, S.Y.; Han, T.C. Ultra-Broadband Wide-Angle Linear Polarization Converter Based on H-Shaped Metasurface. *Opt. Express* **2018**, *26*, 20913–20919. [CrossRef]
16. Zhao, J.; Wang, J.X.; Qiu, W.B.; Zhao, Z.Y. Investigation of Sensing Characteristic of Graphene Metamaterial Based on Fano Resonance. *Laser Optoelectron. Prog.* **2021**, *58*, 6.
17. Bai, L.; Zhang, X.G.; Jiang, W.; Cui, T. Research Progress of Light-Controlled Electromagnetic Metamaterials. *J. Radars* **2021**, *10*, 240–258.
18. Shaltout, A.M.; Shalaev, V.M.; Brongersma, M.L. Spatiotemporal Light Control with Active Metasurfaces. *Science* **2019**, *364*, 6591. [CrossRef] [PubMed]
19. Abdulkarim, Y.I.; Alkurt, F.O.; Awl, H.N.; Muhammadsharif, F.F.; Bakir, M.; Dalgac, S.; Karaaslan, M.; Luo, H. An Ultrathin and Dual Band Metamaterial Perfect Absorber Based on Znse for the Polarization-Independent in Terahertz Range. *Results Phys.* **2021**, *26*, 11. [CrossRef]
20. Le, D.T.; Tong, B.T.; Nguyen, T.K.T.; Cao, T.N.; Nguyen, H.Q.; Tran, M.C.; Truong, C.L.; Bui, X.K.; Vu, D.L.; Nguyen, T.Q.H. Polarization-Insensitive Dual-Band Terahertz Metamaterial Absorber Based on Asymmetric Arrangement of Two Rectangular-Shaped Resonators. *Optik* **2021**, *245*, 8. [CrossRef]
21. Lu, T.G.; Zhang, D.W.; Qiu, P.Z.; Lian, J.Q.; Jing, M.; Yu, B.B.; Wen, J. Ultrathin Terahertz Dual-Band Perfect Metamaterial Absorber Using Asymmetric Double-Split Rings Resonator. *Symmetry* **2018**, *10*, 293. [CrossRef]
22. Schalch, J.; Duan, G.W.; Zhao, X.G.; Zhang, X.; Averitt, R.D. Terahertz Metamaterial Perfect Absorber with Continuously Tunable Air Spacer Layer. *Appl. Phys. Lett.* **2018**, *113*, 5. [CrossRef]
23. Vasic, B.; Gajic, R. Graphene Induced Spectral Tuning of Metamaterial Absorbers at Mid-Infrared Frequencies. *Appl. Phys. Lett.* **2013**, *103*, 4. [CrossRef]
24. Ma, W.; Wen, Y.Z.; Yu, X.M. Broadband Metamaterial Absorber at Mid-Infrared Using Multiplexed Cross Resonators. *Opt. Express* **2013**, *21*, 30724–30730. [CrossRef] [PubMed]
25. Zhu, W.R.; Zhao, X.P.; Gong, B.Y.; Liu, L.H.; Su, B. Optical Metamaterial Absorber Based on Leaf-Shaped Cells. *Appl. Phys. A-Mater. Sci. Process* **2011**, *102*, 147–151. [CrossRef]

26. Zhu, W.R.; Zhao, X.P. Metamaterial Absorber with Dendritic Cells at Infrared Frequencies. *J. Opt. Soc. Am. B-Opt. Phys.* **2009**, *26*, 2382–2385. [CrossRef]
27. Li, M.L.; Muneer, B.; Yi, Z.X.; Zhu, Q. A Broadband Compatible Multispectral Metamaterial Absorber for Visible, near-Infrared, and Microwave Bands. *Adv. Opt. Mater.* **2018**, *6*, 9. [CrossRef]
28. Teperik, T.V.; de Abajo, F.J.G.; Borisov, A.G.; Abdelsalam, M.; Bartlett, P.N.; Sugawara, Y.; Baumberg, J.J. Omnidirectional Absorption in Nanostructured Metal Surfaces. *Nat. Photonics* **2008**, *2*, 299–301. [CrossRef]
29. Feng, R.; Ding, W.Q.; Liu, L.H.; Chen, L.X.; Qiu, J.; Chen, G.Q. Dual-Band Infrared Perfect Absorber Based on Asymmetric T-Shaped Plasmonic Array. *Opt. Express* **2014**, *22*, A335–A343. [CrossRef]
30. Lee, H.S.; Lee, E.H. Analysis of Optical Impedance Matching and Polarization Matching for Integration of Surface Plasmon Polariton Waveguides and Dielectric Waveguides. *J. Korean Phys. Soc.* **2010**, *57*, 1737–1742. [CrossRef]
31. Hedayati, M.K.; Zillohu, A.U.; Strunskus, T.; Faupel, F.; Elbahri, M. Plasmonic Tunable Metamaterial Absorber as Ultraviolet Protection Film. *Appl. Phys. Lett.* **2014**, *104*, 5. [CrossRef]
32. Mattiucci, N.; Bloemer, M.J.; Akozbek, N.; D'Aguanno, G. Impedance Matched Thin Metamaterials Make Metals Absorbing. *Sci. Rep.* **2013**, *3*, 11. [CrossRef] [PubMed]
33. Duan, G.W.; Schalch, J.; Zhao, X.G.; Li, A.B.; Chen, C.X.; Averitt, R.D.; Zhang, X. A Survey of Theoretical Models for Terahertz Electromagnetic Metamaterial Absorbers. *Sens. Actuator A-Phys.* **2019**, *287*, 21–28. [CrossRef]
34. Ghosh, S.; Srivastava, K.V. An Equivalent Circuit Model of Fss-Based Metamaterial Absorber Using Coupled Line Theory. *IEEE Antennas Wirel. Propag. Lett.* **2015**, *14*, 511–514. [CrossRef]
35. Ming, X.S.; Liu, X.Y.; Sun, L.Q.; Padilla, W.J. Degenerate Critical Coupling in All-Dielectric Metasurface Absorbers. *Opt. Express* **2017**, *25*, 24658–24669. [CrossRef]
36. Rothenberg, J.M.; Chen, C.P.; Ackert, J.J.; Dadap, J.I.; Knights, A.P.; Bergman, K.; Osgood, R.M.; Grote, R.R. Experimental Demonstration of Coherent Perfect Absorption in a Silicon Photonic Racetrack Resonator. *Opt. Lett.* **2016**, *41*, 2537–2540. [CrossRef]
37. Dutta-Gupta, S.; Martin, O.J.F.; Gupta, S.D.; Agarwal, G.S. Controllable Coherent Perfect Absorption in a Composite Film. *Opt. Express* **2012**, *20*, 1330–1336. [CrossRef]
38. Luo, X.; Cheng, Z.Q.; Zhai, X.; Liu, Z.M.; Li, S.Q.; Liu, J.P.; Wang, L.L.; Lin, Q.; Zhou, Y.H. A Tunable Dual-Band and Polarization-Insensitive Coherent Perfect Absorber Based on Double-Layers Graphene Hybrid Waveguide. *Nanoscale Res. Lett.* **2019**, *14*, 8. [CrossRef]
39. Tao, H.; Landy, N.I.; Bingham, C.M.; Zhang, X.; Averitt, R.D.; Padilla, W.J. A Metamaterial Absorber for the Terahertz Regime: Design, Fabrication and Characterization. *Opt. Express* **2008**, *16*, 7181–7188. [CrossRef] [PubMed]
40. Ye, Y.Q.; Jin, Y.; He, S.L. Omnidirectional, Polarization-Insensitive and Broadband Thin Absorber in the Terahertz Regime. *J. Opt. Soc. Am. B-Opt. Phys.* **2010**, *27*, 498–504. [CrossRef]
41. Abdulkarim, Y.I.; Awl, H.N.; Alkurt, F.O.; Muhammadsharif, F.F.; Saeed, S.R.; Karaaslan, M.; Bakir, M.; Luo, H. A Thermally Stable and Polarization Insensitive Square-Shaped Water Metamaterial with Ultra-Broadband Absorption. *J. Mater. Res. Technol.-JMRT* **2021**, *13*, 1150–1158. [CrossRef]
42. Astorino, M.D.; Frezza, F.; Tedeschi, N. Ultra-Thin Narrow-Band, Complementary Narrow-Band, and Dual-Band Metamaterial Absorbers for Applications in the Thz Regime. *J. Appl. Phys.* **2017**, *121*, 10. [CrossRef]
43. Cheng, R.J.; Xu, L.; Yu, X.; Zou, L.E.; Shen, Y.; Deng, X.H. High-Sensitivity Biosensor for Identification of Protein Based on Terahertz Fano Resonance Metasurfaces. *Opt. Commun.* **2020**, *473*, 4. [CrossRef]
44. Wang, S.; Wang, Z.; Cao, T. Design of Terahertz Ultra-High Sensitivity Biosensor Based on Flexible Metamaterial. *Electron. Compon. Mater.* **2021**, *40*, 882–887.
45. Fan, K.B.; Suen, J.Y.; Liu, X.Y.; Padilla, W.J. All-Dielectric Metasurface Absorbers for Uncooled Terahertz Imaging. *Optica* **2017**, *4*, 601–604. [CrossRef]
46. Liu, X.Y.; Fan, K.B.; Shadrivov, I.V.; Padilla, W.J. Experimental Realization of a Terahertz All-Dielectric Metasurface Absorber. *Opt. Express* **2017**, *25*, 191–201. [CrossRef]
47. Wen, Q.Y.; Zhang, H.W.; Xie, Y.S.; Yang, Q.H.; Liu, Y.L. Dual Band Terahertz Metamaterial Absorber: Design, Fabrication, and Characterization. *Appl. Phys. Lett.* **2009**, *95*, 3. [CrossRef]
48. Ma, Y.; Chen, Q.; Grant, J.; Saha, S.C.; Khalid, A.; Cumming, D.R.S. A Terahertz Polarization Insensitive Dual Band Metamaterial Absorber. *Opt. Lett.* **2011**, *36*, 945–947. [CrossRef]
49. Shan, Y.; Chen, L.; Shi, C.; Cheng, Z.X.; Zang, X.F.; Xu, B.Q.; Zhu, Y.M. Ultrathin Flexible Dual Band Terahertz Absorber. *Opt. Commun.* **2015**, *350*, 63–70. [CrossRef]
50. Xie, W.L.; Sun, P.; Wang, J.; Feng, H.; Luo, Q.P.; Xie, Q.; Guo, L.H.; Zhang, Z.X.; Sun, J.F.; Zhao, Q.; et al. Polarization-Independent Dual Narrow-Band Perfect Metamaterial Absorber for Optical Communication. *Microw. Opt. Technol. Lett.* **2022**, *64*, 1310–1316. [CrossRef]
51. Wang, Y.; Zhu, D.Y.; Cui, Z.J.; Hou, L.; Lin, L.; Qu, F.F.; Liu, X.X.; Nie, P.C. All-Dielectric Terahertz Plasmonic Metamaterial Absorbers and High-Sensitivity Sensing. *ACS Omega* **2019**, *4*, 18645–18652. [CrossRef]
52. Shen, X.P.; Yang, Y.; Zang, Y.Z.; Gu, J.Q.; Han, J.G.; Zhang, W.L.; Cui, T.J. Triple-Band Terahertz Metamaterial Absorber: Design, Experiment, and Physical Interpretation. *Appl. Phys. Lett.* **2012**, *101*, 4. [CrossRef]

53. Wang, B.X.; Wang, G.Z. Two Compact SRR Resonators Enabling Five-Band Perfect Absorption. *Mater. Lett.* **2016**, *180*, 317–321. [CrossRef]
54. Huang, L.; Chowdhury, D.R.; Ramani, S.; Reiten, M.T.; Luo, S.N.; Taylor, A.J.; Chen, H.T. Experimental Demonstration of Terahertz Metamaterial Absorbers with a Broad and Flat High Absorption Band. *Opt. Lett.* **2012**, *37*, 154–156. [CrossRef] [PubMed]
55. Feng, H.; Li, X.M.; Wang, M.; Xia, F.; Zhang, K.; Kong, W.J.; Dong, L.F.; Yun, M.J. Ultrabroadband Metamaterial Absorbers from Ultraviolet to near-Infrared Based on Multiple Resonances for Harvesting Solar Energy. *Opt. Express* **2021**, *29*, 6000–6010. [CrossRef]
56. Li, X.; Chen, Y.T.; Chen, J.; Jiang, X.P.; He, J.; Jing, Q.; Yang, J.B. Full Spectrum Ultra-Wideband Absorber with Stacked Round Hole Disks. *Optik* **2022**, *249*, 9. [CrossRef]
57. Huang, Y.W.; Kaj, K.; Chen, C.X.; Yang, Z.W.; Ul Haque, S.R.; Zhang, Y.; Zhao, X.G.; Averitt, R.D.; Zhang, X. Broadband Terahertz Silicon Membrane Metasurface Absorber. *ACS Photonics* **2022**, *9*, 1150–1156. [CrossRef]
58. Xu, Z.C.; Gao, R.M.; Ding, C.F.; Wu, L.; Zhang, Y.T.; Yao, J.Q. Photoexcited Broadband Blueshift Tunable Perfect Terahertz Metamaterial Absorber. *Opt. Mater.* **2015**, *42*, 148–151. [CrossRef]
59. Zhao, X.G.; Wang, Y.; Schalch, J.; Duan, G.W.; Crernin, K.; Zhang, J.D.; Chen, C.X.; Averitt, R.D.; Zhang, X. Optically Modulated Ultra-Broadband All-Silicon Metamaterial Terahertz Absorbers. *ACS Photonics* **2019**, *6*, 830–837. [CrossRef]
60. Wang, Y.; Yue, L.S.; Cui, Z.J.; Zhang, X.J.; Zhang, X.; Zhu, Y.Q.; Zhang, K. Optically Tunable Single Narrow Band All-Dielectric Terahertz Metamaterials Absorber. *AIP Adv.* **2020**, *10*, 7. [CrossRef]
61. Lan, J.X.; Zhang, R.X.; Bai, H.; Zhang, C.D.; Zhang, X.; Hu, W.; Wang, L.; Lu, Y.Q. Tunable Broadband Terahertz Absorber Based on Laser-Induced Graphene. *Chin. Opt. Lett.* **2022**, *20*, 4. [CrossRef]
62. Sun, P.; You, C.L.; Mahigir, A.; Liu, T.T.; Xia, F.; Kong, W.J.; Veronis, G.; Dowling, J.P.; Dong, L.F.; Yun, M.J. Graphene-Based Dual-Band Independently Tunable Infrared Absorber. *Nanoscale* **2018**, *10*, 15564–15570. [CrossRef]
63. Wu, D.; Wang, M.; Feng, H.; Xu, Z.X.; Liu, Y.P.; Xia, F.; Zhang, K.; Kong, W.J.; Dong, L.F.; Yun, M.J. Independently Tunable Perfect Absorber Based on the Plasmonic Properties in Double-Layer Graphene. *Carbon* **2019**, *155*, 618–623. [CrossRef]
64. Feng, H.; Xu, Z.X.; Li, K.; Wang, M.; Xie, W.L.; Luo, Q.P.; Chen, B.Y.; Kong, W.J.; Yun, M.J. Tunable Polarization-Independent and Angle-Insensitive Broadband Terahertz Absorber with Graphene Metamaterials. *Opt. Express* **2021**, *29*, 7158–7167. [CrossRef]
65. Zhang, Z.X.; Xie, Q.; Guo, L.H.; Su, C.X.; Wang, M.; Xia, F.; Sun, J.F.; Li, K.; Feng, H.; Yun, M.J. Dual-Controlled Tunable Dual-Band and Ultra-Broadband Coherent Perfect Absorber in the Thz Range. *Opt. Express* **2022**, *30*, 30832–30844. [CrossRef] [PubMed]
66. Huang, X.; He, W.; Yang, F.; Ran, J.; Yang, Q.; Xie, S.Y. Thermally Tunable Metamaterial Absorber Based on Strontium Titanate in the Terahertz Regime. *Opt. Mater. Express* **2019**, *9*, 1377–1385. [CrossRef]
67. Shrekenhamer, D.; Chen, W.C.; Padilla, W.J. Liquid Crystal Tunable Metamaterial Absorber. *Phys. Rev. Lett.* **2013**, *110*, 177403. [CrossRef]
68. Zhang, Y.Q.; Dong, H.X.; Mou, N.L.; Li, H.N.; Yao, X.; Zhang, L. Tunable and Transparent Broadband Metamaterial Absorber with Water-Based Substrate for Optical Window Applications. *Nanoscale* **2021**, *13*, 7831–7837. [CrossRef]

photonics

Communication

Synthetic Mueller Imaging Polarimetry

José J. Gil [1,*] and Ignacio San José [2]

1 Photonic Technologies Group, Universidad de Zaragoza, Pedro Cerbuna 12, 50009 Zaragoza, Spain
2 Instituto Aragonés de Estadística, Gobierno de Aragón, Bernardino Ramazzini 5, 50015 Zaragoza, Spain; isanjose@aragon.es
* Correspondence: ppgil@unizar.es

Abstract: The transformation of the state of polarization of a light beam via its linear interaction with a material medium can be modeled through the Stokes–Mueller formalism. The Mueller matrix associated with a given interaction depends on many aspects of the measurement configuration. In particular, different Mueller matrices can be measured for a fixed material sample depending on the spectral profile of the light probe. For a given light probe and a given sample with inhomogeneous spatial behavior, the polarimetric descriptors of the point-to-point Mueller matrices can be mapped, leading to respective polarimetric images. The procedure can be repeated sequentially using light probes with different central frequencies. In addition, the point-to-point Mueller matrices, consecutively measured, can be combined synthetically through convex sums leading to respective new Mueller matrices, in general with increased polarimetric randomness, thus exhibiting specific values for the associated polarimetric descriptors, including the indices of polarimetric purity, and generating new polarimetric images which are different from those obtained from the original Mueller matrices. In this work, the fundamentals for such synthetic generation of additional polarimetric images are described, providing a new tool that enhances the exploitation of Mueller polarimetry.

Keywords: Mueller matrices; polarimetry; diattenuation; polarizance; depolarization

1. Introduction

Nowadays, Mueller imaging polarimetry constitutes a mature set of technologies and procedures with an enormous variety of applications that, among many others, cover areas like analysis of biological tissues [1,2] and remote sensing [3–5].

Once the point-to-point Mueller matrices of a given sample have been measured, the mathematical treatment of the information obtained is crucial for its exploitation. In fact, the methods used for the serial or parallel decompositions of the measured Mueller matrices as well as for the identification of significant physical descriptors related to the enpolarizing, retarding and depolarizing properties of the sample stand out in polarimetry.

Mueller polarimeters are composed of (see Figure 1) [6–16]

- A collimated light source with a fixed spectral profile.
- A polarization state generator (PSG) that modulates the state of a polarization incident on the material sample.
- A setup for the appropriate placement and relative orientation of the material sample on which the light probe beam impinges.
- A polarization state analyzer (PSA) that modulates the state of polarization of the light probe after its interaction with the sample.
- A detection, record and processing device that allows the measurement of the intensity of the light probe after passing through the PSG, the sample and the PSA.

Citation: Gil, J.J.; San José, I. Synthetic Mueller Imaging Polarimetry. *Photonics* **2023**, *10*, 969. https://doi.org/10.3390/photonics10090969

Received: 24 July 2023
Revised: 16 August 2023
Accepted: 21 August 2023
Published: 24 August 2023

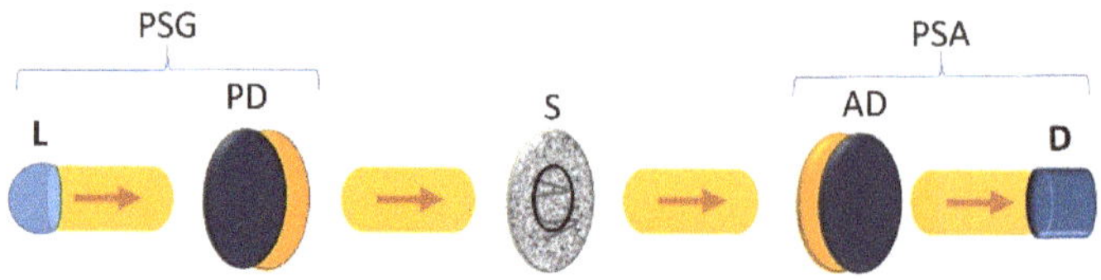

Figure 1. General structure of an imaging Mueller polarimeter. The polarization state generator (PSG) is constituted by the light source (L), which provides a collimated beam, and a polarizing device (PD). After emerging from the PSG, the light probe is incident on a spatial section of the sample (S), which in general exhibits heterogeneous polarimetric behavior in the area covered by the light probe and whose point-to-point Mueller matrices are under measurement. The polarization state analyzer (PSA) is constituted by an analyzing device (AD), and the detection and processing system (D).

Specific Mueller polarimeters are designed for their operation on samples with a homogeneous behavior over the cross-section covered by the light probe, in which case a single Mueller matrix is obtained for each measurement, while the so-called imaging Mueller polarimeters are designed for the measurement of the point-to-point Mueller matrices covering a certain area of the sample with heterogeneous spatial behavior.

Thus, in Mueller imaging polarimetry, the sample can be considered pixelated, in such a manner that a different Mueller matrix is measured for each pixel (Figure 2). Note that we are using the term pixels to refer to small areas of the sample, which in general have a common shape and size, so that a number of pixels cover the entire cross-section of the sample under measurement. Therefore, each pixel is characterized by a specific and fixed Mueller matrix, which is in general different from those of other pixels. Since, in general, a given Mueller matrix can be parameterized in terms of up to sixteen independent parameters, up to sixteen independent images can be obtained for each complete polarimetric measurement. Nevertheless, when appropriate, additional images can be generated from specific polarimetric descriptors.

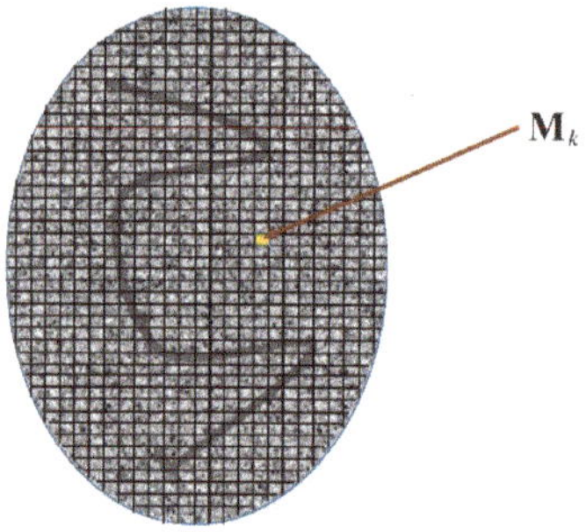

Figure 2. The area of the sample under measurement impinged by the light probe of the polarimeter is composed of a number m of small elements (pixels) whose respective Mueller matrices are measured. For a particular pixel k, its corresponding Mueller matrix is denoted by $\mathbf{M}_k$, so that the subscript k runs the entire set of m pixels $(k = 1, \ldots, m)$. When a series of n independent measurements is performed for a fixed sample, the respective Mueller matrices of the pixel k are denoted as $\mathbf{M}_{kl}$, with $l = 1, \ldots, n$.

Let us now consider a sequence of n independent measurements over a given fixed material sample, which are performed with different light probes with respective spectral profiles (Figure 3), in such a manner that each independent measurement produces a specific set of m point-to point Mueller matrices, which are denoted as $\mathbf{M}_{kl}$ $(k = 1, \ldots, m,$ $l = 1, \ldots, n)$ where indices k and l refer, respectively to the k pixel and the l measurement.

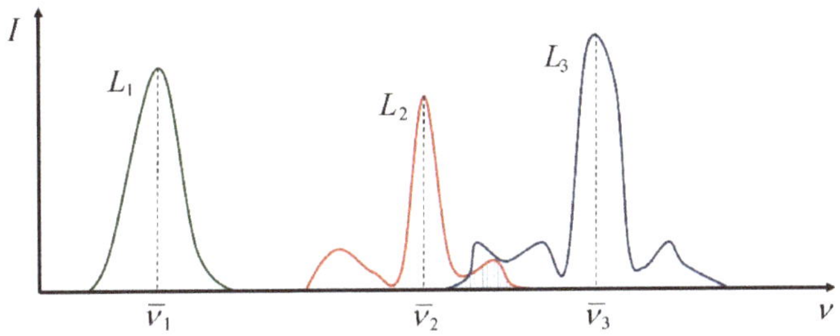

Figure 3. For a given Mueller imaging polarimeter, different complete measurements can be performed by varying the spectral profile of the light probe (L_1, L_2 and L_3 in the figure), leading to respective sets of Mueller matrices associated with the points (pixels) of the sample. The Mueller matrix of a given pixel depends on both the central frequency and the shape of the spectral profile of the light probe. The successive measurements can be performed either with spectral profiles that partially overlap or not. The shaded area represents some overlap of L_2 and L_3.

Depending on the central frequency, shape and width of the spectral profile used in a given measurement l, the Mueller matrix of the k pixel adopts a specific form $\mathbf{M}_{kl}$, which can be either nondepolarizing (or pure, in the sense that it preserves the degree of polarization of incident totally polarized light), or depolarizing. In general, samples exhibiting inhomogeneous behavior over the cross-section covered by the light probe are considered and consequently, the separate representations of the sixteen elements $m_{kl,ij}$ $(i, j = 0, 1, 2, 3)$, where k maps the entire set of m pixels of the sample under measurement, lead to specific images.

It is well known that the elements $m_{kl,ij}$ of $\mathbf{M}_{kl}$ do not provide direct and independent information about the polarimetric properties of the sample for the spectral profile of the light probe used in the corresponding measurement, but are related to the enpolarizing (diattenuating and polarizing), retarding and depolarizing properties in an intricate manner. This is the reason why alternative sets of polarimetric descriptors (other than $m_{kl,ij}$) are frequently used to generate more significant polarimetric images that describe, in a more or less decoupled manner, the spatial variation of specific properties. A possible parameterization of $\mathbf{M}_{kl}$ in terms of sixteen independent parameters directly related to specific polarimetric features is based on the so-called arrow decomposition of Mueller matrices [17–20].

The main argument which will be dealt with in this work relies on the fact that, beyond the polarimetric images obtained for each measurement l, the information already obtained through a set of n independent measurements of $\mathbf{M}_{kl}$ can be used to generate new sets of synthetic images that enhance the contrast or visualization of certain properties or structures of the sample. In particular, for instance, let us build the Mueller matrix

$$\mathbf{M}_k = c_1\mathbf{M}_{k1} + c_2\mathbf{M}_{k2} + \ldots + c_n\mathbf{M}_{kn}, \\ [c_1 + c_2 + \ldots + c_n = 1, c_i > 0, i = 1, \ldots n], \quad (1)$$

which is directly synthesized from the independent n measurements but whose associated enpolarizing, retarding and depolarizing properties are different from those of each $\mathbf{M}_{kl}$. In fact, $\mathbf{M}_k$ carries information about the variation of the polarimetric behavior of the sample as the spectral profile of the probing light is modified. In other words, the set of measured $\mathbf{M}_{kl}$ contains implicit information on certain properties of the sample that can only be represented in the form of polarimetric images obtained from synthetic parallel combinations performed as indicated in Equation (1).

It should be emphasized that the abovementioned Mueller imaging measurements are performed through conventional methods and that the aim of this work is to describe the general theoretical framework used for the synthesis, through a very simple averaging procedure which does not add computational complexity, of significant polarimetric images based on the parallel composition of the Mueller matrices obtained through a set of n (with $n \geq 2$) independent conventional measurements of a given material sample. To achieve this, this communication is organized as follows. The main necessary concepts and notations are summarized in Section 2. Section 3 is devoted to the general formulation of the synthetic Mueller imaging polarimetry approach, whose nature and scope is discussed in Section 4. The main conclusions are summarized in Section 5.

Note that although the term "light" is used in this communication, it should be understood in the wide sense of referring generally to arbitrary frequencies within different areas of the electromagnetic spectrum and not exclusively within the optical range. This is the case, for example, for synthetic aperture radar polarimetry (SAR polarimetry) for which the electromagnetic probe belongs to the microwave range.

2. Theoretical Background

The transformation of polarized light by the action of a linear medium (under fixed interaction conditions) can always be represented mathematically as $\mathbf{s}' = \mathbf{M}\mathbf{s}$, where $\mathbf{s}$ and $\mathbf{s}'$ are the Stokes vectors that represent the polarization states of the incident and emerging light beams, and $\mathbf{M}$ is the Mueller matrix that performs the linear transformation, which can be expressed in the partitioned form [21–23]

$$\mathbf{M} = m_{00}\hat{\mathbf{M}},\ \hat{\mathbf{M}} \equiv \begin{pmatrix} 1 & \mathbf{D}^T \\ \mathbf{P} & \mathbf{m} \end{pmatrix},$$
$$\mathbf{m} \equiv \frac{1}{m_{00}} \begin{pmatrix} m_{11} & m_{12} & m_{13} \\ m_{21} & m_{22} & m_{23} \\ m_{31} & m_{32} & m_{33} \end{pmatrix}, \tag{2}$$
$$\mathbf{D} \equiv \frac{(m_{01}, m_{02}, m_{03})^T}{m_{00}},\ \mathbf{P} \equiv \frac{(m_{10}, m_{20}, m_{30})^T}{m_{00}},$$

where m_{ij} $(i,j = 0,1,2,3)$ denote the elements of $\mathbf{M}$; superscript T indicates transpose; m_{00} represents the mean intensity coefficient (MIC), i.e., the ratio between the intensity of the emerging light and the intensity of incident unpolarized light; $\mathbf{D}$ and $\mathbf{P}$ are the diattenuation and polarizance vectors, with absolute values D (diattenuation) and P (polarizance), and $\mathbf{m}$ is the normalized 3×3 submatrix associated with $\mathbf{M}$.

A proper measure of the ability of $\mathbf{M}$ to preserve the degree of polarization of totally polarized incident light is given by the degree of polarimetric purity of $\mathbf{M}$ (also called depolarization index) [24], P_Δ, which can be expressed as

$$P_\Delta = \sqrt{\frac{D^2 + P^2 + 3P_S^2}{3}} = \sqrt{\frac{2P_P^2}{3} + P_S^2} \quad \left[P_P \equiv \sqrt{\frac{D^2 + P^2}{2}} \right], \tag{3}$$

where P_S is the polarimetric dimension index (also called the degree of spherical purity), defined as

$$P_S \equiv \frac{\|\mathbf{m}\|_2}{\sqrt{3}} \left[\|\mathbf{m}\|_2 \equiv \frac{1}{m_{00}} \sqrt{\sum_{k,l=1}^{3} m_{kl}^2} \right], \tag{4}$$

$\|\mathbf{m}\|_2$ being the Frobenius norm of $\mathbf{m}$, and P_P is the so-called degree of polarizance, or enpolarizance, giving an overall measure of the power of $\mathbf{M}$ to increase the degree of polarization of the interacting light in either forward or reverse interactions [17–19].

Maximum degree of polarimetric purity, $P_\Delta = 1$, is exhibited uniquely by nondepolarizing (or pure) media (i.e., media that do not decrease the degree of polarization of totally polarized incident light), while $P_\Delta = 0$ corresponds to perfect depolarizers, with an associated Mueller matrix $\mathbf{M}_{\Delta 0} = m_{00}\mathrm{diag}(1,0,0,0)$. Maximum P_S, $P_S = 1$, implies $P_\Delta = 1$ with $P_P = 0$ (nondepolarizing and nonenpolarizing media), which corresponds uniquely to retarders; minimal P_S, $P_S = 0$, corresponds to media exhibiting $\mathbf{m} = \mathbf{0}$. Maximal enpolarizance,$P_P = 1$, implies $P_\Delta = 1$ and corresponds to perfect polarizers, while the minimal, $P_P = 0$, is exhibited by nonenpolarizing interactions (either nondepolarizing or depolarizing) [18,19].

In general, two kinds of decompositions of a Mueller matrix can be considered, namely serial decompositions (through product of Mueller matrices) and parallel decompositions (through weighted sums of Mueller matrices) [25,26].

Parallel decompositions consist of representing a Mueller matrix as a weighted sum of Mueller matrices. The physical meaning of parallel decompositions is that the incoming electromagnetic wave splits into a set of pencils that interact, without overlapping, with a number of material components that are spatially distributed in the illuminated area, and the emerging pencils are incoherently recombined into the emerging beam.

Thus, the notion of parallel composition of Mueller matrices underlies the very concept of the Mueller matrix and obeys the rule that the coefficients of the Mueller components in the sum should be positive and sum to one (convex sum) [25,26]. This property is directly linked to the so-called covariance criterion, namely, given a Mueller matrix $\mathbf{M}$, its associated Hermitian coherency matrix $\mathbf{C}(\mathbf{M})$ is positive semidefinite [27], which is equivalent to the fact that any Mueller matrix can be expressed as a sum of nondepolarizing Mueller matrices [28]. The explicit expression of $\mathbf{C}(\mathbf{M})$, in terms of the elements m_{ij} of $\mathbf{M}$, is given by [26,27]

$$\mathbf{C}(\mathbf{M}) = \frac{1}{4}\begin{pmatrix} \begin{matrix} m_{00}+m_{11} \\ +m_{22}+m_{33} \end{matrix} & \begin{matrix} m_{01}+m_{10} \\ -i(m_{23}-m_{32}) \end{matrix} & \begin{matrix} m_{02}+m_{20} \\ +i(m_{13}-m_{31}) \end{matrix} & \begin{matrix} m_{03}+m_{30} \\ -i(m_{12}-m_{21}) \end{matrix} \\ \begin{matrix} m_{01}+m_{10} \\ +i(m_{23}-m_{32}) \end{matrix} & \begin{matrix} m_{00}+m_{11} \\ -m_{22}-m_{33} \end{matrix} & \begin{matrix} m_{12}+m_{21} \\ +i(m_{03}-m_{30}) \end{matrix} & \begin{matrix} m_{13}+m_{31} \\ -i(m_{02}-m_{20}) \end{matrix} \\ \begin{matrix} m_{02}+m_{20} \\ -i(m_{13}-m_{31}) \end{matrix} & \begin{matrix} m_{12}+m_{21} \\ -i(m_{03}-m_{30}) \end{matrix} & \begin{matrix} m_{00}-m_{11} \\ +m_{22}-m_{33} \end{matrix} & \begin{matrix} m_{23}+m_{32} \\ +i(m_{01}-m_{10}) \end{matrix} \\ \begin{matrix} m_{03}+m_{30} \\ +i(m_{12}-m_{21}) \end{matrix} & \begin{matrix} m_{13}+m_{31} \\ +i(m_{02}-m_{20}) \end{matrix} & \begin{matrix} m_{23}+m_{32} \\ -i(m_{01}-m_{10}) \end{matrix} & \begin{matrix} m_{00}-m_{11} \\ -m_{22}+m_{33} \end{matrix} \end{pmatrix}. \tag{5}$$

The general formulation for the synthesis of a Mueller matrix associated with a linear polarimetric interaction from the properties of the parallel components of the medium on which the considered light beam interacts has been dealt with in [26,29–34].

Since $\mathbf{C}$ is a positive semi-definite Hermitian matrix, it can always be diagonalized through a unitary transformation of the form

$$\mathbf{C} = \mathbf{U}\mathrm{diag}(\lambda_0, \lambda_1, \lambda_2, \lambda_3)\mathbf{U}^\dagger. \tag{6}$$

where λ_i are the four ordered $(0 \le \lambda_3 \le \lambda_2 \le \lambda_1 \le \lambda_0)$ nonnegative eigenvalues of $\mathbf{C}$. The columns $\mathbf{u}_i$ $(i = 0,1,2,3)$ of the 4×4 unitary matrix $\mathbf{U}$ are the orthonormal eigenvectors of $\mathbf{C}$.

Nondepolarizing Mueller matrices have the genuine property that they exhibit a single nonzero eigenvalue (i.e., $\lambda_0 \neq 0$, $\lambda_1 = \lambda_2 = \lambda_3 = 0$) and hereafter, wherever appropriate, we will use the subscript J to refer to Mueller and coherency matrices associated with nondepolarizing media, which will be generically denoted as $\mathbf{M}_J$ and $\mathbf{C}_J$. The coherency matrix $\mathbf{C}_J$ associated with a nondepolarizing Mueller matrix $\mathbf{M}_J$ can always be expressed as $\mathbf{C}_J = m_{00}(\mathbf{c} \otimes \mathbf{c}^\dagger)$, where $\mathbf{c}$ (called the coherency vector of $\mathbf{M}_J$) is a unit vector with four complex components [34].

Complete quantitative information on the structure of polarimetric purity-randomness exhibited by the interaction represented by $\mathbf{M}$ is provided by the set of three indices of polarimetric purity (IPP) defined as follows from the normalized eigenvalues of $\mathbf{C}$ [35]

$$\begin{gathered} P_1 \equiv \hat{\lambda}_0 - \hat{\lambda}_1,\ P_2 \equiv \hat{\lambda}_0 + \hat{\lambda}_1 - 2\hat{\lambda}_2,\ P_3 \equiv \hat{\lambda}_0 + \hat{\lambda}_1 + \hat{\lambda}_2 - 3\hat{\lambda}_3, \\ \left[\hat{\lambda}_i \equiv \lambda_i / m_{00}\ (i = 0,1,2,3),\ 0 \le P_1 \le P_2 \le P_3 \le 1\right]. \end{gathered} \tag{7}$$

Given a Mueller matrix $\mathbf{M}$, it can always be submitted to the so-called arbitrary decomposition [25,36]

$$\begin{gathered} \mathbf{M} = \sum_{i=0}^{r-1} k_i\, \mathbf{M}_{Ji}, \\ \mathbf{M}_{Ji} = m_{00i}\hat{\mathbf{M}}_{Ji},\ \hat{\mathbf{M}}_{Ji} \equiv \begin{pmatrix} 1 & \mathbf{D}_i^T \\ \mathbf{P}_i & \mathbf{m}_i \end{pmatrix} \equiv \mathbf{M}_J\,(\mathbf{c}_i), \\ m_{00i}(1 + D_i) \le 1,\ k_i = \frac{m_{00}}{m_{00i}} \frac{1}{\mathbf{c}_i^\dagger\, \mathbf{C}^-\, \mathbf{c}_i},\ \left[\sum_{i=0}^{r-1} k_i = 1\right], \end{gathered} \tag{8}$$

where $r = \mathrm{rank}\mathbf{C}$, with $1 \le r \le 4$, which coincides with the number of independent nondepolarizing parallel components of $\mathbf{M}$; $\mathbf{C}^-$ is the pseudoinverse of $\hat{\mathbf{C}}$, defined as $\mathbf{C}^- = \mathbf{U}\boldsymbol{\Lambda}^-\ \mathbf{U}^\dagger$, with $\boldsymbol{\Lambda}^-$ being the diagonal matrix whose r first diagonal elements are $1/\lambda_0, 1/\lambda_1, \ldots, 1/\lambda_{r-1}$ and whose last $4 - r$ elements are zero [36].

Consider now the following modified singular value decomposition of the submatrix $\mathbf{m}$ of $\mathbf{M}$,

$$\begin{gathered} \mathbf{m}{=}\mathbf{m}_{RO}\mathbf{m}_A\mathbf{m}_{RI}, \\ \begin{bmatrix} \mathbf{m}_{Ri}^{-1} = \mathbf{m}_{Ri}^{\mathrm{T}},\ \det\mathbf{m}_{Ri} = +1 (i = I, O), \\ \mathbf{m}_A \equiv \mathrm{diag}(a_1, a_2, \varepsilon a_3),\ a_1 \ge a_2 \ge a_3 \ge 0,\ \varepsilon \equiv \det\mathbf{m}/|\det\mathbf{m}|, \end{bmatrix} \end{gathered} \tag{9}$$

where the nonnegative parameters (a_1, a_2, a_3) are the singular values of $\mathbf{m}$ (taken in decreasing order), and $\mathbf{m}_{Ri}$ are proper orthogonal 3×3 matrices, and consequently the associated 4×4 matrices of the form

$$\mathbf{M}_{Ri} = \begin{pmatrix} 1 & \mathbf{0}^{\mathrm{T}} \\ \mathbf{0} & \mathbf{m}_{Ri} \end{pmatrix} (i = I, O) \tag{10}$$

are orthogonal Mueller matrices (representing respective transparent retarders). The arrow form $\mathbf{M}_A\ (\mathbf{M})$ of $\mathbf{M}$ is then defined as [17]

$$\begin{gathered} \mathbf{M}_A(\mathbf{M}) \equiv \mathbf{M}_{RO}^{\mathrm{T}}\ \mathbf{M}\mathbf{M}_{RI}^{\mathrm{T}} = m_{00} \begin{pmatrix} 1 & \mathbf{D}_A^{\mathrm{T}} \\ \mathbf{P}_A & \mathbf{m}_A \end{pmatrix} \\ \begin{bmatrix} \mathbf{m}_A \equiv \mathbf{m}_{RO}^{\mathrm{T}}\ \mathbf{m}\mathbf{m}_{RI}^{\mathrm{T}} = \mathrm{diag}(a_1, a_2, \varepsilon a_3) \\ a_1 \ge a_2 \ge a_3 \ge 0,\ \varepsilon \equiv \det\mathbf{m}/|\det\mathbf{m}| \\ \mathbf{D}_A = \mathbf{m}_{RI}\,\mathbf{D},\ \mathbf{P}_A = \mathbf{m}_{RO}^{T}\,\mathbf{P} \end{bmatrix} \end{gathered} \tag{11}$$

and contains up to ten nonzero elements. The corresponding arrow decomposition of $\mathbf{M}$ is defined as

$$\mathbf{M} = \mathbf{M}_{RO}\mathbf{M}_A\,\mathbf{M}_{RI}. \tag{12}$$

Note that the diattenuation and polarizance vectors, $\mathbf{D}$ and $\mathbf{P}$, of $\mathbf{M}$ are recovered from those of $\mathbf{M}_A$ through the respective transformations $\mathbf{D} = \mathbf{m}_{RI}^{\mathrm{T}}\mathbf{D}_A$ and $\mathbf{P} = \mathbf{m}_{RO}\,\mathbf{P}_A$, which preserve the absolute values of the transformed vectors.

The arrow decomposition allows for the parameterization of $\mathbf{M}$ in terms of the following significant sixteen parameters [20,37]:

- The three angular parameters $(\varphi_I, \chi_I, \Delta_I)$ determining the entrance retarder, where φ_I and χ_I are the azimuth and ellipticity of the fast eigenstate and Δ_I is the retardance;
- The three angular parameters $(\varphi_O, \chi_O, \Delta_O)$ determining the exit retarder;

- The three parameters (φ_D, χ_D, D) determining the diattenuation vector $\mathbf{D}$ of $\mathbf{M}$, where φ_D and χ_D are the azimuth and ellipticity of the eigenstate with smaller attenuation; or, alternatively, the three parameters $(\varphi_{DA}, \chi_{DA}, D)$ determining the diattenuation vector $\mathbf{D}_A = \mathbf{m}_{RI}\mathbf{D}$ of $\mathbf{M}_A$;
- The three parameters (φ_P, χ_P, P) determining the polarizance vector $\mathbf{P}$ of $\mathbf{M}$ or, alternatively, the three parameters $(\varphi_{PA}, \chi_{PA}, P)$ determining the polarizance vector $\mathbf{P}_A = \mathbf{m}_{RO}^T\mathbf{D}$ of $\mathbf{M}_A$;
- The three indices of polarimetric purity P_1, P_2, P_3 of $\mathbf{M}$ (which coincide with those of $\mathbf{M}_A$);
- The MIC m_{00} of $\mathbf{M}$ (which coincides with that of $\mathbf{M}_A$).

Consequently, independent mappings of the above sixteen descriptors provide respective polarimetric images that reflect the variations and associated contrast of each parameter. Obviously, many other parameterizations of $\mathbf{M}$ are also possible, but we emphasize the fact that those listed above provide separate information on the enpolarizing ($\mathbf{D}$ and $\mathbf{P}$ vectors), retarding ($\mathbf{M}_{RI}$ and $\mathbf{M}_{RO}$ matrices), depolarizing (P_1, P_2, P_3) and MIC (m_{00}) properties for each spectral profile of the light probe used in the imaging polarimeter.

Furthermore, despite the implicit redundancy of the information held by other descriptors, it can be useful to generate additional images associated with parameters like, for instance, P_Δ, P_S and P_P.

The parallel composition of Mueller matrices that only differ in their MICs does not produce changes in the fifteen remaining parameters described above. Thus, hereafter, the fifteen parameters $\varphi_I, \chi_I, \Delta_I, \varphi_O, \chi_O, \Delta_O, \varphi_D, \chi_D, D, \varphi_P, \chi_P, P, P_1, P_2, P_3$ will be referred to as the polarimetric descriptors of the Mueller matrix to which they correspond.

In the case of nondepolarizing Mueller matrices, the arrow decomposition adopts the symmetric form [26,38,39]

$$\mathbf{M}_J = \mathbf{M}_{RO}\,\mathbf{M}_{DL0}\,\mathbf{M}_{RI} \tag{13}$$

where the central matrix has the general form

$$\mathbf{M}_{DL0} = m_{00}\begin{pmatrix} 1 & D & 0 & 0 \\ D & 1 & 0 & 0 \\ 0 & 0 & \sqrt{1-D^2} & 0 \\ 0 & 0 & 0 & \sqrt{1-D^2} \end{pmatrix}, \tag{14}$$

so that it represents a horizontal-aligned linear diattenuator, while the set of entrance and exit retarders depends on five parameters through the expressions [26]

$$\mathbf{M}_{RI} = \mathbf{M}_{RL0}\,(\Delta/2)\mathbf{M}_{RL1}\,(\varphi_1, \Delta_1),\ \ \mathbf{M}_{RO} = \mathbf{M}_{RL2}(\varphi_2, \Delta_2)\,\mathbf{M}_{RL0}\,(\Delta/2), \tag{15}$$

where $\mathbf{M}_{RL1}\,(\varphi_1, \Delta_1)$, $\mathbf{M}_{RL2}(\varphi_2, \Delta_2)$ and $\mathbf{M}_{RL0}\,(\Delta/2)$ represent respective linear retarders with retardances Δ_1, Δ_2 and $\Delta/2$, and whose fast axes are oriented at angles φ_1, φ_2 and zero. Recall that diattenuation and polarizance necessarily coincide for nondepolarizing Mueller matrices $(P = D)$.

3. Synthetic Mueller Imaging Polarimetry

For the sake of clarity, let us first consider a fixed material sample submitted to two independent point-to-point Mueller measurements which only differ in the spectral profile of the light probe, with respective central frequencies denoted as ν_1. and ν_2. It is also assumed that the shift $\nu_2 - \nu_1$ is big enough for the sample to have different polarimetric behavior, that is to say, some or all of the set of fifteen polarimetric descriptors described in the last paragraph of Section 2 take different values for both light probes. No particular assumptions are made about the shapes and bandwidths of the spectral profiles, although the implications of their possible partial overlap will be discussed in Section 4.

Let $\mathbf{M}_{kl}$ $(k = 1,\ldots,m,\ l = 1,2)$ be the point-to-point Mueller matrices obtained from each independent measurement ($l = 1,2$ corresponding, respectively, to the light probes with central frequencies ν_1. and ν_2), where we are using the notation introduced in Section 1.

For a given pixel k, one can always build the synthetic Mueller matrix

$$\mathbf{M}_k = \frac{1}{2}\mathbf{M}_{k1} + \frac{1}{2}\mathbf{M}_{k2}. \quad (16)$$

As long as a certain property of the sample behaves differently in both measurements, a respective new polarimetric image associated with $\mathbf{M}_k$ arises. Note that other pairs of coefficients c_1, c_2 (with $c_1 + c_2 = 1$) different from $c_1 = c_2 = 1/2$ can be used, but there is no apparent advantage in straying from the simpler option.

To emphasize the power of this simple procedure, let us consider the special (but common) case in which both $\mathbf{M}_{k1}$ and $\mathbf{M}_{k2}$ are nondepolarizing. Then, provided $\hat{\mathbf{M}}_{k1} \neq \hat{\mathbf{M}}_{k2}$ (that is, proportional Mueller matrices are excluded because of their equivalent polarimetric behavior, which only differ in their respective MICs), $\mathbf{M}_k$ is depolarizing, so that its polarimetric descriptors differ from those of both $\mathbf{M}_{k1}$ and $\mathbf{M}_{k2}$. In particular, the inequality $P_1 < 1$ is necessarily satisfied (recall that nondepolarizing Mueller matrices always satisfy $P_1 = P_2 = P_3 = P_\Delta = 1$) and, consequently, the mapping of P_1 associated with each pixel k provides a new and peculiar polarimetric image whose nature is substantially different from the images obtained for $\mathbf{M}_{k1}$ and $\mathbf{M}_{k2}$. In fact, it can be said that certain images obtained from the synthetic $\mathbf{M}_k$ provide information about the differences in the polarimetric behavior of the sample for both spectral profiles considered.

Obviously, when either $\mathbf{M}_{k1}$ or $\mathbf{M}_{k2}$ is depolarizing, the polarimetric descriptors of $\mathbf{M}_k$ are also genuinely different from those of the sole $\mathbf{M}_{k1}$ or $\mathbf{M}_{k2}$.

The power of the synthetic procedure for the generation of images that hold information about the changes in the polarimetric behavior for different spectral profiles of the light probe can be enhanced by using more than two light probes (i.e., $n > 2$). In general, new point-to-point Mueller matrices can be synthesized through the simple procedure indicated in Equation (1). The larger n and the larger $\nu_2 - \nu_1$ are (within the range considered: optical, microwave...), the greater polarimetric randomness is generated in $\mathbf{M}_k$, in such a manner that the point-to-point values of the three IPP of $\mathbf{M}_k$ decrease correspondingly and move away from unity. In general, up to three new respective images for the three IPP are generated, while the images obtained for the remaining twelve polarimetric descriptors adopt specific new features, which can reflect different aspects of the variation of the polarimetric behavior of the sample as different spectral profiles of the light probe are used.

Thus, the above procedure for obtaining synthetic Mueller matrices $\mathbf{M}_k$ for the respective pixels of the sample as well as the associated synthetic polarimetric images determined by a set of polarimetric descriptors (for instance those derived from the arrow decomposition of each $\mathbf{M}_k$), can be straightforwardly generalized for n independent measurements performed with respective light probes (see Figure 4).

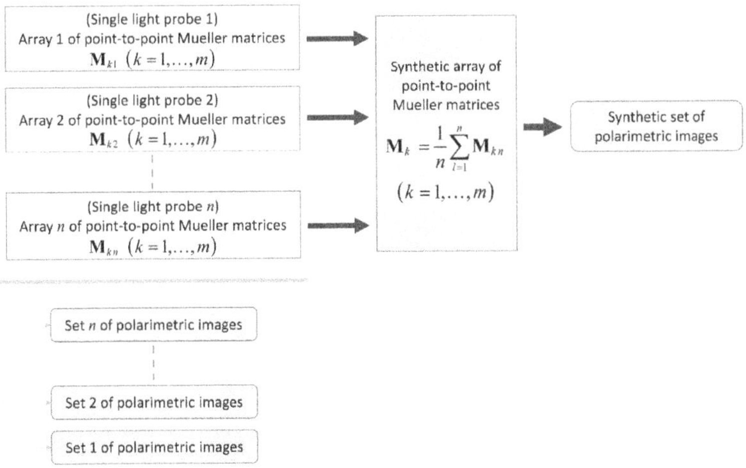

Figure 4. Conventional Mueller imaging polarimetry provides sets of images based on the point-to-point dependence of the polarimetric descriptors derived from the Mueller matrices measured for each pixel in the sample. By maintaining the configuration of the polarimeter and the sample, different conventional measurements can be performed for respective spectral profiles of the light probe. Once a number of n, with $n \geq 2$, independent measurements is performed, new pixel-to-pixel Mueller matrices can be synthesized, without the necessity of additional measurements, by averaging the measured ones, providing new polarimetric images, which are different from those of the original (measured) Mueller matrices and reflect the differences of the polarimetric behavior of the sample for the light probes used. When appropriate, in accordance with the criterion of the polarimeter operator, the coefficients in the convex sum of the measured Mueller matrices can be different and not equal to $1/n$.

4. Discussion

Given a point-to-point Mueller matrix measurement on a given material sample with spatially heterogeneous polarimetric behavior (under fixed interaction conditions; i.e., relative orientation of the sample with respect to the direction of the collimated light probe, transmission, reflection or scattering operation mode, angle of observation, etc.), infinite polarimetric images can be generated through the spatial dependence of parameters defined as functions of the sixteen elements of the measured Mueller matrices. Nevertheless, once the spectral profile of the light probe of the imaging polarimeter is fixed, the images obtained for a set of sixteen well-suited parameters representing independent and significant properties, like, for instance, the one derived from the arrow decomposition of the point-to-point Mueller matrix, constitute a reasonable and sufficient framework for inspecting the polarimetric image features of the sample.

Obviously, for each different spectral profile of the light probe used, a new set of images is generated, extending the power to analyze the sample. The changes from one set of images to another one depend on the specific behavior of the sample for each spectral profile, in such a way that both the central frequency and the bandwidth and shape of the profile influence the measurement. In particular, as the bandwidth increases, the more polarimetric randomness is introduced by the interaction with the sample and, consequently, the lower the corresponding indices of polarimetric purity. It can be said that, as the bandwidth is reduced, the more deterministic the polarimetric behavior of the sample is and vice versa, and that, as the bandwidth is increased, the composition of the polarimetric behaviors of the sample for the different frequencies involved increases the depolarization of the emerging light probe before reaching the polarization state analyzer of the polarimeter.

In addition to the specific images obtained for different spectral profiles of the light probe of the polarimeter, which provide specific information for each corresponding kind of interaction, new sets of images can be generated by the synthetic parallel composition of the point-to-point Mueller matrices obtained in each independent experiment. These new Mueller matrices encompass, in a peculiar manner, information on the differences in the behavior of the sample as the type of the light probe changes.

When two independent measurements are taken using spectral profiles whose frequencies do not overlap, the synthesized Mueller matrices coincide (at least from a theoretical point of view) with those obtained through a simple measurement in which both light probes are used simultaneously (provided the relative weights of the Mueller matrices in the convex sum coincide with the relative weights of the intensities used). Nevertheless, when the spectral profiles overlap to some extent, the synthetic matrices differ from those potentially generated experimentally through a simultaneous combination of both light probes. This in no way detracts from interest of the general synthesis of point-to-point Mueller matrices through parallel composition of the ones obtained from independent measurements. In fact, the synthesis leads to a new arrangement of the polarimetric information already obtained in each measurement, which allows us to obtain new images that are well defined from an algebraic point of view.

It should be emphasized that the potential improvements in the contrast of images strongly depend on the nature of the material sample and are not necessarily realized for each independent parameter, but can be made evident for some descriptors, which are different for each case. Some preliminary analyses of certain samples of biological tissues allow us to say that the new approach enhances the visualization of certain structures. Specific and comprehensive analyses for particular kinds of material samples (thin films, LCD devices, scattering by particles, biological tissues, samples inspected by SAR polarimetry...) require further studies that have a strongly contextual nature and could be the subject of future works that fall outside the scope of this communication, whose main objective is to present the theoretical approach to synthetic imaging Mueller polarimetry.

In general, all the polarimetric descriptors of the synthetic point-to-point Mueller matrix $\mathbf{M}_k$ are different from those of the n measured Mueller matrices $(\mathbf{M}_{k1}, \ldots \mathbf{M}_{kn})$ used as its parallel components of $\mathbf{M}_k$ $[\mathbf{M}_k = (1/n)(\mathbf{M}_{k1} + \mathbf{M}_{k2} + \ldots + \mathbf{M}_{kn})]$, while the integer number r of IPP with values below their maximum (with $P_1 \leq P_2 \leq P_3 \leq 1$) increases as the number of measurements involved increases (recall that $r = \operatorname{rank}\mathbf{C}(\mathbf{M}_k)$). This property is evident when the measured Mueller matrices are nondepolarizing, thus satisfying $1 = r = P_\Delta(\mathbf{M}_{kl}) = P_1(\mathbf{M}_{kl}) = P_2(\mathbf{M}_{kl}) = P_3(\mathbf{M}_{kl})$ $(l = 1, \ldots, n)$, while the synthetic $\mathbf{M}_k$ is necessarily depolarizing, with $2 \leq r(\mathbf{M}_k) \leq 4$. The specific value of r exhibited by $\mathbf{M}_k$ depends strongly on the nature and differences among its components $\mathbf{M}_{kl}$ $(l = 1, \ldots, n)$.

5. Conclusions

The sets of point-to-point Mueller matrices associated with a given material sample and obtained via a number n of independent Mueller imaging measurements provide information that, beyond that held by the measured Mueller matrices themselves, can be rearranged into a set of synthetic point-to-point Mueller matrices whose polarimetric descriptors (enpolarizing, retarding and depolarizing parameters) depend on the changes in the polarimetric behavior of the sample when the spectral profile of the light probe of the polarimeter changes.

The procedure used to generate the point-to-point synthetic Mueller matrices is as simple as calculating the respective mean Mueller matrices, which exhibit polarimetric descriptors that, in general, are different from those of the measured Mueller matrices. This allows us to generate new polarimetric images of the sample that can enhance the visualization of certain structural properties of the sample.

Funding: This research received no external funding.

Informed Consent Statement: Not applicable.

Data Availability Statement: Not applicable.

Conflicts of Interest: The authors declare no conflict of interest.

References

1. Ghosh, N.; Vitkin, I.A. Tissue polarimetry: Concepts, challenges, applications, and outlook. *J. Biomed. Opt.* **2011**, *16*, 110801. [CrossRef] [PubMed]
2. Ignatenko, D.N.; Shkirin, A.V.; Lobachevsky, Y.P.; Gudkov, S.V. Applications of Mueller Matrix Polarimetry to Biological and Agricultural Diagnostics: A Review. *Appl. Sci.* **2022**, *12*, 5258. [CrossRef]
3. Touzi, R.; Boerner, W.M.; Lee, J.S.; Lueneburg, E. A review of polarimetry in the context of synthetic aperture radar: Concepts and information extraction. *Can. J. Remote Sens.* **2004**, *30*, 380–407. [CrossRef]
4. Tyo, J.S.; Goldstein, D.L.; Chenault, D.B.; Shaw, J.A. Review of passive imaging polarimetry for remote sensing applications. *Appl. Opt.* **2006**, *45*, 5453–5469. [CrossRef]
5. Cloude, S.R. *Polarisation: Applications in Remote Sensing*; Oxford University: London, UK, 2009.
6. Azzam, R.M.A. Photopolarimetric measurement of the Mueller matrix by Fourier analysis of a single detected signal. *Opt. Lett.* **1977**, *2*, 148–150. [CrossRef] [PubMed]
7. Hauge, P.S. Mueller matrix ellipsometry with imperfect compensators. *J. Opt. Soc. Am.* **1978**, *68*, 1519–1528. [CrossRef]
8. Gil, J.J. Metodo Dinamico de Determinacion de Matrices de Mueller por Analisis de Fourier. Master's Thesis, University of Zaragoza, Zaragoza, Spain, 1979. Available online: https://www.researchgate.net/publication/235981117_Metodo_dinamico_de_determinacion_de_parametros_de_Stokes_y_matrices_de_Mueller_por_analisis_de_Fourier (accessed on 23 July 2023).
9. Goldstein, D.H. Mueller matrix dual-rotating retarder polarimeter. *Appl. Opt.* **1992**, *31*, 6676–6683. [CrossRef]
10. Pezzaniti, J.L.; Chipman, R.A. Mueller Matrix Imaging Polarimetry. *Opt. Eng.* **1995**, *34*, 1558–1568. [CrossRef]
11. Garcia-Caurel, E.; De Martina, A.; Drevillon, B. Spectroscopic Mueller polarimeter based on liquid crystal devices. *Thin Solid Film.* **2004**, *455–456*, 120–123. [CrossRef]
12. Zallat, J.; Ainouz, S.; Stoll, M.P. Optimal configurations for imaging polarimeters: Impact of image noise and systematic errors. *J. Opt. A Pure Appl. Opt.* **2006**, *8*, 807–814. [CrossRef]
13. Arteaga, O.; Freudenthal, J.; Wang, B.; Kahr, B. Mueller matrix polarimetry with four photoelastic modulators: Theory and calibration. *Appl. Opt.* **2012**, *51*, 6805–6817. [CrossRef]
14. Azzam, R.M.A. Stokes-vector and Mueller-matrix polarimetry. *J. Opt. Soc. Am. A* **2016**, *33*, 1396–1408. [CrossRef] [PubMed]
15. Bian, S.; Cui, C.; Arteaga, O. Mueller matrix ellipsometer based on discrete-angle rotating Fresnel rhomb compensators. *J. Opt. Soc. Am. A* **2016**, *33*, 1396–1408. [CrossRef]
16. Gil, J.J. Absolute Mueller polarimeters based on dual-rotating imperfect retarders and arbitrary ratio of angular velocities. *Dynamics* **2023**, *3*, 250–271. [CrossRef]
17. Gil, J.J. Transmittance constraints in serial decompositions of Mueller matrices. The arrow form of a Mueller matrix. *J. Opt. Soc. Am. A* **2013**, *30*, 701–707. [CrossRef] [PubMed]
18. Gil, J.J. Structure of polarimetric purity of a Mueller matrix and sources of depolarization. *Opt. Commun.* **2016**, *368*, 165–173. [CrossRef]
19. Gil, J.J. Components of purity of a Mueller matrix. *J. Opt. Soc. Am. A* **2011**, *28*, 1578–1585. [CrossRef]
20. Gil, J.J.; San José, I.; Canabal-Carbia, M.; Estévez, I.; González-Arnay, E.; Luque, J.; Garnatje, T.; Campos, J.; Lizana, A. Polarimetric Images of Biological Tissues Based on the Arrow Decomposition of Mueller Matrices. *Photonics* **2023**, *10*, 669. [CrossRef]
21. Robson, B.A. *The Theory of Polarization Phenomena*; Clarendon Press: Oxford, UK, 1975.
22. Xing, Z.-F. On the deterministic and non-deterministic Mueller matrix. *J. Mod. Opt.* **1992**, *39*, 461–484. [CrossRef]
23. Lu, S.-Y.; Chipman, R.A. Interpretation of Mueller matrices based on polar decomposition. *J. Opt. Soc. Am. A* **1996**, *13*, 1106–1113. [CrossRef]
24. Gil, J.J.; Bernabéu, E. Depolarization and polarization indices of an optical system. *Opt. Acta* **1986**, *33*, 185–189. [CrossRef]
25. Gil, J.J. Polarimetric characterization of light and media–Physical quantities involved in polarimetric phenomena. *Eur. Phys. J. Appl. Phys.* **2007**, *40*, 1–47. [CrossRef]
26. Gil, J.J.; Ossikovski, R. *Polarized Light and the Mueller Matrix Approach*, 2nd ed.; CRC Press: Boca Raton, FL, USA, 2022.
27. Cloude, S.R. Group theory and polarisation algebra. *Optik* **1986**, *75*, 26–36.
28. Gil, J.J.; Bernabéu, E. A depolarization criterion in Mueller matrices. *Opt. Acta* **1985**, *32*, 259–261. [CrossRef]
29. Hingerl, K.; Ossikovski, R. General approach for modeling partial coherence in spectroscopic Mueller matrix polarimetry. *Opt. Lett.* **2016**, *41*, 219–222. [CrossRef]
30. Ossikovski, R.; Hingerl, K. General formalism for partial spatial coherence in reflection Mueller matrix polarimetry. *Opt. Lett.* **2016**, *41*, 4044–4047. [CrossRef]
31. Kuntman, E.; Kuntman, M.A.; Arteaga, O. Vector and matrix states for Mueller matrices of nondepolarizing optical media. *J. Opt. Soc. Am. A* **2017**, *34*, 80–86. [CrossRef]

32. Kuntman, E.; Kuntman, M.A.; Sancho-Parramon, J.; Arteaga, O. Formalism of optical coherence and polarization based on material media states. *Phys. Rev. A* **2017**, *95*, 063819. [CrossRef]
33. Ossikovski, R.; Gil, J.J. Basic properties and classification of Mueller matrices derived from their statistical definition. *J. Opt. Soc. Am. A* **2017**, *34*, 1727–1737. [CrossRef]
34. San José, I.; Gil, J.J. Coherency vector formalism for polarimetric transformations. *Opt. Commun.* **2020**, *475*, 126230. [CrossRef]
35. San José, I.; Gil, J.J. Invariant indices of polarimetric purity: Generalized indices of purity for n × n covariance matrices. *Opt. Commun.* **2011**, *284*, 38–47. [CrossRef]
36. Gil, J.J.; San José, I. Arbitrary decomposition of a Mueller matrix. *Opt. Lett.* **2019**, *44*, 5715–5718. [CrossRef] [PubMed]
37. Gil, J.J. Physical Quantities Involved in a Mueller Matrix. In Proceedings of the SPIE 9853, Baltimore, MD, USA, 17–21 April 2016; p. 985302.
38. Gil, J.J. Determination of Polarization Parameters in Matricial Representation. Theoretical Contribution and Development of an Automatic Measurement Device. Ph.D. Thesis, University of Zaragoza, Zaragoza, Spain, 1983. Available online: http://zaguan.unizar.es/record/10680/files (accessed on 20 August 2023).
39. Ossikovski, R. Interpretation of nondepolarizing Mueller matrices based on singular-value decomposition. *J. Opt. Soc. Am. A* **2008**, *25*, 473–482. [CrossRef] [PubMed]

photonics

MDPI

Article

Calibration of Waveplate Retardance Fluctuation Due to Field-of-View Effect in Mueller Matrix Ellipsometer

Zhou Jiang, Song Zhang *, Hao Jiang and Shiyuan Liu

State Key Laboratory for Digital Manufacturing Equipment and Technology, Huazhong University of Science and Technology, Wuhan 430074, China
* Correspondence: songzhang@hust.edu.cn

Abstract: Leveraging their unique phase modulation characteristics, birefringent waveplates have been widely used in various optical systems. With the development of material science and manufacturing techniques, the polarization properties of waveplates have become increasingly complex and diverse. Among these properties, the field-of-view effect of the waveplate caused due to manufacturing defects or improper installation procedures is extremely difficult to calibrate and seriously affects the precision and accuracy of the relevant optical systems. In this paper, a calibration method that can compensate for the field-of-view effect of waveplates installed in the instrument is proposed. Moreover, to approve the fidelity of the proposed calibration method, a series of film thickness measurement experiments are carried out. The results show that under different installation conditions of the waveplates, the precision and accuracy of the film thickness measured with the proposed method significantly improved. This method can be expected to reduce the assembly difficulty of such optical systems, while also improving their accuracy and stability.

Keywords: Mueller matrix ellipsometer; birefringent; waveplate; calibration; field-of-view effect; thickness measurement

1. Introduction

The waveplate is one of the most commonly used optical components in optical systems. It can produce an additional optical path difference (or phase difference) between two mutually perpendicular light components. Thanks to its unique polarization modulation characteristics, the waveplate has been widely used in various optical systems, such as interferometry [1,2], polarimeter/ellipsometry [3–9], birefringent filters [10,11], etc. Meanwhile, with the development of material technologies, the materials of waveplates are no longer limited to traditional quartz, and cover a much wider and richer range, including gypsum, $LiTaO_3$, ZnO, etc. [12–15]. This enables waveplates to exhibit complex and diverse polarization characteristics, while also increasing the difficulty in the characterization and calibration of their polarization properties. Therefore, the accurate polarization property calibration [16–19] of the waveplate plays an important role in improving the accuracy and stability of the relevant optical instruments.

In previous studies, it was observed that the retardance of a birefringent waveplate fluctuated significantly when the incident angle of the light and the azimuth of the waveplate varied at the same time. This phenomenon existing in the waveplates is called the field-of-view effect [20,21]. While the field-of-view effect of waveplates has been utilized for applications such as attitude angle tracking [20], the loss of precision and accuracy it causes in other optical systems that rely on waveplates for phase modulation is unacceptable. Therefore, it is of great importance to calibrate and compensate for the retardance fluctuation caused by the field-of-view effect of waveplates in optical systems.

There are many researchers dedicated to investigating the relationship between the incident angle of light and the retardance of waveplates using ellipsometry [21–23]. West and Smith have comprehensively studied the errors associated with birefringent waveplates,

Citation: Jiang, Z.; Zhang, S.; Jiang, H.; Liu, S. Calibration of Waveplate Retardance Fluctuation Due to Field-of-View Effect in Mueller Matrix Ellipsometer. *Photonics* **2023**, *10*, 1038. https://doi.org/10.3390/photonics10091038

Received: 18 August 2023
Revised: 8 September 2023
Accepted: 9 September 2023
Published: 12 September 2023

including the thickness error, field-of-view errors, optic axis tilt errors and misalignment. They developed a piece of equipment consisting of three Glan–Thompson polarizers, a Soleil–Babinet compensator, a multiline He–Ne laser source and a photomultiplier tube detector to measure the absolute retardance of the waveplate [24]. These studies clearly exhibit the significant influence of the incident light angle on the retardance of waveplates. Ruder et al. [25] used dual continuously rotating anisotropic mirrors to construct a single-wavelength Mueller matrix ellipsometer in a normal transmission configuration. However, few studies have investigated how to calibrate the system when the incident light is tilted with respect to the waveplate. This may be a limitation to improving the precision and accuracy of high-precision optical systems.

In this paper, a calibration method for compensating for the field-of-view effect of waveplates in optical systems is proposed. Firstly, a characterization model of the field-of-view effect in waveplates is proposed. Subsequently, a series of tilt angle measurement experiments are carried out. The consistency between the simulated attitude angles of waveplates and the measured tilt angles demonstrate the correctness and the effectiveness of the proposed method. In addition, the proposed characterization model is applied into the calibration of a single-wavelength Mueller matrix ellipsometer (SWE). Compared to the measured thicknesses on a set of standard SiO_2 thin films given using a commercial MME, the deviations decreased from 6.5% to 1.8% with the field-of-view error considered. It is expected that the proposed calibration method can improve the accuracy and precision of the instrument, while also reducing the difficulty of the instrument assembly.

2. Characterization of the Waveplate

The polarization state of the light passing through the compensator can deviate from the theoretical expectation due to the design and manufacturing defects, as well as unsatisfactory installations. Generally speaking, the waveplates exhibit small depolarizations due to manufacturing defects, so the characterization of practical waveplates at the azimuth θ can be expressed using the flowing Mueller matrix formalism:

$$\boldsymbol{M}_{\mathrm{C}}(\delta,\theta,b,c) = \boldsymbol{R}(-\theta)\boldsymbol{M}_{\mathrm{C}}^{\mathrm{ideal}}(\delta)\boldsymbol{M}_{\mathrm{Dep}}(b,c)\boldsymbol{R}(\theta), \tag{1}$$

where $\boldsymbol{M}_{\mathrm{C}}^{\mathrm{ideal}}$ is the Mueller matrix of the ideal compensator, and can be represented as [26]:

$$\boldsymbol{M}_{\mathrm{C}}^{\mathrm{ideal}} = \begin{bmatrix} 1 & 0 & 0 & 0 \\ 0 & 1 & 0 & 0 \\ 0 & 0 & \cos(\delta) & \sin(\delta) \\ 0 & 0 & -\sin(\delta) & \cos(\delta) \end{bmatrix}, \tag{2}$$

where δ denotes the retardance of the compensator. $\boldsymbol{M}_{\mathrm{Dep}}$ is the Mueller matrix of the depolarization effect in the waveplates, and can be expressed as [27]:

$$\boldsymbol{M}_{\mathrm{Dep}} = \begin{bmatrix} 1 & 0 & 0 & 0 \\ 0 & 1-c & 0 & 0 \\ 0 & 0 & 1-b & 0 \\ 0 & 0 & 0 & 1-b \end{bmatrix}, \tag{3}$$

where b and c are the linear depolarization parameters of the compensator. $\boldsymbol{R}(\theta)$ is the Chandrasekhar matrix that can unify the optical axis direction of each optical component to the incident plane reference system, and its matrix form can be denoted as:

$$\boldsymbol{R}(\theta) = \begin{bmatrix} 1 & 0 & 0 & 0 \\ 0 & \cos(2\theta) & \sin(2\theta) & 0 \\ 0 & -\sin(2\theta) & \cos(2\theta) & 0 \\ 0 & 0 & 0 & 1 \end{bmatrix}, \tag{4}$$

In this paper, a compound zero-order waveplate was selected as the compensator to study the retardance fluctuation introduced as a result of the field-of-view effect. The compound zero-order waveplate was composed of two multiorder single waveplates composed of quartz, whose optical axes were oriented perpendicular to each other, as shown in Figure 1. Without losing generality, we assumed that the optical axis of the thicker multiorder single waveplate was parallel with the x-axis, and that of the thinner one was parallel with the y-axis. According to the above descriptions and derivations, the retardance of the compound zero-order waveplate under an arbitrary incidence and azimuth could be calculated using [21,28,29].

$$\begin{aligned}\delta(\theta_{\text{tilt}},\beta) &= \tfrac{2\pi}{\lambda}L = \tfrac{2\pi}{\lambda}\sum_{i=1}^{2} d_i\left(\sqrt{n_{yi}^2-\sin^2\theta_{\text{tilt}}}-\sqrt{n_{zi}^2-\sin^2\theta_{\text{tilt}}}\right) \\ &= \tfrac{2\pi}{\lambda}d_1\left(\sqrt{n_e^2-\tfrac{n_e^2\cos^2\beta+n_o^2\sin^2\beta}{n_o^2}\sin^2\theta_{\text{tilt}}}-\sqrt{n_o^2-\sin^2\theta_{\text{tilt}}}\right) \\ &\quad -\tfrac{2\pi}{\lambda}d_2\left(\sqrt{n_e^2-\tfrac{n_e^2\sin^2\beta+n_o^2\cos^2\beta}{n_o^2}\sin^2\theta_{\text{tilt}}}-\sqrt{n_o^2-\sin^2\theta_{\text{tilt}}}\right)\end{aligned}, \quad (5)$$

where θ_{tilt} is the tilt angle of the waveplate, which is defined as the angle between the incident light and the normal direction of the waveplate's surface. Additionally, β is the fast axis azimuth angle, d_1 and d_2 are the thicknesses of the thicker single waveplate and the thinner single waveplate, respectively, and n_e and n_o are the extraordinary index and the ordinary index of the quartz, respectively.

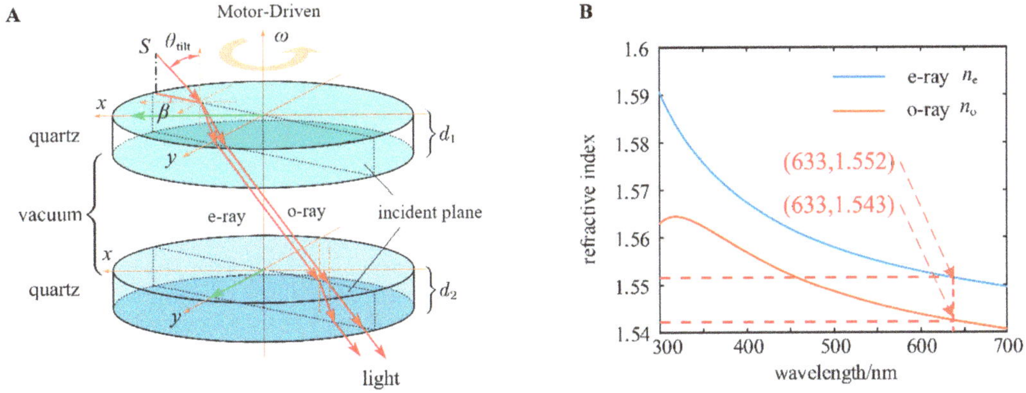

Figure 1. (**A**) Schematic of the wave normal propagation in a waveplate under an arbitrary incidence and an arbitrary azimuth of the incident light. (**B**) Refractive index of quartz.

The dispersion equation (Schott dispersion formula) was:

$$\begin{cases} n_o^2 = a_{0o}+a_{1o}\lambda^2+a_{2o}\lambda^3+a_{3o}\lambda^4+a_{4o}\lambda^5+a_{5o}\lambda^6 \\ n_e^2 = a_{0e}+a_{1e}\lambda^2+a_{2e}\lambda^3+a_{3e}\lambda^4+a_{4e}\lambda^5+a_{5e}\lambda^6 \end{cases}, \quad (6)$$

where a_{io} and a_{ie} (i = 1, . . ., 5) are the ordinary dispersion coefficient and the extraordinary dispersion coefficient, respectively, which can be found in the material libraries of most manufacturers. λ is the wavelength of the incident light. It could be calculated that n_e = 1.5517, n_o = 1.5426 when the material was quartz and the wavelength was 632.8 nm. It could be seen from the data provided by the manufacturer that the total thickness of the quartz biplate (including the air gap ≈ 203.200 μm) was approximately 2177.990 μm. According to Equation (5), we could obtain the effective thickness (i.e., the difference between the thicknesses of the two multiorder single waveplates for a compound zero-order biplate) of the biplate, amounting to 17.480 μm, and the designed thicknesses of the two single waveplates were d_1 = 996.135 μm and d_2 = 978.655 μm. Therefore, the initial field-of-view parameters could be determined with these conditions. Figure 2 shows the simulation results of the retardance fluctuation versus the azimuth at different waveplate tilt angles.

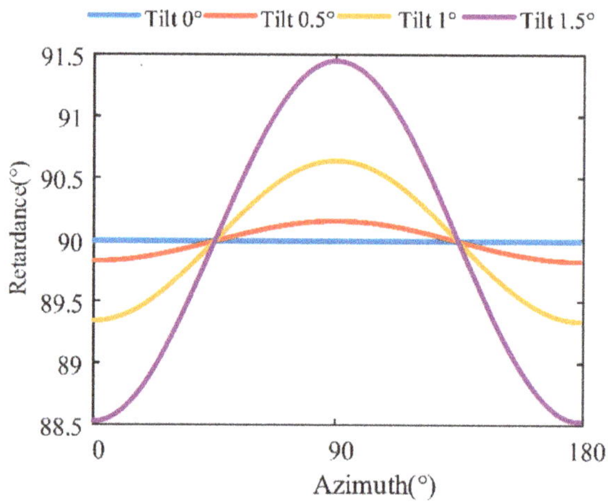

Figure 2. Retardance oscillations versus the azimuth at different waveplate tilt angles.

3. System and Calibration Method

3.1. Single-Wavelength Mueller Matrix Ellipsometer System

The instrument is a self-developed SWE, which could measure 16 Mueller matrix elements simultaneously. As schematically shown in Figure 3, the SWE consisted of three parts: a CW He–Ne laser (HRS015B 100-240VAC, Thorlabs, Newton, NJ, USA), a polarization state generator (PSG) and a polarization state analyzer (PSA). The laser first transmitted through an optical isolator (IO-2D-633-VLP, Thorlabs, Newton, NJ, USA), which prevented the interference of the reflected light. A beam (BS025, Thorlabs, Newton, NJ, USA) split the laser into two beams with a 1:9 intensity ratio. One beam entered detector1 (PDA36A2, Thorlabs, Newton, NJ, USA) directly to monitor the intensity fluctuation of the light source, while the other entered the main optical path. After passing through a bandpass filter (FLH633-5, Thorlabs, Newton, NJ, USA) and being reflected on a mirror (64-013, Edmund, Barrington, NJ, USA), the light would incident on a sample through the PSG at an angle of 65 degrees. The PSG consisted of a polarizer (LPVISC100-MP2, Thorlabs, Newton, NJ, USA) and a waveplate (WPQ10M-633, Thorlabs, Newton, NJ, USA). The PSA then modulated the sample-reflected light, which would eventually be captured by detector2 (PDA36A2, Thorlabs, Newton, NJ, USA). The PSA consisted of a waveplate and a polarizer. The two detachable focus lenses installed in the PSG and PSA could reduce the size of the light spot when the size of the measured sample was very small. The self-developed instrument could obtain the full Mueller matrix of the sample using the above configuration. A high-precision data acquisition card (USB6281, NI, Austin, TX, USA) was required to meet the criteria of the high-precision real-time measurement.

Generally speaking, the detected light intensity matrix $\boldsymbol{I}_{\text{dec}}$ could be modeled as the product of the modulation matrix $\boldsymbol{G}$ of the PSG, the Mueller matrix $\boldsymbol{M}_{\text{S}}$ of the sample and the demodulation matrix $\boldsymbol{A}$ of the PSA [30]:

$$I_{\text{dec}} = A \cdot M_{\text{S}} \cdot G, \tag{7}$$

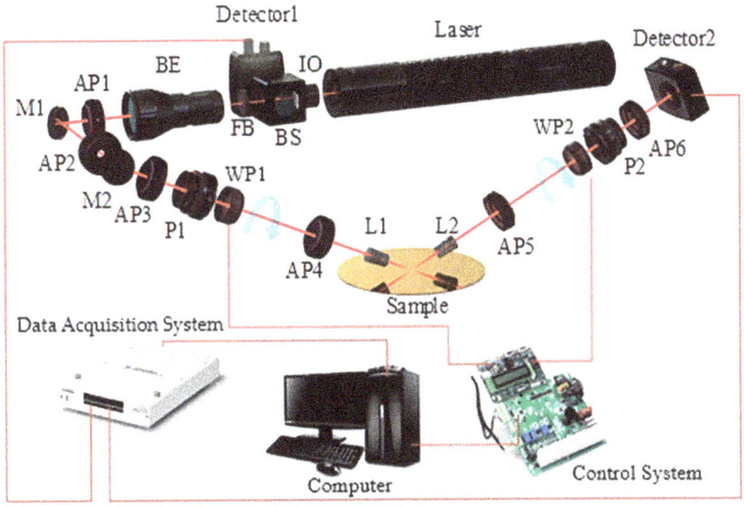

Figure 3. Critical components and beam path of the SWE. The ellipsometer was composed of a He–Ne laser light source (laser), an optical isolator (IO), a beam splitter (BS), two detectors (detector1 and detector2), a narrowband filter (FB), a beam expander (BE) (GBE-03A, Thorlabs, Newton, NJ, USA), six apertures (AP1–AP6) (SM1D12CZ, Thorlabs, Newton, NJ, USA), two mirrors (M1, M2), two polarizers (P1 and P2) (LPVISC100-MP2, Thorlabs, Newton, NJ, USA), two continuously rotating waveplates (WP1 and WP2) (WPQ10M-633, Thorlabs, Newton, NJ, USA) and two focus lens (L1 and L2). Incident and reflected beams are denoted in red.

Let us denote the 4 × 4 Mueller matrix of the sample as:

$$M_S = [m_{i,j}]_{4\times4}, (i, j = 1, 2, 3, 4), \quad (8)$$

The modulation matrix G and demodulation matrix A could be represented as:

$$G = \begin{bmatrix} S^1_{\text{PSG}} & S^2_{\text{PSG}} & \cdots & S^k_{\text{PSG}} & \cdots & S^K_{\text{PSG}} \end{bmatrix}, \quad (9)$$

$$A = \begin{bmatrix} H^1_{\text{PSA}} & H^2_{\text{PSA}} & \cdots & H^k_{\text{PSA}} & \cdots & H^K_{\text{PSA}} \end{bmatrix}, \quad (10)$$

where S^k_{PSG} and H^k_{PSA} are the kth Stokes vector of the polarized light output from the PSG and PSA, respectively. K is the number of the total sampling point in an optical cycle.

According to the optical path of the SWE, the kth Stokes vector S^k_{PSG} and H^k_{PSA} could be calculated with Equations (11) and (12), respectively [30]:

$$S^k_{\text{PSG}} = \left\{R\left(-C^k_1\right)\cdot M_C(\delta^k_1, b_1, c_1)\cdot R\left(C^k_1\right)\right\}\cdot\{R(-P)\cdot M_P(Dt_1, \delta_P)\cdot R(P)\}\cdot S_{\text{in}}, \quad (11)$$

$$H^k_{\text{PSA}} = \begin{bmatrix}1 & 0 & 0 & 0\end{bmatrix}\cdot\{R(-A)\cdot M_A(Dt_2, \delta_A)\cdot R(A)\}\cdot\left\{R\left(-C^k_2\right)\cdot M_C(\delta^k_2, b_2, c_2)\cdot R\left(C^k_2\right)\right\}, \quad (12)$$

where P and A are the azimuth angle of the polarizer and the analyzer in the PSG and PSA, respectively. The δ_P and δ_A are the weak birefringence retardances of the polarizer and analyzer, respectively. The first and second compensators were driven using two servo hollow motors (AgilityRH, Applimotion, Loomis, CA, USA), and their fast axis azimuth was changed according to the following relations: $C^k_1 = \omega_1 t_k + C^{\text{initial}}_1$ and $C^k_2 = \omega_2 t_k + C^{\text{initial}}_2$. C^{initial}_1 and C^{initial}_2 are the initial azimuths of the compensators. ω_1 and ω_2 are the rotation speed of the first and second compensators, respectively. Additionally, the rotation ratio of ω_1 and ω_2 was set to 1:5 in our instrument.

The actual Mueller matrix of the polarizer and analyzer could be shown as [31,32]:

$$M_{P/A}(Dt, \delta_{P/A}) = \begin{bmatrix} 1 & Dt & 0 & 0 \\ Dt & 1 & 0 & 0 \\ 0 & 0 & 2\cos(\delta_{P/A})\sqrt{1-Dt^2} & 2\sin(\delta_{P/A})\sqrt{1-Dt^2} \\ 0 & 0 & -2\sin(\delta_{P/A})\sqrt{1-Dt^2} & 2\cos(\delta_{P/A})\sqrt{1-Dt^2} \end{bmatrix}, \quad (13)$$

where Dt is the extinction parameter, which represents the ratio of the difference between the transmittance of the p-polarized light and the s-polarized light passing through the polarizer to the sum of the transmittances of the two polarization components:

$$Dt = \frac{I_p - I_s}{I_p + I_s}, \quad (14)$$

3.2. Calibration of the Waveplate Retardance Fluctuation

As observed in Section 2, the retardance of the compound zero-order waveplate varied sinusoidally with the azimuth, and the fluctuation amplitude increased with a greater tilt angle. To prevent an inaccurate calibration caused due to systematic parameter coupling, Equations (15) and (16) could substitute the waveplate characterization model from Section 2 when performing the regression calibration using the Levenberg–Marquardt algorithm.

$$\delta_1^k = A_{C1} \cdot \sin(\omega_1 t_k + \phi_{C1}) + \delta_{C1}^{center}, \quad (15)$$

$$\delta_2^k = A_{C2} \cdot \sin(\omega_2 t_k + \phi_{C2}) + \delta_{C2}^{center}, \quad (16)$$

where A_{C1} and A_{C2} are the retardance amplitudes of the first and second compensators, respectively. ϕ_{C1} and ϕ_{C2} are the azimuth angles of the waveplates. The δ center C1 and δ center C2 are the central retardances of the first and second waveplate, respectively.

The instrument needed to be carefully calibrated to maintain high performance [33]. The instrument was calibrated with a series of standard SiO_2 film samples. The theoretical Mueller matrices of the samples could be calculated from the refractive indices (n and k), the thicknesses d and the incidence angles θ of the measurements, while the measured Mueller matrices could be obtained with the SWE. In this case, the system parameters of the SWE that needed to be calibrated involved the azimuthal angles of polarizer P, analyzer A, the weak birefringence retardance δ_P and δ_A of the polarizer and analyzer, the extinction parameters Dt_1 and Dt_2, the initial azimuths $C_1^{initial}$ and $C_2^{initial}$ of the compensators, the retardance of the first and second compensators δ center C1 and δ center C2, the rotation speed of first and second compensators ω_1 and ω_2, the retardance amplitude of the first and second compensators A_{C1} and A_{C2}, the azimuth angles of waveplates ϕ_{C1} and ϕ_{C2}, the thicknesses d and the incidence angle θ, as well as the depolarization parameters of the first and second compensators b_1, c_1 and b_2, c_2. Moreover, the nonlinear parameters of the detectors a_0, a_1, a_2 and a_3 had to be considered, which could be defined with the characterization model of the detector's nonlinear response [34]:

$$I_{out} = \alpha_0 + \alpha_1 I_{in} + \alpha_2 I_{in}^2 + \alpha_3 I_{in}^3, \quad (17)$$

where I_{in} is the input light intensity of the detector and I_{out} is the output signal value of the detector.

Therefore, the system parameter p_{sys} could be written in the vector form as:

$$p_{sys} = [P,\ Dt_1,\ \delta_P C_1^{initial},\ \omega_1,\ \delta \text{ center C1},\ A_{C1},\ \phi_{C1},\ b_1,\ c_1,\ C_2^{initial},\ \omega_1,\ \delta \text{ center C2},\ A_{C2},\ \phi_{C2},\ b_2,\ c_2,\ P,\ Dt_1,\ \delta_P,\ \theta_{incident},\ d,\ a_0,\ a_1,\ a_2,\ a_3], \quad (18)$$

With the cost function defined as Equation (19), the system parameter p_{sys} could be obtained from the measured light intensity I^{meas} by using the Levenberg–Marquardt algorithm [35].

$$p_{sys} = \underset{p_{sys} \in \Omega_p}{\operatorname{argmin}} \left[I^{meas} - I^{calc}(p_{sys}) \right]^T \Gamma^+_{I^{meas}} \left[I^{meas} - I^{calc}(p_{sys}) \right], \tag{19}$$

where I^{meas} is the actual measurement intensity matrix, I^{calc} is the theoretical intensity matrix, Ω_p indicates the value range of the system parameters, $\Gamma^+_{I^{meas}}$ is the Moore–Penrose pseudoinverse of the covariance matrix of the measured intensity matrix. In addition, $\Gamma^+_{I^{meas}} = \left(\Gamma^+_{I^{meas}} \cdot \Gamma^+_{I^{meas}}\right)^{-1}$. The tilting-induced retardance error of the polarizer was ignored, since it usually varied within $\pm 0.05°$ when the tilt angle was less than 5°, which was quite small compared with the field-of-view effect of the waveplate. The relative optical parameters in the corresponding defined ranges could be decoupled and extracted. It was noted that Equations (18) and (19) yielded values of P, Dt_1, $C_1^{initial}$, ω_1, δ center C1, A_{C1}, ϕ_{C1}, b_1, c_1, $C_2^{initial}$, ω_2, δ center C2, A_{C2}, ϕ_{C2}, b_2, c_2, A and Dt_2 in the ranges $-180° \leq P, A, C_1^{initial}, C_2^{initial} \leq 180°$, $0.95 \leq Dt_1, Dt_2 \leq 1$, $80° \leq \delta$ center C1, δ center C2 $\leq 100°$, $1438°/\mathrm{s} \leq \omega_1 \leq 1442°/\mathrm{s}$, $7198°/\mathrm{s} \leq \omega_2 \leq 7202°/\mathrm{s}$, $-1 \leq b_1, c_1, b_2, c_2 \leq 1$, $60° \leq \theta_{incident} \leq 70°$, $0\ \mathrm{nm} \leq d \leq 100\ \mathrm{nm}$, $-180° \leq \phi_{C1}, \phi_{C2} \leq 180°$, $-10 \leq A_{C1}, A_{C2} \leq 10$, $0 \leq a_1 \leq 2$ and $-1 \leq a_1, a_2, a_3 \leq 1$, respectively.

When the calibration was completed, the retardance sequence in an optical cycle could be obtained by substituting the system parameters A_{C1}, A_{C2}, ω_1, ω_2, δ center C1, δ center C2, ϕ_{C1} and ϕ_{C2} into Equations (15) and (16). The retardance sequence of C1 and C2 could be denoted as:

$$\delta_{1/2} = \begin{bmatrix} \delta_{1/2}^1 & \delta_{1/2}^2 & \delta_{1/2}^3 & \cdots & \delta_{1/2}^K \end{bmatrix}, \tag{20}$$

Similarly, utilizing the nonlinear regression fitting method, the parameters of the field-of-view errors $h_C = [\theta_{tilt}, \beta, d_1, d_2]$ could be accurately determined.

$$h_C = \underset{h_C \in \Omega_h}{\operatorname{argmin}} \left[\delta^{meas} - \delta^{calc}(h_C) \right]^T \Gamma^+_{\delta^{meas}} \left[\delta^{meas} - \delta^{calc}(h_C) \right], \tag{21}$$

where δ^{meas} is the actual fitted retardance sequence of C1 and C2 calculated with Equation (21) and δ^{calc} is the theoretical retardance sequence calculated with Equation (1); Ω_h indicates the value range of the field-of-view error parameter h_C, $\Gamma^+_{\delta^{meas}}$ is the Moore–Penrose pseudoinverse of the covariance matrix of the actual fitted retardance sequence and $\Gamma^+_{\delta^{meas}} = \left(\Gamma^+_{\delta^{meas}} \cdot \Gamma_{\delta^{meas}}\right)^{-1} \Gamma^T_{\delta^{meas}}$. Then, the system parameter M_S could be obtained from the following:

$$M_S = \underset{M_S \in \Omega_M}{\operatorname{argmin}} \left[I^{meas} - I^{calc}(p_{sys}, M_S) \right]^T \Gamma^+_{I^{meas}} \left[I^{meas} - I^{calc}(p_{sys}, M_S) \right], \tag{22}$$

where Ω_M indicates the value range of the system Mueller matrix. Then, the thickness d of the sample could be obtained from the following:

$$d = \underset{d \in \Omega_d}{\operatorname{argmin}} \left[M^{meas} - M^{calc}(a, d) \right]^T \Gamma^+_{M^{meas}} \left[M^{meas} - M^{calc}(a, d) \right], \tag{23}$$

where Ω_d indicates the value range of the thickness, a denotes the priori value of the reconstruction, M^{meas} is the measurement Mueller matrix, M^{calc} is the theoretical Mueller matrix and $\Gamma^+_{M^{meas}}$ is the Moore–Penrose pseudoinverse of the covariance matrix of the measured Mueller matrix, as well as $\Gamma^+_{M^{meas}} = \left(\Gamma^+_{M^{meas}} \cdot \Gamma_{M^{meas}}\right)^{-1} \cdot \Gamma^T_{M^{meas}}$.

4. Experiments and Results

In this section, the validity of the characterization method was first examined through an offline experiment of the field-of-view effect. Further, the feasibility and effectiveness of the proposed characterization model was demonstrated through the measurement experiments on a set of standard SiO_2 thin films with different thicknesses.

As shown in Figure 4, the method for determining the tilt angle of the beam was proposed and an offline validation experiment was carried out to ensure that the proposed method was useful. The pitching of the laser could be adjusted accurately by rotating Mirror2. A dual-size adjustable aperture and dual-reflecting mirror were introduced to ensure the accurate alignment of the laser. By adjusting the attitude angles of the reflecting mirrors to guide the laser through the small apertures, the accuracy of the alignment could be evaluated by observing the shape of the laser spot through the Cameron Beam Profiler (BC106N-VIS/M, Thorlabs, Newton, NJ, USA) (CBP). When the optical path was perfectly aligned, a small round spot on the screen of the CBP would be achieved and the position of the spot intensity peak on the screen would be nearly consistent regardless of how the CBP moved along the rail.

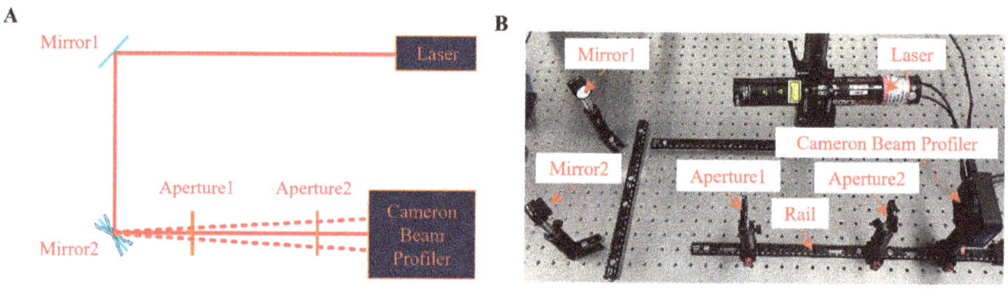

Figure 4. An offline validation experiment to measure the deflection angle of a beam: (**A**) schematic diagram; (**B**) experiment setup.

Then, as illustrated in Figure 5, a simple method based on a geometric principle was employed to determine the tilt angle of the beam. The relationship between the tilt angle θ_{tilt} and the spatial distance x and a was

$$a = x \cdot \tan(\theta_{tilt}), \tag{24}$$

where a is the moving distance of the small round spot on the screen and x is the moving distance of the CBP along the rail. Therefore, we could use Mirror2 to slightly adjust the deflection angle of the laser. For example, when the moving distance x of the CBP along the rail was 25,000 μm, a 0.5° deflection angle of the laser would be obtained if the moving distance of the small round spot on the screen between Position1 and Position2 was $a = x \cdot \tan(\theta_{tilt})$ = 218 μm. Similarly, when the movement distance of the small round spot on the screen was 436 μm and 655 μm, 1° and 1.5° deflection angles could be produced, respectively.

As shown in Figure 6, a tilt angle adjustment experiment setup on the SWE was built. Firstly, in the same way as Figure 4, a dual-size adjustable aperture and dual-reflecting mirror were introduced to ensure the accurate alignment of the laser. Moreover, a CBP was mounted onto an oblique moving stage (PHS-662C-YG, SIGMAKOKI, Sumida-ku, Tokyo, Japan), which could ensure the photosensitive screen was perpendicular to the optical axis and provided a 25,000 μm stroke along the optical axis. With the measurement configuration described above, the allowable tilt angle range of the waveplate in the instrument was 0°~1.5°. Meanwhile, according to the geometric relationship between the tilt angles θ_{tilt1} and θ_{tilt2} of the waveplates in the PSG and PSA shown in Figure 7, it could be assumed $\theta_{tilt1} \approx \theta_{tilt2}$.

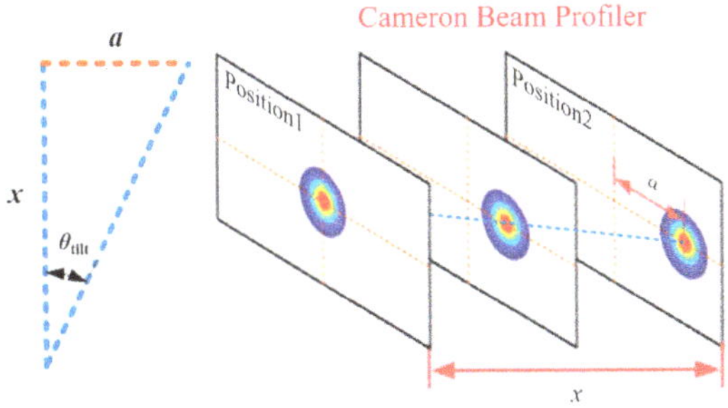

Figure 5. Geometric principles of the determination of the deflection angle of the beam.

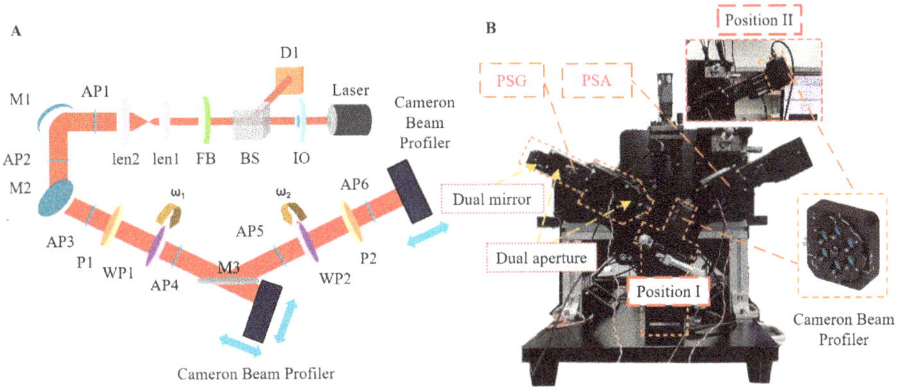

Figure 6. A deflection angle adjustment experiment: (**A**) schematic diagram; (**B**) experiment setup.

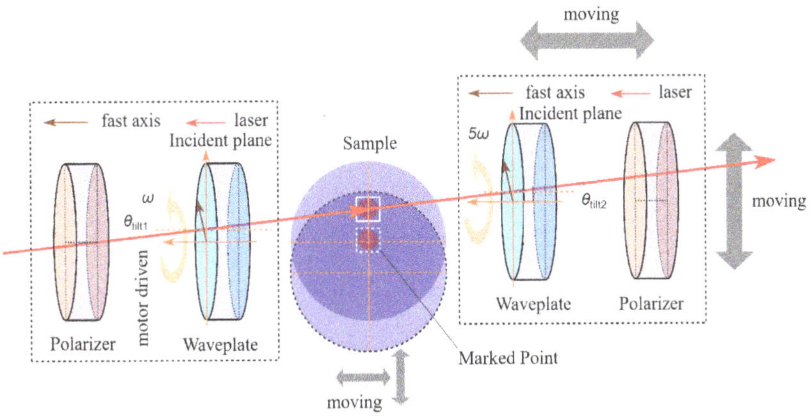

Figure 7. Schematic of the laser propagation in the constructed SWE and the field-of-view effect.

From Figure 7, we could find that the incident positions of the laser at different optical elements were uncertain due to the mechanical installation and adjustment errors. Therefore, it was very difficult to ensure that the laser passed through the center of the polarizers and the waveplates, which meant that more system errors would be introduced into the measurement system.

To ensure the accuracy of the standard SiO_2 film thickness used for the instrument calibration, the film thicknesses were measured with a commercial spectroscopic ellipsometer (RC2 Ellipsometer, J.A. Woollam, Lincoln, NE, USA). To guarantee the measurements were carried out at the same location, a tag was attached to the center of the sample surface, whose edge was parallel to the locating edge of the sample, as shown in Figure 8. The measurements were carried out 30 times on the point next to the left edge of the tag. During the measurement, the sample was held with a vac-sorb pump installed on the sample stage to ensure no movement was introduced during the test.

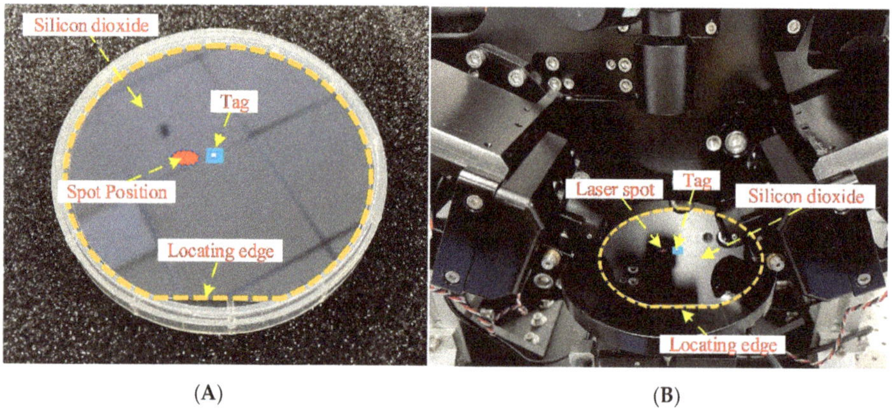

Figure 8. The diagram of the point positioning. (**A**) The silicon wafer; (**B**) area of measurement.

During the calibration process of the instrument, the SiO_2 thin film sample with a nominal thickness of 14.87 nm at a wavelength of 633 nm was used as the standard calibration sample. To demonstrate the successful application of the proposed method in the instrument parameter calibration, the comparison between the measured light intensity and the simulated light intensity at the waveplate tilt angle of 0° was chosen as an example. It could be observed from Figure 9 that the field-of-view effect of the waveplate was well characterized and the measured curves and fitting curves matched well. The results showed that the parameters and polarization effects of each component in the system were accurately calibrated.

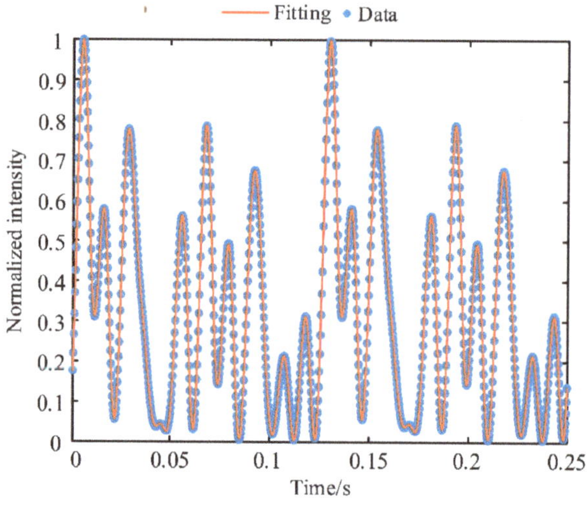

Figure 9. Measured and fitting results of the SiO_2 thin film sample with thickness of 14.87 nm at the wavelength of 633 nm, the instrument incident angle of 65° and the waveplate tilt angle of 0°.

Table 1 summarizes the system parameter calibration results at different waveplate tilt angles. The SiO_2 thin film sample with a nominal thickness of 14.87 nm was used as the calibration sample. As Table 1 shows, when changing the waveplate tilt angle, the system parameters had to be recalibrated. When the calibration procedure was complete, most of the system parameters were fixed, except for ω_1, ω_2, C, C_2^{initial}, A_{C1}, A_{C2}, ϕ_{C1}, ϕ_{C2} and a_1. The ranges of the unfixed system parameters were determined according to the actual experiment: $1438^\circ/\text{s} \le \omega_1 \le 1442^\circ/\text{s}$, $7198^\circ/\text{s} \le \omega_2 \le 7202^\circ/\text{s}$, calibration value $- 5^\circ \le C_1^{\text{initial}}$, $C_2^{\text{initial}} \le$ calibration value $+ 5^\circ$, calibration value $- 0.2 \le A_{C1}$, $A_{C2} \le$ calibration value $+ 0.2$, calibration value $- 5^\circ \le \phi_{C1}$, $\phi_{C2} \le$ calibration value $+ 5^\circ$ and $0 \le a_1 \le 1.5$.

Table 1. System parameters extracted from the calibration of waveplates at different waveplate tilt angles.

Tilt Angle (°) (Absolute Value)	System Parameters		System Parameters		System Parameters	
0	P (°)	33.530	ω_2 (°/s)	7200.006	b_2	0.012
	A (°)	28.650	A_{C1}	−0.011	c_2	0.003
	Dt_1	0.995	A_{C2}	0.034	θ_{incident} (°)	64.944
	Dt_2	0.989	ϕ_{C1} (°/s)	−42.173	a_0	−0.003
	δ_P (°)	0	ϕ_{C2} (°/s)	42.717	a_1	1.010
	δ_A (°)	0	$\delta_{C1}^{\text{center}}$ (°/s)	90.343	a_2	−0.037
	C_1^{initial} (°)	−41.663	$\delta_{C2}^{\text{center}}$ (°/s)	90.490	a_3	0.029
	C_2^{initial} (°)	−62.177	b_1	−0.001		
	ω_1 (°/s)	1440.136	c_1	−0.005		
0.5	P (°)	33.561	ω_2 (°/s)	7200.059	b_2	0.010
	A (°)	28.547	A_{C1}	−0.142	c_2	0.003
	Dt_1	0.995	A_{C2}	0.275	θ_{incident} (°)	64.944
	Dt_2	0.989	ϕ_{C1} (°/s)	−68.006	a_0	−0.003
	δ_P (°)	0	ϕ_{C2} (°/s)	46.404	a_1	1.012
	δ_A (°)	0	$\delta_{C1}^{\text{center}}$ (°/s)	90.341	a_2	−0.037
	C_1^{initial} (°)	−41.829	$\delta_{C2}^{\text{center}}$ (°/s)	90.706	a_3	0.029
	C_2^{initial} (°)	−63.398	b_1	−0.005		
	ω_1 (°/s)	1440.382	c_1	−0.003		
1	P (°)	33.655	ω_2 (°/s)	7200.350	b_2	0.019
	A (°)	28.238	A_{C1}	−0.553	c_2	0.003
	Dt_1	0.990	A_{C2}	−1.193	θ_{incident} (°)	65.360
	Dt_2	0.989	ϕ_{C1} (°/s)	−53.970	a_0	−0.003
	δ_P (°)	0	ϕ_{C2} (°/s)	−91.971	a_1	1.004
	δ_A (°)	0	$\delta_{C1}^{\text{center}}$ (°/s)	89.875	a_2	−0.037
	C_1^{initial} (°)	−41.474	$\delta_{C2}^{\text{center}}$ (°/s)	91.329	a_3	0.029
	C_2^{initial} (°)	−62.944	b_1	0.006		
	ω_1 (°/s)	1441.193	c_1	−0.018		
1.5	P (°)	33.806	ω_2 (°/s)	7199.663	b_2	0.051
	A (°)	27.392	A_{C1}	−1.445	c_2	−0.016
	Dt_1	0.987	A_{C2}	−2.006	θ_{incident} (°)	65.111
	Dt_2	0.989	ϕ_{C1} (°/s)	21.634	a_0	−0.003
	δ_P (°)	0	ϕ_{C2} (°/s)	−49.868	a_1	0.994
	δ_A (°)	0	$\delta_{C1}^{\text{center}}$ (°/s)	88.207	a_2	−0.037
	C_1^{initial} (°)	−25.577	$\delta_{C2}^{\text{center}}$ (°/s)	89.573	a_3	0.029
	C_2^{initial} (°)	−67.841	b_1	0.007		
	ω_1 (°/s)	1441.522	c_1	−0.025		

The SiO_2 thin film samples with nominal thicknesses of 14.87 nm, 26.62 nm, 30.70 nm, 53.84 nm and 57.04 nm were measured at different waveplate tilt angles with the SWE. Then, the field-of-view error parameters could be calculated according to Equation (21). Five sets of field-of-view error parameters could be obtained after measuring the five SiO_2

thin film samples, of which we took the average. Table 2 shows that the waveplate tilt angle θ_{tilt} in the PSG was almost equal to the value set using the CBP. Moreover, it could be observed that the waveplate tilt angle θ_{tilt} in the PSG was close to the waveplate tilt angle θ_{tilt} in the PSA, and this phenomenon conformed to the prediction above. The eight field-of-view error parameters could be obtained with the proposed method. The calculated d_1 and d_2 in the PSG were different from d_1 and d_2 in the PSA. We deduced the reason to be that the incident positions of the laser at the waveplates were different, which meant that manufacture and installation errors (thickness error, optic axis tilt errors and fast axis misalignment [21]) were introduced.

Table 2. Parameters of field-of-view effect extracted from the calibration of waveplates at different waveplate tilt angles.

Tilt Angle (°) (Absolute Value)	Field-of-View Error Parameters							
	PSG				PSA			
	θ_{tilt} (°)	β (°)	d_1 (μm)	d_2 (μm)	θ_{tilt} (°)	β (°)	d_1 (μm)	d_2 (μm)
0	−0.111	−66.074	995.912	978.363	−0.232	66.368	980.360	962.783
0.5	−0.465	101.020	995.913	978.365	−0.655	68.215	980.348	962.728
1	−0.924	108.080	995.798	978.340	−1.359	89.045	991.744	974.004
1.5	−1.491	−33.977	996.408	979.274	−1.756	109.951	998.445	981.046

The retardance fluctuation of the waveplates could be observed in Figure 10. The amplitude of the retardance fluctuation increased with the waveplate tilt angle increasing. The central retardance of the waveplates was different at different waveplate tilt angles.

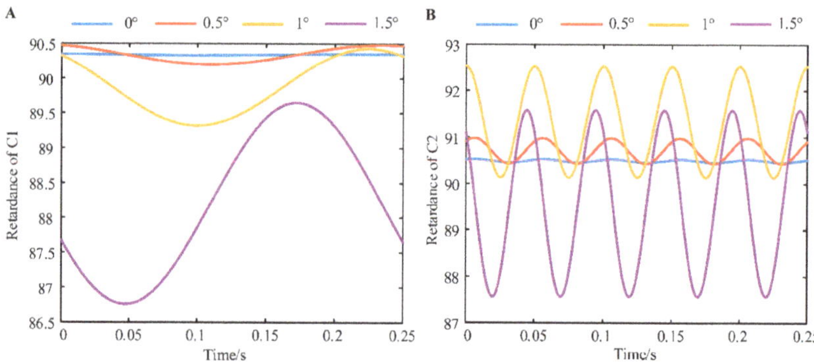

Figure 10. Retardance oscillation calibration results versus the time (the azimuth) at different waveplate tilt angles: (**A**) PSG; (**B**)PSA.

To evaluate the performance of the SWE calibrated with the proposed method, we conducted measurement experiments on standard SiO_2 films of varying thicknesses at different waveplate tilt angles. Since comparing the Mueller matrix results for the same sample was the most direct and reliable way to evaluate the SWE versus the commercial MME, a 57.04 nm SiO_2 film was chosen as the test sample. Figure 11 summarizes the Mueller matrices measured with the different methods. The matrix elements showed good agreement between the proposed methods and with the values reported using the commercial MME. Meanwhile, the proposed method showed smaller errors in the Mueller matrix elements compared to the conventional methods. In addition, ΔM_{con} increased with the waveplate tilt angle, while ΔM_{pro} did not vary with the waveplate tilt angle.

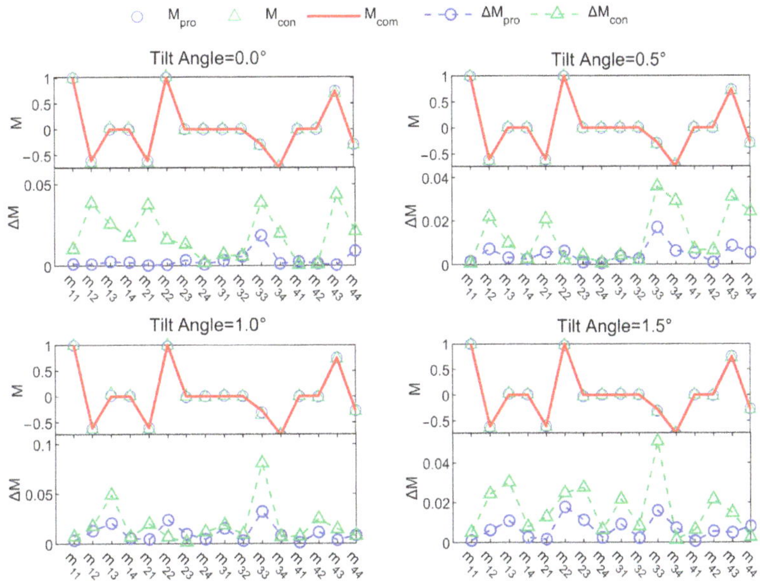

Figure 11. Mueller matrix comparison of SiO_2 thin films with thickness of 57.04 nm at different waveplate tilt angles; $\boldsymbol{M}_{\text{pro}}$ represent the Mueller matrix measured with the proposed method, $\boldsymbol{M}_{\text{con}}$ represents the Mueller matrix measured with the conventional method and $\boldsymbol{M}_{\text{com}}$ represents the Mueller matrix measured with the commercial MME; $\Delta\boldsymbol{M}_{\text{pro}}$ represents the absolute value of the difference between $\boldsymbol{M}_{\text{pro}}$ and $\boldsymbol{M}_{\text{com}}$, while $\Delta\boldsymbol{M}_{\text{con}}$ represents the absolute value of the difference between $\boldsymbol{M}_{\text{con}}$ and $\boldsymbol{M}_{\text{com}}$.

Moreover, the thicknesses of the SiO_2 thin films extracted from the measured Mueller matrix at the waveplate tilt angles of 0°, 0.5°, 1° and 1.5° are summarized in Table 3. It should be noted in advance that the baseline for the thin film thickness deviation was the results reported with the commercially MME. As shown in Table 3, when the waveplate tilt angle was 0°, both the conventional method and the proposed method exhibited good performance in the thin film thickness measurement, and deviations in the thickness measurement were within 1.5%. However, the deviations in the thin film thickness obtained using the conventional method increased significantly as the waveplate tilt angle increased. In contrast, the deviations in the thin film thickness obtained with the proposed method remained stable within 1.8%, barely increasing with the rise in the waveplate tilt angle. Based on the above analysis of the measurement results, it could be concluded that the proposed calibration method could improve the accuracy and precision of the instrument and reduce the difficulty of the instrument assembly.

Table 3. Thickness measurement results of the five SiO_2 samples at different waveplate tilt angles.

Tilt Angle (°)	Silicon Dioxide	RC2 *d* (nm)	Conventional Method *d* (nm)	Conventional Method |Δ (nm)|	Deviation	Proposed Method *d* (nm)	Proposed Method |Δ (nm)|	Deviation
0°	Sample1	14.97	14.75	0.22	1.470%	14.75	0.22	0.1470%
	Sample2	25.38	25.42	0.04	0.158%	25.38	0.00	0.000%
	Sample3	30.33	30.07	0.26	0.857%	30.05	0.28	0.890%
	Sample4	53.01	52.48	0.53	1.000%	52.44	0.57	1.075%
	Sample5	57.04	56.62	0.42	0.736%	56.56	0.48	0.842%
0.5°	Sample1	14.97	14.94	0.03	0.200%	14.95	0.02	0.134%
	Sample2	25.38	24.66	0.72	2.837%	25.33	0.05	0.197%
	Sample3	30.33	29.06	1.27	4.187%	30.01	0.32	1.055%
	Sample4	53.01	49.92	3.09	5.829%	52.22	0.79	1.490%
	Sample5	57.04	53.73	3.31	5.803%	56.27	0.77	1.350%
1.0°	Sample1	14.97	15.06	0.09	0.601%	15.08	0.11	0.735%
	Sample2	25.38	26.49	1.11	4.622%	25.74	0.36	1.418%
	Sample3	30.33	30.98	0.65	2.143%	29.95	0.38	1.253%
	Sample4	53.01	51.15	1.86	3.509%	53.57	0.56	1.056%
	Sample5	57.04	55.01	2.03	3.559%	57.68	0.63	1.104%
1.5°	Sample1	14.97	16.01	1.04	6.496%	15.11	0.14	0.935%
	Sample2	25.38	25.53	0.15	0.59%	25.78	0.40	1.576%
	Sample3	30.33	29.78	0.55	1.813%	30.26	0.07	0.231%
	Sample4	53.01	50.54	2.46	4.641%	52.11	0.90	1.698%
	Sample5	57.04	54.32	2.72	4.769%	56.06	0.98	1.718%

5. Conclusions

In this work, a waveplate characterization model and calibration method were proposed for a self-developed SWE. To compensate for the field-of-view effect in rotating waveplates with tilted incidences, we proposed a calibration method that could obtain the attitude angles of waveplates installed in the instrument and enable a decoupled extraction of all field-of-view error parameters, so that the systematic error could be evaluated reasonably. The consistency between the waveplate tilt angle from the calibrated results and offline measurement demonstrated the correctness and effectiveness of the proposed method. With the proposed method applied, the deviations in the thickness measurement on the SiO_2 thin film samples were within 1.8% compared to the results reported with the commercial MME, when the waveplate tilt angle varied as much as 0°, 0.5°, 1° or 1.5°. The proposed calibration method could not only improve the accuracy and precision of the instrument, but also provide theoretical guidance for the installation and commissioning of the relevant optical systems.

Author Contributions: Conceptualization, Z.J., S.Z. and H.J.; methodology, Z.J., S.Z. and H.J.; software, Z.J. and S.Z.; validation, Z.J. and S.Z.; formal analysis, Z.J. and S.Z.; investigation, Z.J. and S.Z.; resources, S.L. and H.J.; data curation, Z.J. and S.Z.; writing—original draft preparation, Z.J. and S.Z.; writing—review and editing, Z.J., S.Z., H.J. and S.L.; visualization, Z.J. and S.Z.; supervision, S.L. and H.J.; project administration, S.L. and H.J.; funding acquisition, S.L. and H.J. All authors have read and agreed to the published version of the manuscript.

Funding: This work was funded by the National Natural Science Foundation of China (51975232 and 52205592) and the National Key Research and Development Plan (grant no. 2017YFF0204705).

Institutional Review Board Statement: Not applicable.

Informed Consent Statement: Not applicable.

Data Availability Statement: Not applicable.

Conflicts of Interest: The authors declare no conflict of interest.

References

1. Lee, S.H.; Kim, M.Y.; Ser, J.-I.; Park, J. Asymmetric polarization-based frequency scanning interferometer. *Opt. Express* **2015**, *23*, 7333–7344. [CrossRef]
2. Liu, L.; Zeng, A.; Zhu, L.; Huang, H. Lateral shearing interferometer with variable shearing for measurement of a small beam. *Opt. Lett.* **2014**, *39*, 1992–1995. [CrossRef]
3. Sato, T.; Araki, T.; Sasaki, Y.; Tsuru, T.; Tadokoro, T.; Kawakami, S. Compact ellipsometer employing a static polarimeter module with arrayed polarizer and wave-plate elements. *Appl. Opt.* **2007**, *46*, 4963–4967. [CrossRef] [PubMed]
4. Aspnes, D.E. Spectroscopic ellipsometry—Past, present, and future. *Thin Solid Films* **2014**, *571*, 334–344. [CrossRef]
5. Aas, L.M.S.; Ellingsen, P.G.; Fladmark, B.E.; Letnes, P.A.; Kildemo, M. Overdetermined broadband spectroscopic Mueller matrix polarimeter designed by genetic algorithms. *Opt. Express* **2013**, *2021*, 8753–8762. [CrossRef] [PubMed]
6. De Martino, A.; Kim, Y.-K.; Garcia-Caurel, E.; Laude, B.; Drévillon, B. Optimized Mueller polarimeter with liquid crystals. *Opt. Lett.* **2003**, *28*, 616–618. [CrossRef]
7. Williams, P.A. Rotating-wave-plate Stokes polarimeter for differential group delay measurements of polarization-mode dispersion. *Appl. Opt.* **1999**, *38*, 6508–6515. [CrossRef]
8. Aspnes, D.E.; Studna, A.A. High Precision Scanning Ellipsometer. *Appl. Opt.* **1975**, *14*, 220–228. [CrossRef]
9. Azzam, R.M.A. Photopolarimetric measurement of the Mueller matrix by Fourier analysis of a single detected signal. *Opt. Lett.* **1978**, *2*, 148–150. [CrossRef]
10. Sun, C.; Ye, S.; Jian, S. Solc-Like Tunable Birefringent Fiber Filter Based on Elliptical-Core Spun Fiber. *IEEE Photonics Technol. Lett.* **2017**, *29*, 1031–1034. [CrossRef]
11. Demirbas, U.; Uecker, R.; Fujimoto, J.G.; Leitenstorfer, A. Multicolor lasers using birefringent filters: Experimental demonstration with Cr:Nd:GSGG and Cr:LiSAF. *Opt. Express* **2017**, *25*, 2594–2607. [CrossRef] [PubMed]
12. Hagen, N.; Yoshida, K.; Shan, Y.; Otani, Y. Apophyllite waveplates. *Appl. Opt.* **2022**, *61*, 6518–6526. [CrossRef] [PubMed]
13. Emam-Ismail, M. Experimental realization of short-wavelength infrared half wave retarder fabricated from ZnO single crystal plates. *Opt. Mater.* **2023**, *135*, 113297. [CrossRef]
14. Emam-Ismail, M. Wide spectral range multiple orders and half-wave achromatic phase retarders fabricated from two lithium tantalite single crystal plates. *Opt. Laser Technol.* **2015**, *74*, 108–118. [CrossRef]
15. Emam-Ismail, M. Experimental realization of multi, zero, dual order and achromatic gypsum wave plate in a wavelength range 400–1000 nm. *Opt. Commun.* **2013**, *292*, 18–24. [CrossRef]
16. Boulbry, B.; Bousquet, B.; Jeune, B.L.; Guern, Y.; Lotrian, J. Polarization errors associated with zero-order achromatic quarter-wave plates in the whole visible spectral range. *Opt. Express* **2001**, *9*, 225–235. [CrossRef]
17. Dai, H.; Yan, C. Measurement errors resulted from misalignment errors of the retarder in a rotating-retarder complete Stokes polarimeter. *Opt. Express* **2014**, *22*, 11869–11883. [CrossRef]
18. Dong, H.; Tang, M.; Gong, Y. Measurement errors induced by deformation of optical axes of achromatic waveplate retarders in RRFP Stokes polarimeters. *Opt. Express* **2012**, *20*, 26649–26666. [CrossRef]
19. Lee, J.; Rovira, P.I.; An, I.; Collins, R.W. Alignment and calibration of the MgF_2 biplate compensator for applications in rotating-compensator multichannel ellipsometry. *J. Opt. Soc. Am. A* **2001**, *18*, 1980–1985. [CrossRef]
20. Zhang, S.; Jiang, H.; Gu, H.; Chen, X.; Liu, S. Attitude metrology based on the field-of-view effect of birefringence using high-speed polarimetry. *Opt. Lett.* **2020**, *45*, 2074–2077. [CrossRef]
21. Gu, H.; Chen, X.; Zhang, C.; Jiang, H.; Liu, S. Study of the retardance of a birefringent waveplate at tilt incidence by Mueller matrix ellipsometer. *J. Opt.* **2018**, *20*, 015401. [CrossRef]
22. Hsieh, C.H.; Tsai, C.C.; Wei, H.C.; Yu, L.P.; Wu, J.S.; Chou, C. Determination of retardation parameters of multiple-order wave plate using a phase-sensitive heterodyne ellipsometer. *Appl. Opt.* **2007**, *46*, 5944–5950. [CrossRef] [PubMed]
23. Arteaga, O.; Canillas, A.; Jellison, G.E., Jr. Determination of the components of the gyration tensor of quartz by oblique incidence transmission two-modulator generalized ellipsometry. *Appl. Opt.* **2009**, *48*, 5307–5317. [CrossRef] [PubMed]
24. West, E.A.; Smith, M.H. Polarization errors associated with birefringent waveplates. *Opt. Eng.* **1995**, *2265*, 1574–1580. [CrossRef]
25. Ruder, A.; Wright, B.; Peev, D.; Feder, R.; Kilic, U.; Hilfiker, M.; Schubert, E.; Herzinger, C.M.; Schubert, M. Mueller matrix ellipsometer using dual continuously rotating anisotropic mirrors. *Opt. Lett.* **2020**, *45*, 3541–3544. [CrossRef]
26. Fujiwara, H. *Spectroscopic Ellipsometry: Principles and Applications*; John Wiley & Sons: Chichester, UK, 2007; pp. 52–59.
27. Liao, C.C.; Lo, Y.L. Extraction of anisotropic parameters of turbid media using hybrid model comprising differential-and decomposition-based Mueller matrices. *Opt. Express* **2013**, *21*, 16831–16853. [CrossRef]
28. Veiras, F.E.; Perez, L.I.; Garea, M.T. Phase shift formulas in uniaxial media: An application to waveplates. *Appl. Opt.* **2010**, *49*, 2769–2777. [CrossRef]
29. Avendaño-Alejo, M.; Rosete-Aguilar, M. Optical path difference in a plane-parallel uniaxial plate. *J. Opt. Soc. Am. A* **2006**, *234*, 926–932. [CrossRef]
30. Bass, M.; Van Stryland, E.W.; Williams, D.R.; Wolfe, W.L. *Handbook of Optics*; McGraw-Hill: New York, NY, USA, 1995; pp. 15.1–15.100.
31. Pezzaniti, J.L.; Chipman, R.A. Angular dependence of polarizing beam-splitter cubes. *Appl. Opt.* **1994**, *33*, 1916–1929. [CrossRef]
32. Lo, Y.-L.; Pham, T.-T.-H.; Chen, P.-C. Characterization on five effective parameters of anisotropic optical material using Stokes parameters—Demonstration by a fiber-type polarimeter. *Opt. Express* **2010**, *18*, 9133–9150. [CrossRef]

33. El-Agez, T.M.; Taya, S.A. Development and construction of rotating polarizer analyzer ellipsometer. *Opt. Lasers Eng.* **2011**, *49*, 507–513. [CrossRef]
34. Argenti, F.; Torricelli, G.; Alparone, L. MMSE filtering of generalised signal-dependent noise in spatial and shift-invariant wavelet domains. *Signal Process.* **2006**, *86*, 2056–2066. [CrossRef]
35. Novikova, T.; Bulkin, P. Inverse problem of Mueller polarimetry for metrological applications. *J. Inverse Ill-Posed Probl.* **2021**, *29*, 759–774. [CrossRef]

photonics

MDPI

Article

High-Performance Polarization Imaging Reconstruction in Scattering System under Natural Light Conditions with an Improved U-Net

Bing Lin, Xueqiang Fan, Dekui Li and Zhongyi Guo *

School of Computer and Information, Hefei University of Technology, Hefei 230009, China
* Correspondence: guozhongyi@hfut.edu.cn; Tel.: +86-18655151981

Abstract: Imaging through scattering media faces great challenges. Object information will be seriously degraded by scattering media, and the final imaging quality will be poor. In order to improve imaging quality, we propose using the transmitting characteristics of an object's polarization information, to achieve imaging through scattering media under natural light using an improved U-net. In this paper, we choose ground glass as the scattering medium and capture the polarization images of targets through the scattering medium by a polarization camera. Experimental results show that the proposed model can reconstruct target information from highly damaged images, and for the same material object, the trained network model has a superior generalization without considering its structural shapes. Meanwhile, we have also investigated the effect of the distance between the target and the ground glass on the reconstructing performance, in which, and although the mismatch distance between the training set and the testing sample expands to 1 cm, the modified U-net can also efficaciously reconstruct the targets.

Keywords: polarization; polarization imaging; U-net

Citation: Lin, B.; Fan, X.; Li, D.; Guo, Z. High-Performance Polarization Imaging Reconstruction in Scattering System under Natural Light Conditions with an Improved U-Net. *Photonics* **2023**, *10*, 204. https://doi.org/10.3390/photonics10020204

Received: 18 January 2023
Revised: 5 February 2023
Accepted: 6 February 2023
Published: 13 February 2023

1. Introduction

Scattering media, such as the atmosphere [1–3], underwater environments [4,5], and biological tissues [6,7], are among the important factors affecting imaging quality in reality. When light passes through a scattering medium, the ballistic light decays rapidly, and target information will be severely corrupted. In order to get imaging results that are as clear as possible, many typical imaging techniques have been proposed, including transmission matrices [8,9], wavefront shaping [10], light storage effects [11,12], and ghost imaging [13–16]. However, these methods have certain limitations and so do not work better in complex scattering media situations. Moreover, they cost a lot of time and money.

Following developments in polarization transmission theory in recent years [17,18], polarization technology now plays an important role in solving target imaging from scattering media [19–21]. In recent years, physical models and image processing methods based on polarization information have been proposed to improve the clarity of imaging in scattering media [22,23]. In 1996, J.S. Tyo et al., proposed the Polarization Difference (PD) method for imaging through scattering media [24]. In 2001, Y.Y. Schechner et al., added polarization effects to the atmospheric defogging model [25]. Liang et al. proposed that the estimated parameters of the angel of polarization (AoP) can be used in defogging [26], which not only significantly improves the clarity of blurry images but also can be applied to dense fog environments [27]; they also tried to fuse visible and infrared polarized images together to defog and improve target recognition efficiency [28]. Hu et al., proposed a recovery algorithm based on the estimation of polarization differential imaging that takes into account the previously overlooked polarized light radiated by the target itself and proves that it is feasible to improve and optimize the quality of the recovered image [29]. In addition, they also proposed a method based on corrected transmittance to clearly improve the quality of the underwater image [30]. Shao et al. developed an active polarization imaging technology based on wavelength selection [31], which uses the dependence of scattering light at different wavelengths in a turbid underwater situation. In addition, Guo et al. obtained the Muller matrix (MM) of a scattering medium based on the Monte Carlo (MC) algorithm [3] and proposed a polarization inversion method to study polarization transmission characteristics in layered dispersion systems [15,17,32–34], a layered atmosphere [19], and underwater environments [5].

At the same time, deep-learning (DL) techniques have been verified to be a very effective method for damaged image recovery by researchers who used DL to find a mapping relationship between speckled images caused by scattering and original targets [35]. A "one to all" convolutional neural network (CNN) can learn the characteristic information in speckle patterns obtained in the same scattering medium [36]. Li et al. established the "IDiffNet" network structure, which is made up of a tightly connected CNN architecture, to learn the characteristics of the scattering medium and proved that the network's superior generalization capability through the network still works in spite of input data from other scattering media [37]. Lyu et al., proposed a hybrid neural network based on computational imaging in thick scattering media to reconstruct target information hidden in the scattering medium [38]. Sun et al., reconstructed the scattered spot image using the DL algorithm in the low-light environment, which cannot be imaged using a traditional imaging method because the resulting spot contains limited information and has the influence of Poisson noise [39]. Zhu et al., used the autocorrelation imaging of scattered spots to learn the generalized statistical invariants of the scattering medium using DL networks, which improves the applicability of the network model [40]. The combination of polarization information and the DL method has also become an important direction of imaging reconstruction. Li et al. used Q information to train the network and prove that the model-Q has superior generalization and robustness in different aspects [41]. Li et al. proposed the PDRDN to achieve the removing of the underwater fog effect using four angle polarization pictures (0°, 45°, 90°, 135°) [42]. In addition, DL based on polarization is applied to target detection [43–45], underwater imaging [46], image denoising [47], and image fusion [48], etc., which can get higher detection accuracy, significant noise suppression, and effective removal of the scattered light, and can obtain more detailed target information. However, data-driven network models depend too much on the data, resulting in limited generalization capabilities, which is also a major difficulty in applying deep learning to reality. On the one hand, training the network with stable target features will improve the stability of the network. Even if the external environment changes within a certain range, it will not affect the reconstruction results of the trained model. Therefore, in order to improve the stability of the model, we use the polarization information of the target as the training set, which carries stable target features during transmission. The stable target feature carried by the polarization information is capable of adapting to many changes of environment, thereby improving the generalization ability of the network model.

Effective physical priors can prompt networks to find an optimal solution for different situations. The degree of polarization (DoP) is the ratio of the polarization to the total light intensity, and it can be considered as the most intense polarization state. Thus, in this paper, we use DoP to focus the polarization characteristics of the scattering system and then utilize the powerful DL to obtain the polarization characteristics from the scattering system, which can solve the generalization problem of single material objects in scattering scenes and reduce the dependence of deep learning on data. Experimental results demonstrate that the network model trained by DoP has a better recovery performance, and for targets that are not in the training set, the model can still recover them with high accuracy. What is more, the model can still work when there is a mismatch distance between the training set and the testing sample. Moreover, the influence of polarization characteristics also provides a certain basis for the application of deep learning in polarization information-based remote sensing. Finally, we present the quantitative-evaluation results with multiple indicators, which show the accuracy and robustness of the scheme, and reflect the great potential of combining physical knowledge and deep learning technology.

2. Materials and Methods

2.1. Physical Foundation

Light can be represented by the Stokes vector $S = (I, Q, U, V)^T$ whether it is polarized or non-polarized [49]. Elements in a Stokes vector can be obtained from the intensity of four angles (0°, 45°, 90°, 135°):

$$S = \begin{bmatrix} I \\ Q \\ U \\ V \end{bmatrix} = \begin{bmatrix} \langle E_{0x}E_{0x}^* + E_{0y}E_{0y}^* \rangle \\ \langle E_{0x}E_{0x}^* - E_{0y}E_{0y}^* \rangle \\ \langle E_{0x}E_{0y}^* + E_{0y}E_{0x}^* \rangle \\ i\langle E_{0x}E_{0y}^* - E_{0y}E_{0x}^* \rangle \end{bmatrix} = \begin{bmatrix} I_{0^\circ} + I_{90^\circ} \\ I_{0^\circ} - I_{90^\circ} \\ I_{45^\circ} - I_{135^\circ} \\ I_{R^\circ} - I_{L^\circ} \end{bmatrix} \tag{1}$$

where I is the total light intensity, Q is the difference between horizontal and vertical components, U is the difference between 45° and 135° components, and V represents the difference between right-handed and left-handed components. The components in Stokes vector satisfy:

$$I^2 \geq Q^2 + U^2 + V^2 \tag{2}$$

The Stokes vector is relative to the light intensity. An existing focal-plane polarization camera can directly obtain polarization pictures of four angles (0°, 45°, 90°, 135°). Therefore, we can easily get three elements of (I, Q, U), minus the V component.

Polarization information of light can be destroyed by scattering media during the transmission process, and the process can be expressed as:

$$S_{out} = MS_{obj} \tag{3}$$

where M is the Muller matrix (MM) of the scattering media, S_{out} represents the Stokes vector of output light, and S_{obj} represents the object's Stokes vector in the incident light. The aim is to reconstruct targets using the S_{obj}; therefore, Equation (3) is transformed and expressed as follows:

$$S_{obj} = M^{-1}S_{out} \tag{4}$$

where M^{-1} is the inverse of M, which contains the polarization characteristics of the scattering media. For scattering media, the larger optical thickness (OT) becomes, the more damaging target polarization information will be; therefore, the detector can only capture spots which contain limited information from the target. For targets, when the difference of polarization characteristics between target and background is slight, the receiver cannot completely distinguish them.

Reconstructing the target can be regarded as an inverse process of imaging in scattering media. The DL as an excellent method can be used to solve the inverse process. Inspired by this, we utilized the powerful fitting capacity of DL to obtain the map between speckles and the original images. In order to solve the inverse problem better, it is necessary to make full use of the polarization physical priors. Specifically, the learning framework consists of the pre-physical step and post-neural network step based on a physical prior, which can be seen in Figure 1. Firstly, the pre-physical step is used to acquire the linear-polarization images, and the *DoLP* can be expressed as:

$$DoLP = \frac{\sqrt{Q^2 + U^2}}{I} \tag{5}$$

As the ratio of the linear-polarization component to the total light intensity, *DoLP* is a common polarization parameter and can be used to describe the polarization characteristics of the scattering systems. Therefore, when we use *DoLP* images as a training set to train the network, we can filter out redundant information with more effective characteristics for training the network.

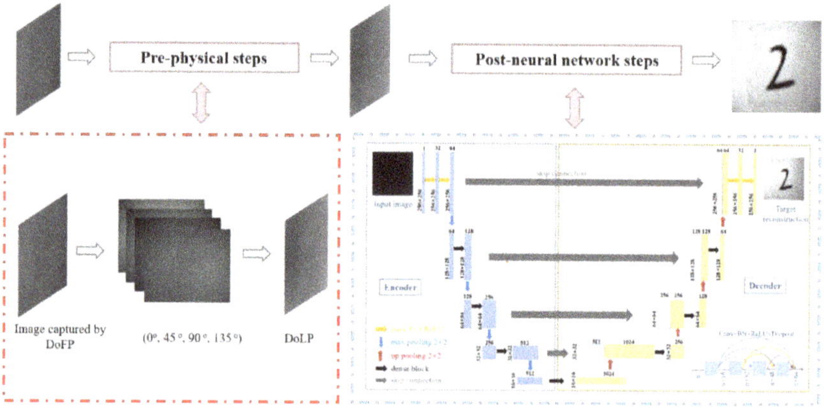

Figure 1. Learning framework.

In addition, the polarization information is very sensitive to the material and the structure of the targets. Therefore, the generalization performance of the model must be most closely related to the scattering medium and the polarization properties of targets. The model trained by the target with the same material will have a broader generalization about materials.

2.2. Measurement System

To get the dataset, we set up a polarization scattering imaging scene in experiments; the schematic of the experimental setup is shown in Figure 2. In order to capture more target information, we placed a polarizer in the front of the LED light source to provide polarization illuminance. The light of $S = (1, 1, 0, 0)^T$ can be modulated by the polarizer, which facilitates the implementation of the polarization algorithm [50]. Then, the polarized light irradiates to the target and is reflected from it. Finally, the reflective light transmits through the ground glass and is captured by the polarization camera (DoFP). In our experiments, the targets are a series of handwritten digits with the ink on the white paper. We put the target at a certain distance behind the ground glass of 5 mm and define the distance between the target and the ground glass as "d".

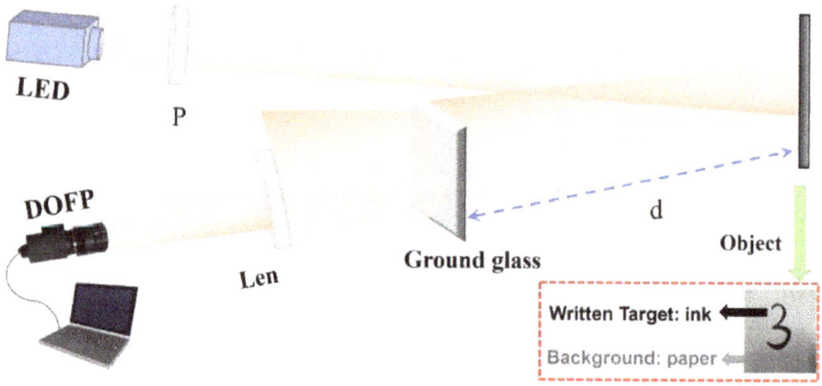

Figure 2. Schematic of the experimental setup.

The polarization camera in our experiment is a commercial DoFP (division of focal plane) polarization camera (LUCID, PHX055S-PC) with pixel counts of 2048 × 2448, whose pixel array surface is covered with a polarization array consisting of four micro-polarizers with four different polarization orientations of 0°, 45°, 90°, and 135°, respectively. The polarization image of the four angles can be used to calculate the image of DoLP. Here, we have captured 200 images with the DoFP cameras and expanded the dataset to 1000 training sets by data enhancement, such as rotation, clipping, etc.; of them, 900 are used for training and 100 for verification.

2.3. Neural Network Design

With the developments of DL technology, many excellent network structures have been built in the field of imaging reconstructions. U-Net, as a fully convolutional neural network structure, has been also proposed for semantic segmentation of medical images. Now it has been showing its superior effects on image reconstruction. The principle of U-Net is similar to that of the self-coder model. Our goal is to extract and reconstruct the target information from the polarization speckles. This process can be regarded as the process of encoding and decoding. Moreover, the skip-connection structure contained in the U-Net solves the problem of gradient explosion and gradient disappearance during training in deeper networks, which is one of the reasons for its excellent performance. DenseNet is a network structure proposed in 2017 [51], and it is a composite layer composed of multiple dense blocks, each of which is connected to the next layer by means of a connection operation. That makes the transmission of features and gradients more efficient and the training process of the network easier [52].

In our scheme, we change the number of convolutional layers and channels of the original U-Net network to form an improved U-Net based DL network, as shown in Figure 1. We replace a single convolutional layer with dense blocks for feature extraction, which will improve the network performance. In the dense block, we use a 3 × 3 convolutional kernel and a circle of padding to ensure that the input and output feature map size is unchanged. Each dense block is connected to the batch normalization and linear activation functions. As the number of network layers and filters increase, the max pooling layer with a step size of 2 × 2 is used to reduce the image length and height to half of the original. In addition, the decoder acts as the inverse of the encoder, and the last layer of each decoder is an up-pooling layer. Throughout our network model, the activation function uses a rectified linear unit (ReLU) that enables fast and efficient training of the network. Meanwhile, in order to reduce the occurrence of overfitting, we add a dropout layer. After that, the images with 256 × 256 pixels can be reconstructed by convolutional layers. In addition, we calculate parameters and floating-point operations (FLOPs) to assess the complexity of the network, which are 53.86 M and 68136.58 M, respectively. During training, the loss function reflects the model's ability to fit the data. Here, we use MAE as the loss function:

$$MAE = \frac{1}{M \times N} \sum_{i=1}^{M} \sum_{j=1}^{N} \| X(i,j) - Y(i,j) \| \tag{6}$$

where $X(i, j)$ and $Y(i, j)$ represent the values of (i, j) pixel in the reconstruction image and in the ground truth, respectively, and M and N are the size of the image.

We trained the model in an image processing unit (NVIDIA RTX 3080) using a Pytorch framework with Python 3.6, training 200 epochs. The optimizer is the Adam (Add Momentum Stochastic Gradient Descent) with a learning rate of 0.001.

3. Results

The polarization characteristics of the target are not easily affected by the scattering media during transmission. Therefore, the model trained with the polarization information of the target is more stable. Therefore, in this section, we designed different test experiments to verify the stability of the trained model with the polarization information of the target.

3.1. Subsection

Unlike the speckled images obtained by laser irradiations, the images obtained by emitting natural light do not have obvious light and dark distributions, and the whole of them is cloudy. Moreover, the greater the distance between the ground glass and the target, the more blurred the outline of the target. At the same time, the spectral width of light reduces the associated length of scattered light and the FoV of the imaging system in a real-world experiment [53,54]. The experiment is set up without ambient light, and we get images only by irradiating the targets with a white-light LED. The results for the circumstance of d = 4.0 cm are shown in Figure 3.

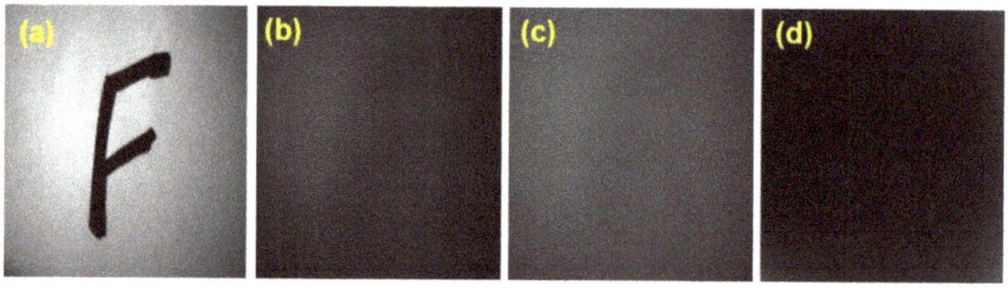

Figure 3. (**a**) Original target; (**b**) Imaging by I_X; (**c**) Imaging by I; (**d**) Imaging by DoP.

With increasing d, the energy of light reaching the ground glass will decrease; therefore, the target information passing through the scattering medium will also be decreasing. We collected the data at a distance of d = 4 cm, where the target profile was completely obscured by the noise from the ground glass, and the calculated DoP image also cannot distinguish the target from the background. Under this condition, we chose the DoP images of the targets as the training set and used the images of targets without the scattering medium as the related labels. The sizes of the scattering images and labels are the same, 256 × 256. After collecting and classifying the data, the proposed methods can be used for training and testing.

In the case of d = 4 cm, we prepared 200 scattering images used as the training sets, which are the DoP imaging results of different structural targets (10 handwriting digits: 0~9) transmitting through the ground glass. Original images without scattering served as the respective labels. We also expanded 200 scattering DoP images to 1000 DoP images, in which 900 and 100 images served as training set and validation set, respectively; the trained network can be called the Model-DoP.

3.2. The Results of Reconstructing Untrained Structural Targets with DoP

In this section, we set targets with different structures, which have not been trained. If the trained Model-DoP can reconstruct those untrained targets, it proves that our proposed method has superior stability on the structure of targets. As shown in Figure 4a, the scattering images are not the samples used to train the Model-DoP, and after transmitting through the ground glass, the corresponding scattering DoP images cannot be distinguished, as depicted in Figure 4b. However, they can be reconstructed well by the trained Model-DoP (as shown in Figure 4), in which the edge of the targets can be identified accurately. The results reflect that the scattering DoP images as training sets can effectively drive the network to learn the polarization characteristics of the different targets, which is helpful to achieve the target reconstruction.

In order to further verify the generalization of the Model-DoP, we changed the structure of the target to test the Model-DoP trained by the digit target. First, we replaced digit targets to English alphabet targets while the background remained unchanged. The reconstructed results are shown in Figure 5. Figure 5b shows the scattering DoP images, and Figure 5c the corresponding reconstructed results. Moreover, we also used some graphics as the targets to further demonstrate the diversity and complexity of the generalization of the Model-DoP. The ground truth, results of the scattering DoP images and reconstructed images are shown in Figure 6. In the case of the limited number of training data, the Model-DoP can reconstruct the untrained targets, including both English alphabets and graphical data, which reflects that the Model-DoP studies not only the mapping relationship between pixels but also the polarization characteristic of different materials. Therefore, the targets with different shapes can be also reconstructed as long as they are made of the same material.

Figure 4. The test results of untrained targets: (**a**) Ground truth; (**b**) Scattering DoP images; (**c**) Images reconstructed by the Model-DoP.

Figure 5. The reconstructed results for untrained alphabetical targets: (**a**) Ground truth; (**b**) Scattering DoP images; (**c**) Images reconstructed by the Model-DoP.

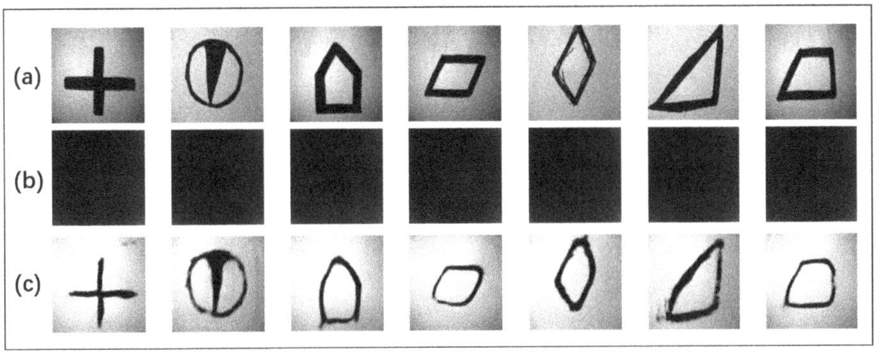

Figure 6. The reconstructed results for untrained graphic targets: (**a**) Ground truth; (**b**) Scattering DoP images; (**c**) Images reconstructed by the Model-DoP.

The Structural Similarity Index (*SSIM*) is a common indicator to evaluate the image quality and measure the similarity of images [55]. Here, we also use the *SSIM* to evaluate the quality of the reconstructed targets, for quantitatively describing the results of the reconstruction and performance of our network. The *SSIM* consists of three parts: brightness, contrast, and structure. Given the original image and the predicted image (*X*, *Y*), the *SSIM* of them can be calculated as follows:

$$SSIM(X,Y) = \frac{(2\mu_X\mu_Y + C_1)(2\sigma_{XY} + C_2)}{(\mu_X^2 + \mu_Y^2 + C_1)(\sigma_X^2 + \sigma_Y^2 + C_2)} \tag{7}$$

where μ_x is the mean of X, μ_Y is the mean of Y, σx is the variance of X, σ_Y is the variance of Y, σ_{XY} is the covariance of X and Y, and C_1 and C_2 are small normal numbers used to avoid the zero denominator. The *SSIM* value range is 0 to 1. The higher the *SSIM* value, the more similar the image.

The *SSIM* of three graphs with different complexity and diversity do not have much difference, which can be seen from Table 1. Although the *SSIMs* have a downward trend with increasing complexity and diversity, the overall fluctuation is not too large. The graphic targets, whose relevance to the target in the training set is the weakest, still have more than 70% similarity.

Table 1. The average SSIM of the target reconstructions for Figures 5–7.

Targets	Digital	Alphabetical	Graphic
SSIM	0.7850	0.7701	0.7688

3.3. The Performance of Model-DoP on the Different Polarization Characteristics

To further investigate the sensitivity of the model-DoP to the polarization properties of the target, we conducted a test using targets composed of other materials that had not been trained. Firstly, the target material was set to steel, and other conditions were unchanged with the background being paper. Therefore, the targets can be called "Steel-Paper" targets, as depicted in Figure 7a. Then, scattering DoP images under natural light condition were obtained, as shown in Figure 7b, and images were entered into the original "Ink–Paper" trained model. The specific reconstruction results are shown in Figure 7c. Due to the high reflectivity and low deflection characteristics of steel, the image obtained through the scattering medium retains a large amount of target information. It can also be seen that the polarization characteristics of steel and paper are quite different from Table 2 [19,56]. Therefore, the model-DoP can also identify the outline of the target.

Table 2. Mueller matrix elements of different materials [19,56].

Material	m22	m33
Paper	0.265	0.247
Wood	0.215	0.16
Ink	0.892	0.921
Steel	0.980	0.977

In addition, the targets can also be set as "Ink–Wood" targets, as shown in Figure 8a, in which the background material is set as wood. The reconstruction results are demonstrated in Figure 8c. From Table 2, because the value of corresponding elements of wood and paper are similar, the model-DoP trained by "Ink–Paper" can distinguish the target and the background. Moreover, the difference between wood and paper impacts the result of the model trained by "Ink–Wood", but it cannot affect the identification and recovery of the target globally. Although the model does not recover well for letter patterns, this problem should be solved by enriching target structures and materials in the training sets.

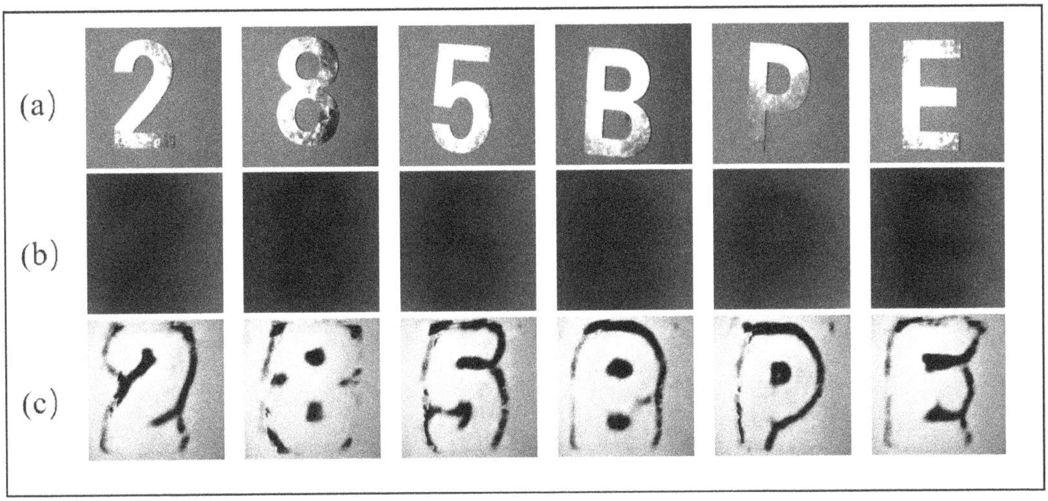

Figure 7. The test results of model-DoP for the untrained target materials. (**a**) Ground truth with target-background as Steel-Paper; (**b**) Scattering DoP images; (**c**) Reconstructed images by the Model-DoP.

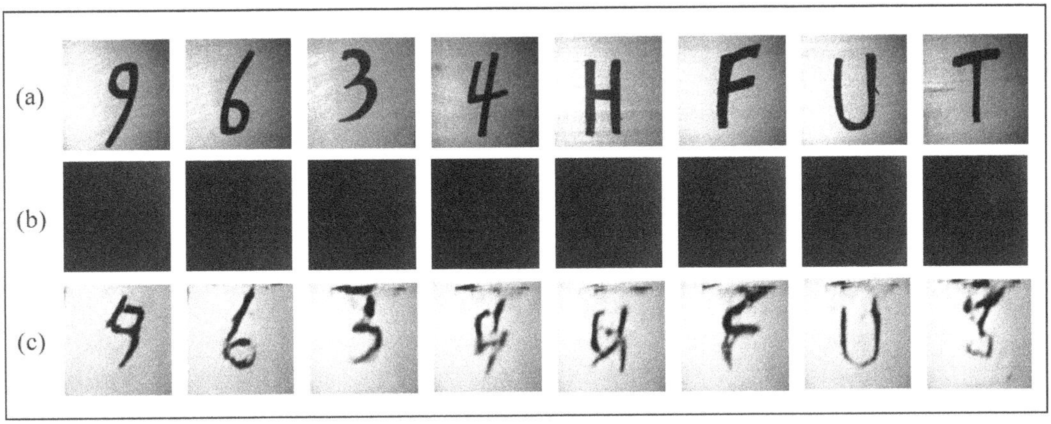

Figure 8. The test results of model-DoP for the untrained background materials. (**a**) Ground truth with target-background as Ink–Wood; (**b**) Scattering DoP images; (**c**) Images reconstructed by the Model-DoP.

Finally, the targets have been set as "Steel-Wood" targets, as shown in Figure 9a, in which the materials of target and background are set as steel and wood, respectively. The model reconstruction results are demonstrated in Figure 9c. The wood background can be distinguished, but the texture cannot be restored. The steel target cannot be recovered with the complete structural information, but the difference in polarization characteristics of the edges can be captured. From Table 2, it can be seen that the difference of corresponding elements of ink and steel is very large. Therefore, the target cannot be recovered very well because of that. When the material is not trained by the DL net, the performance of the target reconstruction will be reduced. The effect of the reconstruction is related to the difference of the polarization properties of the test material and the training material. Therefore, based on the sensitivity of the polarization characteristics of the target, the model, which is trained by the same material target, has a certain cross-material generalization for targets with similar polarization characteristics. It should be noted here that if we train more materials in the DL net, the reconstruction performances would be enhanced for different materials' targets and backgrounds.

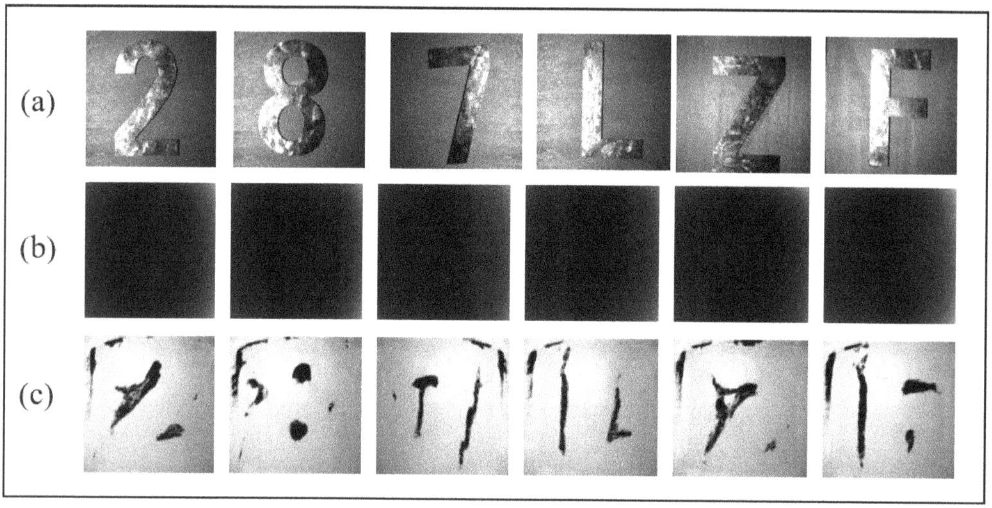

Figure 9. The test results of model-DoP for the untrained target and background materials. (**a**) Ground truth with target-background as Steel–Wood; (**b**) Scattering DoP images; (**c**) Reconstructed images by the Model-DoP.

3.4. *The Performance of Model-DoP on the Generalization of the Imaging Distance*

Different materials have different polarization characters which can be described by a 4×4 matrix called MM. At the same time, when targets and scattering media are determined in a system, the MM of those will not change. Therefore, the trained Model-DoP is still able to reconstruct the targets with different imaging distances (the targets move within a certain range). Therefore, we have also explored the influence of targets at different locations by changing the imaging distances between the ground glass and the targets. We capture the scattering DoP images at the distances of d = 3.5 cm, 4.0 cm, 4.25 cm, 4.5 cm, 5.0 cm and 5.5 cm, and reconstruct the target images through the Model-DoP trained in the imaging distance of d = 4 cm. The results are shown in Figure 10.

It can be seen that the Model-DoP can reconstruct targets at different imaging distances. When d = 3.5 cm, the information of images is enough to provide features for the Model-DoP, and the good retention of the target polarization information strongly improves the imaging quality. Besides, when d is longer than 4.0 cm, the Model-DoP still has a certain generalization ability which is because the model can still obtain some part of the targets' polarization characteristics, allowing the target hidden behind the noise to still be reconstructed until d = 5.0. However, at the imaging distance of d = 5.5 cm, the model cannot reconstruct the target details, though it can still distinguish between the background and the target.

The trained Model-DoP by polarization information is less affected by scattering media because DoP carries stable target features. Therefore, when the target moves within a range, the Model-DoP can still reconstruct it, proving that our proposed method can be adapted to imaging with telescopic distance. Table 3 shows the SSIM of recovered images with increasing imaging distances, where the SSIM is gradually decreasing; however, the magnitude of the decrease is relatively small, which verifies that the DoP can retain the transmitting polarization information in scattering media to improve the stability of the network.

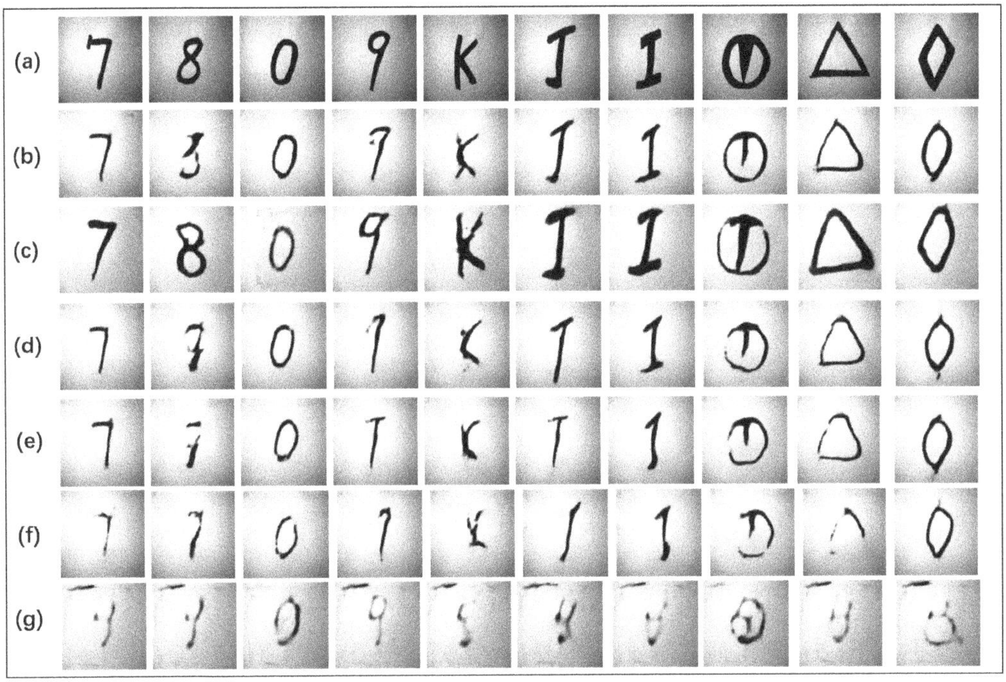

Figure 10. The reconstructed results at different imaging distances by the Model-DoP trained in the imaging distance of d = 4 cm. (**a**) Ground truth; (**b**) d = 3.5 cm; (**c**) d = 4.0 cm; (**d**) d = 4.25 cm; (**e**) d = 4.5 cm; (**f**) d = 5.0 cm; (**g**) d = 5.5 cm.

Table 3. The average SSIM for the reconstructed targets in Figure 10.

Different d	d = 3.5	d = 4.0	d = 4.25	d = 4.5	d = 5.0	d = 5.5
SSIM	0.7557	0.7782	0.7443	0.7339	0.7205	0.6321

3.5. Compared with the Model-I, Model-I_X and Model-Q

DoP can filter out redundant information to a certain extent and focus on the polarization characteristics of targets. Then, the model with both accuracy and stability can be obtained with a small number of datasets. In order to prove that the DoP images as training data are better than I, I_X and Q images, we trained the network to obtain Model-I, Model-I_X, Model-Q and Model-DoP, respectively. Unlike before, we needed to exclude the compensatory effect of emitted polarized light on polarized images, making the difference between different results of the model more obvious. We took the polarizer off and got a series of data directly in natural light conditions. The compared results between those of Model-I, Model-I_X, Model-Q and Model-DoP have been investigated and are demonstrated in Figure 11.

From Figure 11b, it can be seen that targets and backgrounds obtained from Model-I_X can be distinguished; however, the contrast of recovered images is low, and the target structure is distorted, especially letter and graph targets. The I_X component also carries the polarization information of targets, but it also has too much redundant information, making the useful polarization information of targets less prominent. Therefore, in the case of the same amount of data, the network cannot efficiently capture the target polarization information for model building.

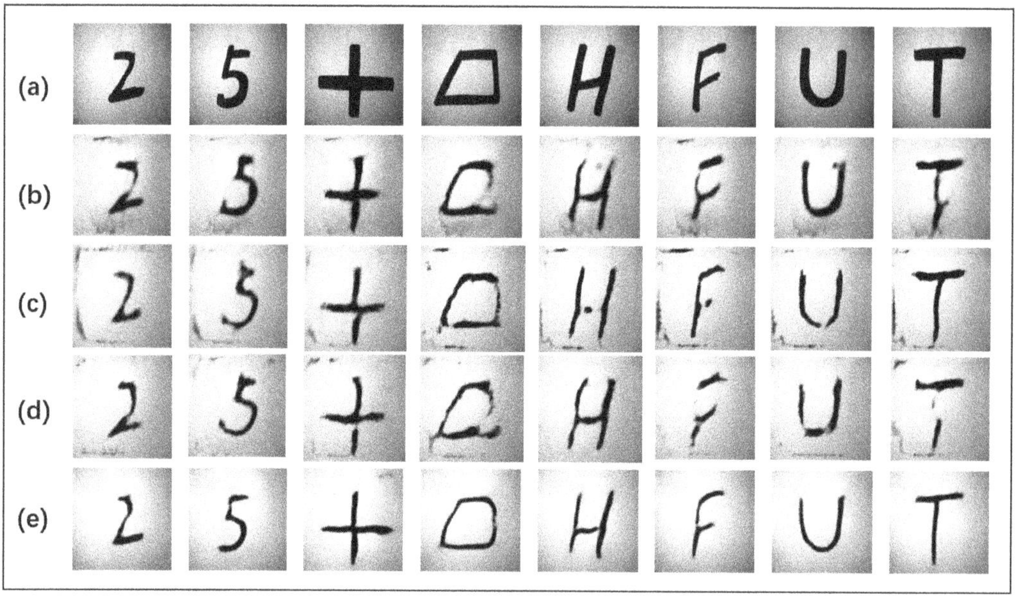

Figure 11. The comparison of Model-I, Model- I_X, model-Q and Model-DoP. (**a**) Ground truth; (**b**) Model-I_X; (**c**) Model-I; (**d**) Model-Q; (**e**) Model-DoP.

The contrast of the result from Model-I is better than that from Model-I_X from Figure 11c, because the intensity is obtained by adding I_X and I_Y, which has more information than I_X alone. However, the background of the result of the Model-I has some noise, especially the edge section. It is precisely because the network trained by intensity cannot accurately distinguish different polarization characteristics.

Thanks to the Q component, which is the difference between I_X and I_Y, it will eliminate some effect of scattering. Therefore, in Figure 11d, the background of the result from the Model-Q has less noise than that from model-I. However, part of the target may not be recovered completely when the gap between the test target and the training target is large, which may contribute to that the Q component may cancel out some target information when the polarization characteristics of the target are not very strong. Therefore, the model-Q has certain restrictions on the material. In Figure 11e, the results of Model-DoP not only recover the goal, but also accurately distinguish the part with different polarization characteristics, although it is not a full reflection. Meanwhile, there is no need to consider the offsets of the targets' polarization information in the DoP information. The quantitative comparison of the four models is shown in Table 4, which further confirms the above information and discussion.

Table 4. The average SSIM for the reconstructed targets in Figure 11.

Different Model	Model-IX	Model-I	Model-Q	Model-DoP
SSIM	0.7498	0.7501	0.7541	0.7746

4. Conclusions

In this article, we combine polarization theory and DL technology to propose a novel method for reconstructing targets. The neural network trained by the DoP can effectively learn the polarization characteristics of different targets and demonstrate certain capability of generalization. Moreover, using the DoP as the polarization information stream reveals a better performance in the reconstruction results, which can provide more target details. Our explorations demonstrate that the polarization information combining the improved U-net shows promise in solving the problem of the information extraction and target identification under strong scattering environments. What is more, for the target with different imaging distances, it can still be reconstructed. It should be further noted that if we use the coherent laser as the illumination source, the much higher performance should be expected, because the speckle effect will be enhanced, as well the characteristics of the scattering fields. In the following work to improve the performance of polarization scattering imaging, we will focus on the following points: (i) extracting usable polarization information from multi-material target information to target reconstruction for more scenarios; (ii) because there is more than one physical quantity that can express the polarization characteristics of the target and they can reflect the different aspects and characteristics of the target, the multi-dimensional polarization information can be used to improve the expression of the target features for improving the performance of target reconstruction.

Author Contributions: Conceptualization, B.L. and Z.G.; methodology, B.L.; software, B.L.; validation, B.L, X.F., D.L. and Z.G.; formal analysis, B.L., X.F. and D.L.; investigation, B.L., X.F. and D.L.; resources, B.L.; writing—original draft preparation, B.L.; writing—review and editing, Z.G.; supervision, Z.G.; funding acquisition, Z.G. All authors have read and agreed to the published version of the manuscript.

Funding: National Natural Science Foundation of China (61775050).

Institutional Review Board Statement: Not applicable.

Informed Consent Statement: Not applicable.

Data Availability Statement: Data underlying the results presented in this paper are not publicly available at this time but may be obtained from the authors upon reasonable request.

Conflicts of Interest: The authors declare no conflict of interest.

References

1. El-Wakeel, A.S.; Mohammed, N.A.; Aly, M.H. Free space optical communications system performance under atmospheric scattering and turbulence for 850 and 1550 nm operation. *Appl. Opt.* **2016**, *55*, 7276–7286. [CrossRef] [PubMed]
2. Xu, Q.; Guo, Z.; Tao, Q.; Jiao, W.; Qu, S.; Gao, J. Multi-spectral characteristics of polarization retrieve in various atmospheric conditions. *Opt. Commun.* **2015**, *339*, 167–170. [CrossRef]
3. Hu, T.W.; Shen, F.; Wang, K.P.; Guo, K.; Liu, X.; Wang, F.; Peng, Z.Y.; Cui, Y.M.; Sun, R.; Ding, Z.Z.; et al. Broad-band transmission characteristics of Polarizations in foggy environments. *Atmosphere* **2019**, *10*, 342. [CrossRef]
4. Purohit, K.; Mandal, S.; Rajagoplan, A.N. Multilevel weighted enhancement for underwater image dehazing. *J. Opt. Soc. Am. A* **2019**, *36*, 1098–1108. [CrossRef] [PubMed]
5. Xu, Q.; Guo, Z.; Tao, Q.; Jiao, W.; Wang, X.; Qu, S.; Gao, J. Transmitting characteristics of the polarization information under seawater. *Appl. Opt.* **2015**, *54*, 6584–6588. [CrossRef] [PubMed]
6. Shen, F.; Zhang, B.; Guo, K.; Guo, Z. The Depolarization Performances of the Polarized Light in Different Scattering Media Systems. *IEEE Photonics J.* **2018**, *10*, 3900212. [CrossRef]
7. Horstmeyer, R.; Ruan, H.; Yang, C. Guidestar-assisted wavefront-shaping methods for focusing light into biological tissue. *Nat. Photonics* **2015**, *9*, 563–571. [CrossRef] [PubMed]
8. Yoon, J.; Lee, K.; Park, J.C.; Park, Y.K. Measuring optical transmission matrices by wavefront shaping. *Opt. Express* **2015**, *23*, 10158–10167. [CrossRef] [PubMed]
9. Shen, F.; Wang, K.P.; Tao, Q.Q.; Xu, X.; Wu, R.W.; Guo, K.; Zhou, H.P.; Yin, Z.P.; Guo, Z.Y. Polarization imaging performances based on different retrieving Mueller matrixes. *Optik* **2018**, *153*, 50–57. [CrossRef]
10. Osnabrugge, G.; Amitonova, L.V.; Vellekoop, I.M. Blind focusing through strongly scattering media using wavefront shaping with nonlinear feedback. *Opt. Express* **2019**, *27*, 11673–11688. [CrossRef] [PubMed]
11. Osnabrugge, G.; Horstmeyer, R.; PapaDoPoulos, I.N.; Judkewitz, B.; Vellekoop, I.M. Generalized optical memory effect. *Optica* **2017**, *4*, 886–892. [CrossRef]
12. Haskel, M.; Stern, A. Modeling optical memory effects with phase screens. *Opt. Express* **2018**, *26*, 29231–29243. [CrossRef]
13. Bromberg, Y.; Katz, O.; Silberberg, Y. Ghost imaging with a single detector. *Phys. Rev. A* **2009**, *79*, 1050–2947. [CrossRef]
14. Ferri, F.; Magatti, D.; Lugiato, L.A.; Gatti, A. Differential Ghost Imaging. *Phys. Rev. Lett.* **2010**, *104*, 253603. [CrossRef] [PubMed]
15. Li, D.; Xu, C.; Yan, L.; Guo, Z. High-Performance Scanning-mode Polarization based Computational Ghost Imaging (SPCGI). *Opt. Express* **2022**, *30*, 17909–17921. [CrossRef] [PubMed]

16. Xu, C.; Li, D.; Guo, K.; Yin, Z.; Guo, Z. Computation ghost imaging with key-patterns for image encryption. *Opt. Commun.* **2023**, *in press*.
17. Li, D.; Xu, C.; Zhang, M.; Wang, X.; Guo, K.; Sun, Y.; Gao, J.; Guo, Z. Measuring glucose concentration in a solution based on the indices of polarimetric purity. *Biomed. Opt. Express* **2021**, *12*, 2447–2459. [CrossRef] [PubMed]
18. Shen, F.; Zhang, M.; Guo, K.; Zhou, H.P.; Peng, Z.Y.; Cui, Y.M.; Wang, F.; Gao, J.; Guo, Z.Y. The Depolarization Performances of Scattering Systems Based on Indices of Polarimetric Purity. *Opt. Express* **2019**, *27*, 28337–28349. [CrossRef]
19. Wang, X.Y.; Hu, T.W.; Li, D.K.; Guo, K.; Gao, J.; Guo, Z.Y. Performances of polarization-retrieve imaging stratified dispersion media. *Remote Sens.* **2020**, *12*, 2895. [CrossRef]
20. Tao, Q.Q.; Sun, Y.X.; Shen, F.; Xu, Q.; Gao, J.; Guo, Z.Y. Active imaging with the AIDS of polarization retrieve in turbid media system. *Opt. Commun.* **2016**, *359*, 405–410. [CrossRef]
21. Wang, P.; Li, D.; Wang, X.; Guo, K.; Sun, Y.; Gao, J.; Guo, Z. Analyzing polarization transmission characteristics in foggy environments based on the indices of polarimetric purity (Is). *IEEE Access* **2020**, *8*, 227703–227709. [CrossRef]
22. Liu, S.; Chen, J.; Xun, Y.; Zhao, X.; Chang, C.H. A New Polarization Image Demosaicking Algorithm by Exploiting Inter-Channel Correlations with Guided Filtering. *IEEE Trans. Image Process.* **2020**, *29*, 7076–7089. [CrossRef]
23. Li, N.; Teurnier, B.L.; Boffety, M.; Goudail, F.; Zhao, Y.; Pan, Q. No-Reference Physics-Based Quality Assessment of Polarization Images and Its Alication to Demosaicking. *IEEE Trans. Image Process.* **2021**, *30*, 8983–8998. [CrossRef] [PubMed]
24. Tyo, J.S.; Rowe, M.P.; Pugh, E.N. Target detection in optically scattering media by polarization-difference imaging. *Appl. Opt.* **1996**, *35*, 1855–1870. [CrossRef]
25. Schechner, Y.Y.; Narasimhan, S.G.; Nayar, S.K. Instant dehazing of images using polarization. In Proceedings of the IEEE Computer Society Conference on Computer Vision and Pattern Recognition (CVPR 2001), Kauai, HI, USA, 8–14 December 2001. [CrossRef]
26. Liang, J.; Ren, L.Y.; Ju, H.J. Visibility enhancement of hazy images based on a universal polarimetric imaging method. *J. Appl. Phys.* **2014**, *116*, 173107. [CrossRef]
27. Liang, J.; Ren, L.; Ju, H. Polarimetric dehazing method for dense haze removal based on distribution analysis of angle of polarization. *Opt. Express* **2015**, *23*, 26146–26157. [CrossRef]
28. Liang, J.; Zhang, W.; Ren, L. Polarimetric dehazing method for visibility improvement based on visible and infrared image fusion. *Appl. Opt.* **2016**, *55*, 8221–8226. [CrossRef]
29. Huang, B.; Liu, T.; Hu, H. Underwater image recovery considering polarization effects of objects. *Opt. Express* **2016**, *24*, 9826–9838. [CrossRef] [PubMed]
30. Hu, H.; Zhao, L.; Huang, B.; Li, X.; Wang, H.; Liu, T. Enhancing Visibility of Polarimetric Underwater Image by Transmittance Correction. *IEEE Photonics J.* **2017**, *9*, 6802310. [CrossRef]
31. Liu, F.; Han, P.L.; Wei, Y.; Yang, K.; Huang, S.Z.; Li, X.; Zhang, G.; Bai, L.; Shao, X.P. Deeply seeing through highly turbid water by active polarization imaging. *Opt. Lett.* **2018**, *43*, 4903–4906. [CrossRef]
32. Xu, Q.; Guo, Z.; Tao, Q.; Jiao, W.; Qu, S.; Gao, J. A novel method of retrieving the polarization qubits after being transmitted in turbid media. *J. Opt.* **2015**, *17*, 035606. [CrossRef]
33. Wang, C.; Gao, J.; Yao, T.; Wang, L.; Sun, Y.; Xie, Z.; Guo, Z. Acquiring reflective polarization from arbitrary multi-layer surface based on Monte Carlo simulation. *Opt. Express* **2016**, *24*, 9397–9411. [CrossRef] [PubMed]
34. Tao, Q.Q.; Guo, Z.Y.; Xu, Q.; Jiao, W.Y.; Wang, X.S.; Qu, S.L.; Gao, J. Retrieving the polarization information for satellite-to-ground light communication. *J. Opt.* **2015**, *17*, 085701. [CrossRef]
35. Jin, K.H.; McCann, M.T.; Froustey, E.; Unser, M. Deep Conutional Neural Network for Inverse Problems in Imaging. *IEEE Trans. Image Process.* **2017**, *26*, 4509–4522. [CrossRef]
36. Li, Y.; Xue, Y.; Tian, L. Deep speckle correlation: A deep learning aroach toward scalable imaging through scattering media. *Optica* **2018**, *5*, 1181–1190. [CrossRef]
37. Li, S.; Deng, M.; Lee, J.; Sinha, A.; Barbastathis, G. Imaging through glass diffusers using densely connected conutional networks. *Optica* **2018**, *5*, 803–813. [CrossRef]
38. Lyu, M.; Wang, H.; Li, G.W.; Zheng, S.S.; Situ, G.H. Learning-based lensless imaging through optically thick scattering media. *Adv. Photon.* **2019**, *1*, 036002. [CrossRef]
39. Sun, L.; Shi, J.H.; Wu, X.Y.; Sun, Y.W.; Zeng, G.H. Photon-limited imaging through scattering medium based on deep learning. *Opt. Express* **2019**, *27*, 33120–33134. [CrossRef]
40. Zhu, S.; Guo, E.L.; Gu, J.; Bai, L.F.; Han, J. Imaging through unknown scattering media based on physics-informed learning. *Photon. Res.* **2021**, *9*, B210–B219. [CrossRef]
41. Li, D.; Lin, B.; Wang, X.; Guo, Z. High-Performance Polarization Remote Sensing with the Modified U-Net Based Deep-Learning Network. *IEEE Trans. Geosci. Remote Sens.* **2022**, *60*, 5621110. [CrossRef]
42. Li, X.; Li, H.; Lin, Y.; Guo, J.; Yang, J.; Yue, H.; Li, K.; Li, C.; Cheng, Z.; Hu, H.; et al. Learning-based denoising for polarimetric images. *Opt. Express* **2020**, *28*, 16309–16321. [CrossRef] [PubMed]
43. Sun, R.; Sun, X.; Chen, F.; Song, Q.; Pan, H. Polarimetric imaging detection using a conutional neural network with three-dimensional and two-dimensional conutional layers. *Appl. Opt.* **2020**, *59*, 151–155. [CrossRef]
44. Wang, Y.; Liu, Q.; Zu, H.; Liu, X.; Xie, R.C.; Wang, F. An end-to-end CNN framework for polarimetric vision tasks based on polarization-parameter-constructing network. *arXiv* **2020**, arXiv:2004.08740.

45. Lin, B.; Fan, X.; Guo, Z. Self-attention module in multi-scale improved U-net (SAM-MIU-net) motivating high-performance polarization scattering imaging. *Optics Express* **2023**, *31*, 3046–3058. [CrossRef]
46. Hu, H.F.; Zhang, Y.B.; Li, X.B.; Lin, Y.; Cheng, Z.Z.; Liu, T. Polarimetric underwater image recovery via deep learning. *Opt. Lasers Eng.* **2020**, *133*, 106152. [CrossRef]
47. Wen, S.J.; Zheng, Y.Q.; Lu, F.; Zhao, Q.P. Conutional demosaicing network for joint chromatic and polarimetric imagery. *Opt. Lett.* **2019**, *44*, 5646–5649. [CrossRef] [PubMed]
48. Zhang, J.; Shao, J.; Chen, J.; Yang, D.; Liang, B.; Liang, R. PFNet: An unsupervised deep network for polarization image fusion. *Opt. Lett.* **2020**, *45*, 1507–1510. [CrossRef]
49. Stokes, S.G.G. *Mathematical and Physical Papers*; Cambridge University Press: Cambridge, UK, 1901.
50. Treibitz, T.; Schechner, Y.Y. Active Polarization Descattering. *IEEE Trans. Pattern Anal. Mach. Intell.* **2008**, *31*, 385–399. [CrossRef]
51. Huang, G.; Liu, Z.; Maaten, L.; Weinberger, K.Q. Densely Connected Conutional Networks. In Proceedings of the IEEE Conference on Computer Vision and Pattern Recognition (CVPR), Honolulu, HI, USA, 21–26 July 2017; IEEE: Piscataway, NJ, USA, 2017; pp. 2261–2269. [CrossRef]
52. Orhan, A.E.; Pitkow, X. Skip connections eliminate singularities. *arXiv* **2017**, arXiv:1701.09175. [CrossRef]
53. Katz, O.; Heidmann, P.; Fink, M.; Gigan, S. Non-invasive single-shot imaging through scattering layers and around corners via speckle correlations. *Nat. Photon.* **2014**, *8*, 784–790. [CrossRef]
54. Zheng, S.S.; Wang, H.; Dong, S.; Wang, F.; Situ, G.H. Incoherent imaging through highly nonstatic and optically thick turbid media based on neural network. *Photon. Res.* **2021**, *9*, B220–B228. [CrossRef]
55. Wang, Z.; Simoncelli, E.P.; Bovik, A.C. Multiscale structural similarity for image quality assessment. In Proceedings of the Thrity-Seventh Asilomar Conference on Signals, Systems & Computers 2003, Pacific Grove, CA, USA, 9–12 November 2003; pp. 1398–1402. [CrossRef]
56. Zhao, Y.Z.; Li, Y.H.; He, W.J.; Liu, Y.; Fu, Y.G. Polarization scattering imaging experiment based on Mueller matrix. *Opt. Commun.* **2021**, *490*, 126892. [CrossRef]

photonics

MDPI

Article

Numerical Study of the Spectrum of TE-Polarized Electromagnetic Waves of a Goubau Line Coated with Graphene

Eugen Smolkin [1] and Yury Smirnov [2,*]

[1] Academy of Technology and Environment, University of Gävle, 801 76 Gävle, Sweden; eugen.smolkin@hig.se
[2] Department of Mathematics and Supercomputing, Penza State University, 440026 Penza, Russia
* Correspondence: smirnovyug@mail.ru

Abstract: The problem of TE-polarized waves in a Goubau line (a perfectly conducting cylinder covered by a concentric layer) coated with graphene is studied. The classification of the waves existing in a Goubau line is carried out. The physical problem is reduced to solving a transmission eigenvalue problem for an ordinary differential equation. The conjugation conditions contain the conductivity of graphene. In this work, we take into account the nonlinearity of graphene. Spectral parameters of the problem are the propagation constants of the waveguide. The article proposes a numerical method for calculating the propagation constants of such waves. A number of numerical experiments were carried out with a Goubau line filled with a dielectric, inhomogeneous dielectric, dielectric with losses, and metamaterial.

Keywords: surface wave; leaky wave; complex wave; Goubau line; Maxwell's equations; spectrum; numerical method

Citation: Smolkin, E.; Smirnov, Y. Numerical Study of the Spectrum of TE-Polarized Electromagnetic Waves of a Goubau Line Coated with Graphene. *Photonics* **2023**, *10*, 1297. https://doi.org/10.3390/photonics10121297

Received: 28 September 2023
Revised: 13 November 2023
Accepted: 21 November 2023
Published: 24 November 2023

1. Introduction

After the discovery of graphene by A. Geim and K. Novoselov [1], the so-called two-dimensional materials, which are considered infinitely thin in a mathematical model and are modeled using special conditions on the surface of a substrate, began to be actively studied. Graphene is the most popular and promising two-dimensional material due to a number of unique properties. It is a layer that is one atom thick and forms a hexagonal crystal lattice of carbon atoms. Graphene is already being used in photonics and optoelectronics. In addition, graphene can also be used in waveguide structures to improve their characteristics. In particular, graphene can be used in various optical devices, such as photodetectors, modulators, and polarizers [2].

A large number of papers have been published on the propagation of electromagnetic waves in waveguide devices containing graphene. Most often, TE and TM waves propagating in a dielectric layer coated with graphene are studied [3–5]. Moreover, graphene can be located on both sides of the dielectric layer and is able to support plasmon modes [6–8]. Plane structures consisting of several alternating layers of dielectric and graphene have also been considered [9].

Graphene nonlinearity was predicted in [10,11], and, later, in [12], it received experimental confirmation. A third-order nonlinearity was found in the terahertz and infrared frequency ranges, similar to the well known Kerr's law. Attempts have been made to use the effect of graphene nonlinearity in photonic devices [13].

This article is devoted to the development of a numerical method for determining the propagation constants of surface, complex, and leaky modes of monochromatic TE-polarized waves in an inhomogeneous Goubau line coated on one side with graphene. At the same time, we take into account the nonlinearity of graphene, which is expressed in changing the standard interface conditions at the interface of media to special conditions containing a nonlinear dependence of conductivity on the intensity of the surface electric field.

Such a statement of the boundary value problem for the system of Maxwell's equations leads to the need to develop new numerical methods for solving the problem, since nonlinear problems of this type have not been studied before. In our formulation of the problem, there are a number of restrictions on the parameters of the waveguide structure. First of all, we do not take into account the absorption in either the dielectric layer or graphene. This assumption is often implemented in practice, since the absorption in the terahertz range is small compared to the strong plasmon response of graphene. In addition, we assume that the conductivity of the graphene depends only on the tangential components of the electric field, more precisely, on its intensity. In this case, the nonlinear part of the conductivity is determined by the amplitude of the TE wave and does not depend on the longitudinal coordinate. This allows us to consider arbitrary, including complex, constant propagations and to study not only surface modes but also complex and leaky modes as well. At the same time, graphene nonlinearity is taken into account in all cases.

From a mathematical point of view, we deal with eigenvalue problems for one differential equation in the case of TE waves with "non-classical" boundary conditions: on the inner boundary, we have one homogeneous condition of the first kind and an additional nonhomogeneous condition of the second kind, which is required in order to determine a discrete set of solutions, and on the external boundary, we have a condition that is nonlinear with respect to a sought-for function.

We will obtain a dispersion equation with the property that its solution is an eigenvalue of the corresponding problem and, conversely, any eigenvalue of the problem is a solution to the dispersion equation. By studying characteristic equations, we will obtain sufficient conditions ensuring that the considered eigenvalue problem has solutions. From a physical point of view, this means that if these conditions are satisfied, then the waveguiding structure studied in the paper is able to support TE-eigenwaves.

An analysis of wave propagation in open metal–dielectric waveguides constitutes an important class of electromagnetic problems. A conducting cylinder covered by a concentric dielectric layer, the Goubau line, is the simplest type of such guiding structures. A complete mathematical investigation of the spectrum of symmetric surface modes in a Goubau line is performed in [14–16].

A large number of papers have been devoted to the study of electromagnetic properties of structures containing graphene layers. Many properties of such structures have been studied quite fully, and there is a significant number of their applications in practice (see, for example, [17–22]).

The classification of waves existing in the Goubau line is an urgent task in the study of waveguides. In this work, a general formulation of the problem is proposed, covering a wide class of TE-polarized waves and various materials filling the waveguide. A numerical method that allows one to calculate the wave propagation constants is proposed. The numerical method is based on the solution to an auxiliary Cauchy problem. Based on this method, it is possible to classify electromagnetic TE-polarized waves existing in a Goubau line: leaky and surface waves (depending on the condition at infinity) and evanescent, propagating and complex waves (depending on the propagation constant).

In this paper, the problem of the propagation of a TE wave in a Goubau line coated with graphene is considered. The graphene coating is considered infinitely thin and leads to a change in one conjugation condition when setting the problem. Such a model is determined by the fact that the graphene layer has a thickness of the order of one atom. Since graphene exhibits nonlinearity, in this article, we will take nonlinearity into account.

The article presents the calculations of the Goubau line filled with a dielectric, inhomogeneous dielectric, dielectric with losses and metamaterial. The calculations were made at different frequencies. Thus, we numerically found complex, evanescent, and propagating leaky waves and complex, evanescent, and propagating surface waves.

2. Statement of the Problem

Consider the three-dimensional space $\mathbb{R}^3$ equipped with the cylindrical coordinate system $O\rho\varphi z$ and filled with a isotropic source-free medium with permittivity $\varepsilon_c\varepsilon_0 \equiv const$ and permeability $\mu_0 \equiv const$, where ε_0 and μ_0 are the permittivity and permeability of a vacuum. We consider electromagnetic waves propagating through a Goubau line:

$$\Sigma := \{(\rho, \varphi) : r_0 \leqslant \rho \leqslant r, 0 \leqslant \varphi < 2\pi\}$$

and a generating line parallel to the axis Oz is placed in $\mathbb{R}^3$.

The cross section of the Goubau line consists of two concentric circles with radii r_0 and r (see Figure 1): r_0 is the radius of the internal (perfectly conducting) cylinder, and $r_0 - r$ is the thickness of the external cylindrical shell.

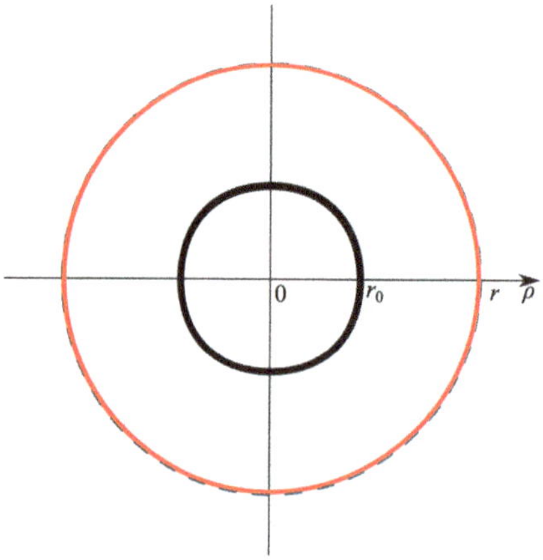

Figure 1. Goubau line coated with graphene.

We assume that the fields depend harmonically on time as $\exp(-i\omega t)$, where $\omega > 0$ is the circular frequency.

The determination of TE-polarized waves reduces to finding nontrivial running-wave solutions of the homogeneous system of Maxwell equations depending on the coordinate z along which the structure is regular in the form $e^{i\gamma z}$,

$$\begin{cases} \nabla \times \mathbf{H} = -i\omega\varepsilon_0\tilde{\varepsilon}\mathbf{E}, \\ \nabla \times \mathbf{E} = i\omega\mu_0\mathbf{H}, \end{cases} \tag{1}$$

$$\mathbf{E} = \left(0,\ E_\varphi(\rho)e^{i\gamma z},\ 0\right),\ \mathbf{H} = \left(H_\rho(\rho)e^{i\gamma z},\ 0,\ H_z(\rho)e^{i\gamma z}\right),$$

with boundary conditions for a tangential electric field component on a perfectly conducting screen

$$E_\varphi\big|_{\rho=r_0} = 0, \tag{2}$$

and the transmission conditions for the tangential electric and magnetic field components on the permittivity discontinuity surface ($\rho = r$)

$$E_\varphi\big|_{r+0} - E_\varphi\big|_{r-0} = 0,\ H_z\big|_{r+0} - H_z\big|_{r-0} = \left(\sigma_1 + \sigma_3\big|E_\varphi\big|_{r+0}\big|^2\right)E_\varphi\big|_{r+0}, \tag{3}$$

where $[f]|_{x_0} = \lim\limits_{x\to x_0-0} f(x) - \lim\limits_{x\to x_0+0} f(x)$ and $\sigma_1 = i\sigma'$, $\sigma_3 = -i\sigma''$, $\sigma' > 0$ and $\sigma'' > 0$.

We will not fix the radiation condition at infinity because we want to consider the problem for arbitrary γ.

We assume that the relative permittivity in the entire space has the form

$$\widetilde{\varepsilon} = \begin{cases} \varepsilon(\rho), & r_0 \leqslant \rho \leqslant r, \\ \varepsilon_c, & \rho > r, \end{cases}$$

We also assume that $\varepsilon(\rho)$ is a continuous function on the segment $[r_0, r]$, i.e., $\varepsilon(\rho) \in C[r_0, r]$.

The problem of normal waves is an eigenvalue problem for the Maxwell equations with spectral parameter γ, which is the propagation constant of waveguiding structure.

The normal wave field in the waveguide can be represented using one scalar function

$$u := E_{\varphi}(\rho).$$

Thus, the problem is reduced to finding the tangential component of the electric field u. Throughout the text below, $(\,\cdot\,)'$ stands for differentiation with respect to ρ.

Definition 1. *The propagating wave is characterized by the real parameter* γ.

Definition 2. *The evanescent wave is characterized by the pure imaginary parameter* γ.

Definition 3. *The complex wave is characterized by the complex parameter* γ *such that* $\Re\gamma\Im\gamma \neq 0$.

Definition 4. *The surface wave satisfies the condition* $u(\rho) \to 0$, $\rho \to \infty$.

Definition 5. *The leaky wave satisfies the condition* $u(\rho) \to \infty$, $\rho \to \infty$.

Remark 1. *The Propagation constant* γ *characterises the behavior of a wave (propagating, evanescent, or complex) in the z direction. The classification of waves as surface or leaky depends on the behavior in the* ρ*-direction.*

We have the following eigenvalue problem for the tangential electric field component u: find $\gamma \in \mathbb{C}$ such that there exist nontrivial solutions of the differential equation

$$u'' + \rho^{-1}u' - \rho^{-2}u + \left(\omega^2\mu_0\varepsilon_0\widetilde{\varepsilon} - \gamma^2\right)u = 0, \tag{4}$$

satisfying the boundary conditions

$$u(r_0) = 0, \tag{5}$$

and the transmission conditions

$$u(r-0) = u(r+0),\ u'(r-0) = u'(r+0) + \varsigma u(r+0), \tag{6}$$

where $\varsigma = \omega\mu_0\left(\sigma' - \sigma''|u(r+0)|^2\right)$.

Thus the resulting field $(\mathbf{E}, \mathbf{H})$ will satisfy all conditions (1)–(3).

For $\rho > r$, we have $\widetilde{\varepsilon} = \varepsilon_c$; then, from (4) , we obtain the equation

$$u'' + \rho^{-1}u' - \rho^{-2}u - \kappa^2 u = 0. \tag{7}$$

In accordance with the condition at infinity (see Definition 4 and 5), we choose a solution to the last equation for surface waves in the form

$$u = C_1 K_1(\kappa\rho),\ \rho > r, \tag{8}$$

and for leaky waves

$$u = C_2 I_1(\kappa\rho),\ \rho > r, \tag{9}$$

where $\kappa = \sqrt{\gamma^2 - \omega^2 \varepsilon_c \varepsilon_0 \mu_0}$ and $\Re\kappa > 0$, C_1 and C_2 are the constants, and K_m and I_m are the modified Bessel functions (Macdonald and Infeld functions) [23].

For $r_0 < \rho < r$, we have $\widetilde{\varepsilon} = \varepsilon(\rho)$; thus, from (4), we obtain the equation

$$u'' + \rho^{-1}u' - \rho^{-2}u + \left(\epsilon - \gamma^2\right)u = 0, \tag{10}$$

where

$$\epsilon(\rho;\omega) = \omega^2 \varepsilon_0 \mu_0 \varepsilon(\rho). \tag{11}$$

Definition 6. *$\gamma \in \mathbb{C}$ is called a propagation constant of the problem if there exists a nontrivial solution u to Equation* (10) *for $r_0 < \rho < r$, satisfying, for $\rho > r$, solutions* (8) *for surface waves and* (9) *for leaky waves, respectively, the boundary condition* (5), *and the transmission conditions* (6).

3. Numerical Method

Let us consider the Cauchy problem for Equation (10) with the initial conditions

$$u(r_0) := 0, \quad u'(r_0) := A, \tag{12}$$

where A is a known constant.

We assume that the Cauchy problem (10), (12) is globally and uniquely solvable on the segment $[r_0, r]$ for given values r_0, r, and its solution continuously depends on the parameter γ. Using the transmission condition on the boundary $\rho = r$ (6), one has the dispersion equation for surface waves

$$\Delta(\gamma) \equiv K_1(\kappa r)u'(r) - \varsigma K_1(\kappa r)u(r) + \left(\kappa K_0(\kappa r) + r^{-1}K_1(\kappa r)\right)u(r) = 0, \tag{13}$$

and the dispersion equation for leaky waves

$$\Delta(\gamma) \equiv I_1(\kappa r)u'(r) - \varsigma I_1(\kappa r)u(r) - \left(\kappa I_0(\kappa r) - r^{-1}I_1(\kappa r)\right)u(r) = 0, \tag{14}$$

where quantities $u(r)$ and $u'(r)$ are obtained from the solution to the Cauchy problem (10), (12).

Let $\gamma = \alpha + i\beta$, where $\alpha, \beta \in \mathbb{R}$. Then, equating to zero the real and imaginary parts of $\Delta(\gamma)$, one obtains a system of real equations for determining the real and imaginary parts of the complex parameter γ:

$$\begin{cases} \Delta_1(\alpha,\beta) := \Re\Delta(\gamma) = 0, \\ \Delta_2(\alpha,\beta) := \Im\Delta(\gamma) = 0. \end{cases} \tag{15}$$

We will numerically solve a system of Equation (15) to determine the pair $(\alpha,\ \beta)$. The solution to each equation of the system (15) is a curve in the plane $O\alpha\beta$. Then, we determine points of intersections of the curves; these points are approximate eigenvalues of the problem. Let us introduce a grid

$$\Big\{\left(\alpha^{(i)}, \beta^{(j)}\right) : \alpha^{(i)} = a_1 + i\tau_1,\ \beta^{(j)} = b_1 + i\tau_2,$$
$$i = \overline{0,n},\ \tau_1 = \frac{a_2 - a_1}{n},\ j = \overline{0,m},\ \tau_2 = \frac{b_2 - b_1}{m}\Big\}$$

with a step $\tau_1 > 0$ in α and a step $\tau_2 > 0$ in β, where a_1, a_2, b_1, b_2 are real fixed constants. We use grid points for the shooting method.

Solving the Cauchy problem (10), (12) for each grid point, one obtains $u(i,j)(r)$ and $u'(i,j)(r)$, $i = \overline{0,n}$, $j = \overline{0,m}$. Since the solution $u(r;\alpha,\beta)$ is continuously dependent on the parameters α and β, it follows that there exists a point $\left(\alpha^{(i)}, \widehat{\beta}^{(j)}\right)$ in the plane $O\alpha\beta$, where $\widehat{\beta}^{(j)} \in \left(\beta^{(j)}, \beta^{(j+1)}\right)$, such that $\Delta_1\left(\alpha^{(i)}, \widehat{\beta}^{(j)}\right) = 0$. The less τ_1 and τ_2, the more exact a solution. Continuing in the same way, one finds a set of pairs $(\alpha^{(p)}, \beta^{(q)})$ that are a curve in the plane $O\alpha\beta$ (the blue curve in Figure 2).

Applying the same approach to the second equation of (15), one obtains another curve in the plane $O\alpha\beta$ (the red curve in Figure 2). This curve is an approximate solution to the equation $\Delta_2(\alpha, \beta) = 0$. It is clear that the intersection points of the curves (the yellow point in Figure 2) are approximate solutions of the problem. By decreasing the steps τ_1 and τ_2, we can obtain arbitrarily accurate solutions.

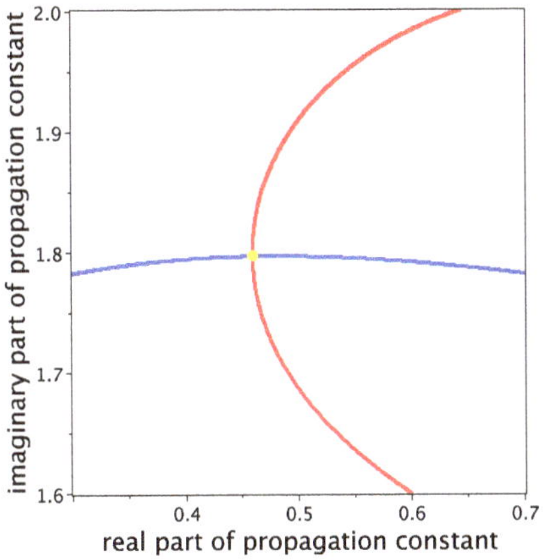

Figure 2. Numerical solution to the system (15): the blue curve is a solution to the first equation of (15); the red curve is a solution to the second equation of (15); the yellow point is a solution to the system (15).

4. Numerical Results

In Figures 3–10, the results of calculating the propagation constants for the problem of the propagation of TE-polarized waves in a Goubau line filled with a dielectric, inhomogeneous dielectric, dielectric with losses, and metamaterial are presented. We have carried out numerical experiments for four frequency values. Propagating, evanescent, complex surface TE waves, and propagating, evanescent, and complex leaky TE waves are found.

The following values of parameters are used for calculations: $\sigma' = 4.67 \times 10^{-4}$ S $\cdot$ m$^2 \cdot$ V^{-2}, $\sigma'' = 2.15 \times 10^{-17}$ S $\cdot$ m$^2 \cdot$ V^{-2}, $\varepsilon_c = 1$, $\varepsilon_0 = 1$, $\mu_0 = 1$, $A = 1 \cdot 10^7$ V$\cdot$m^{-1}, $r_0 = 0.02$ m, $r = 0.04$ m, $a_1 = -3$, $a_2 = 3$, $\tau_1 = 0.025$, $b_1 = -3$, $b_2 = 3$, $\tau_2 = 0.025$.

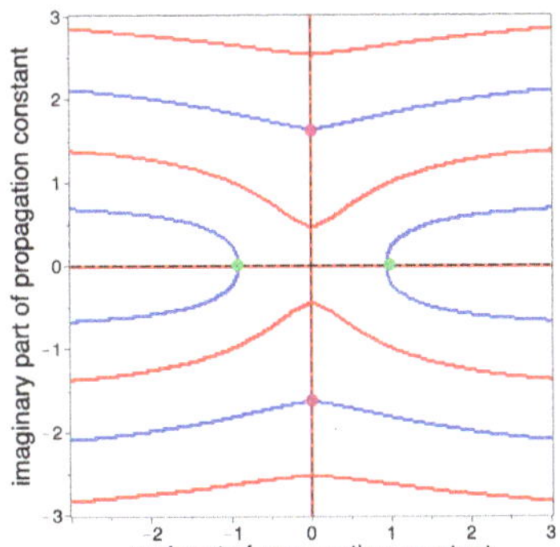

$\omega = 2\pi \cdot 0.25 \cdot 10^{12} Hz$

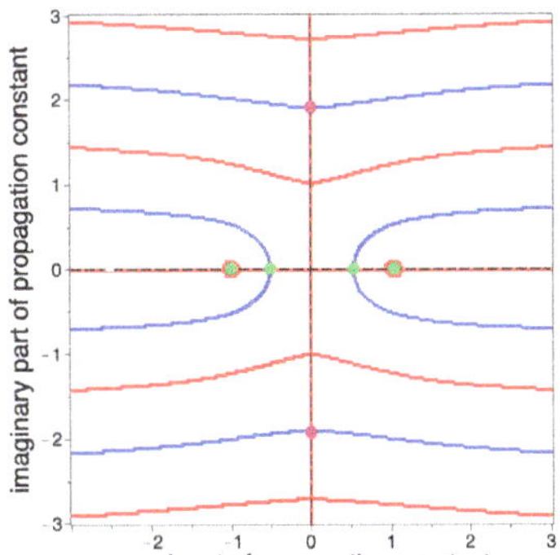

$\omega = 2\pi \cdot 0.5 \cdot 10^{12} Hz$

Figure 3. *Cont.*

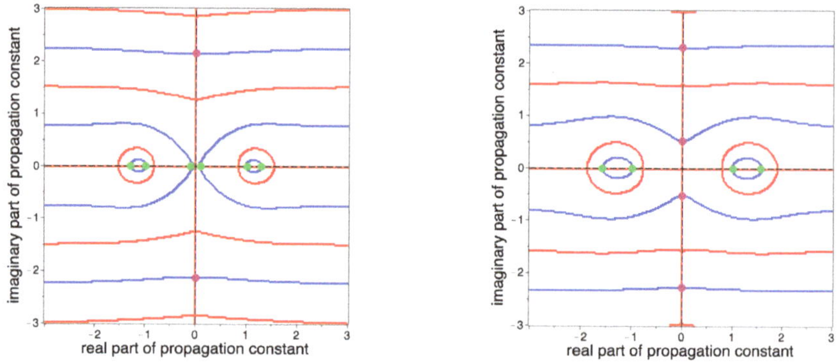

$\omega = 2\pi \cdot 0.75 \cdot 10^{12} Hz$ $\omega = 2\pi \cdot 1.0 \cdot 10^{12} Hz$

Figure 3. Surface waves. Homogeneous dielectric with $\varepsilon = 4$**.** Numerical solution to the system (15): blue curves are a solution to the first equation of (15); red curves are a solution to the second equation of (15); green intersection points correspond to propagating surface TE waves; purple intersection points correspond to evanescent surface TE waves.

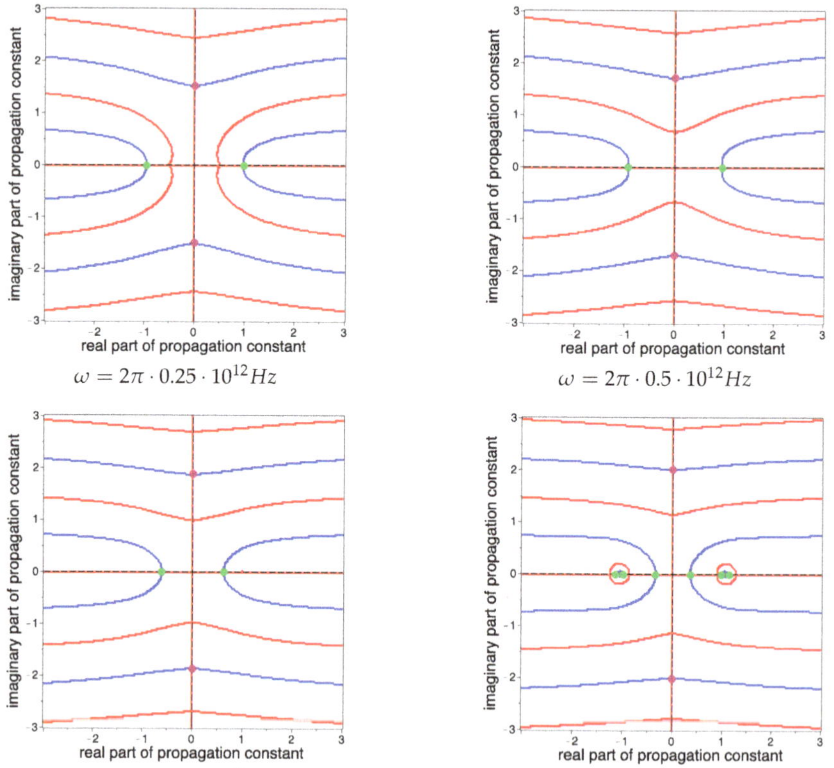

$\omega = 2\pi \cdot 0.25 \cdot 10^{12} Hz$ $\omega = 2\pi \cdot 0.5 \cdot 10^{12} Hz$

$\omega = 2\pi \cdot 0.75 \cdot 10^{12} Hz$ $\omega = 2\pi \cdot 1.0 \cdot 10^{12} Hz$

Figure 4. Surface waves. Inhomogeneous dielectric with $\varepsilon(\rho) = 4 + \frac{\rho}{r - r_0}$**.** Numerical solution to the system (15): blue curves are a solution to the first equation of (15); red curves are a solution to the second equation of (15); green intersection points correspond to propagating surface TE waves; purple intersection points correspond to evanescent surface TE waves.

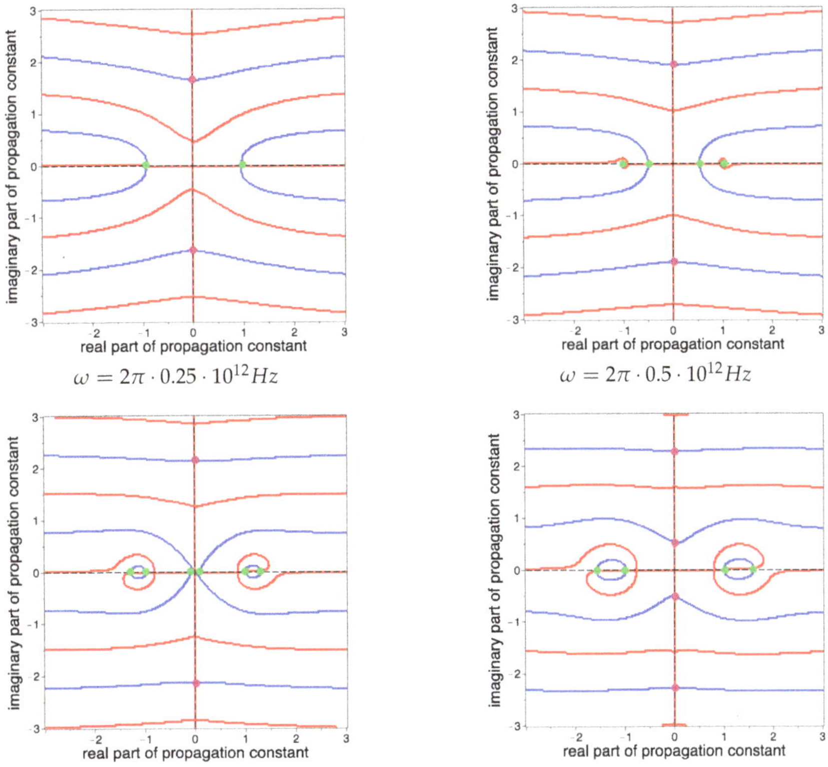

$\omega = 2\pi \cdot 0.25 \cdot 10^{12} Hz$ $\omega = 2\pi \cdot 0.5 \cdot 10^{12} Hz$

$\omega = 2\pi \cdot 0.75 \cdot 10^{12} Hz$ $\omega = 2\pi \cdot 1.0 \cdot 10^{12} Hz$

Figure 5. Surface waves. Dielectric with losses with $\varepsilon = 4 + 0.1\omega i$. Numerical solution to the system (15): blue curves are a solution to the first equation of (15); red curves are a solution to the second equation of (15); green intersection points correspond to propagating surface TE waves; purple intersection points correspond to evanescent surface TE waves.

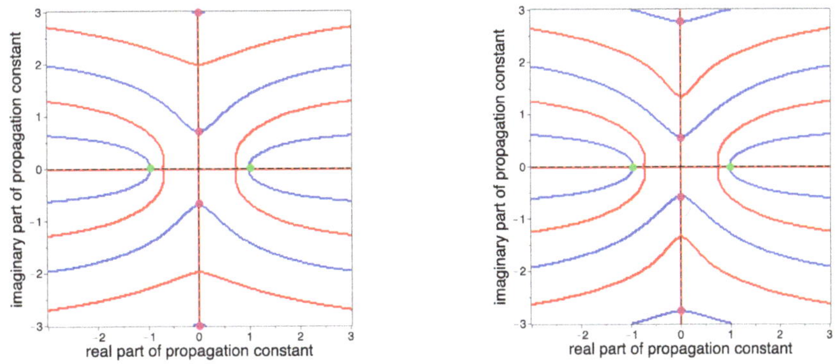

$\omega = 2\pi \cdot 0.25 \cdot 10^{12} Hz$ $\omega = 2\pi \cdot 0.5 \cdot 10^{12} Hz$

Figure 6. *Cont.*

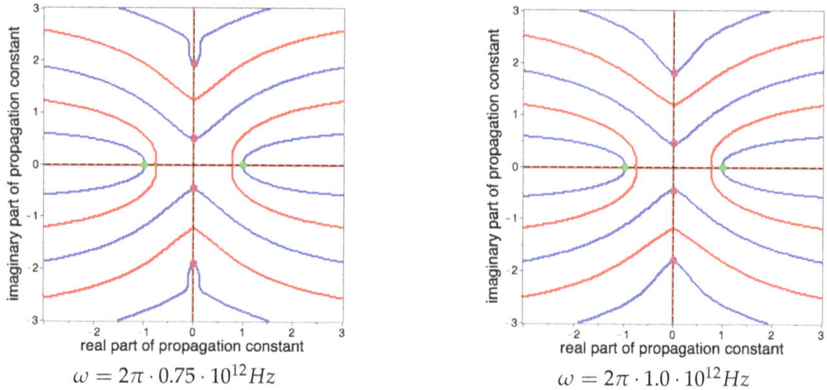

Figure 6. Surface waves. Metamaterial with $\varepsilon = -4$. Numerical solution to the system (15): blue curves are a solution to the first equation of (15); red curves are a solution to the second equation of (15); green intersection points correspond to propagating surface TE waves; purple intersection points correspond to evanescent surface TE waves.

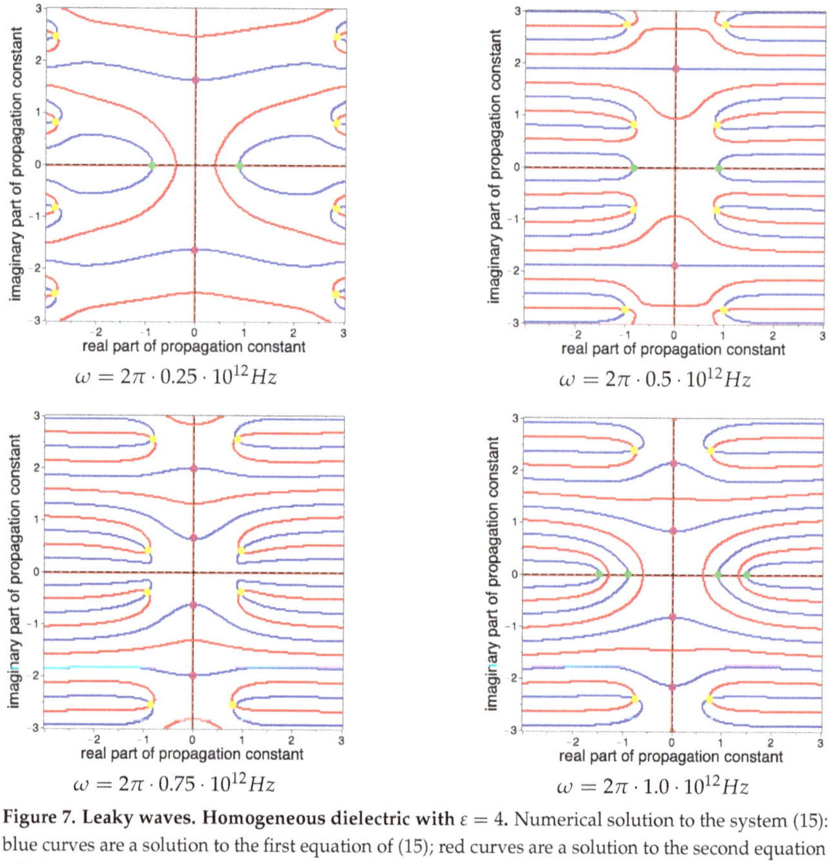

Figure 7. Leaky waves. Homogeneous dielectric with $\varepsilon = 4$. Numerical solution to the system (15): blue curves are a solution to the first equation of (15); red curves are a solution to the second equation of (15); green intersection points correspond to propagating leaky TE waves; purple intersection points correspond to evanescent leaky TE waves; yellow intersection points correspond to complex leaky TE waves.

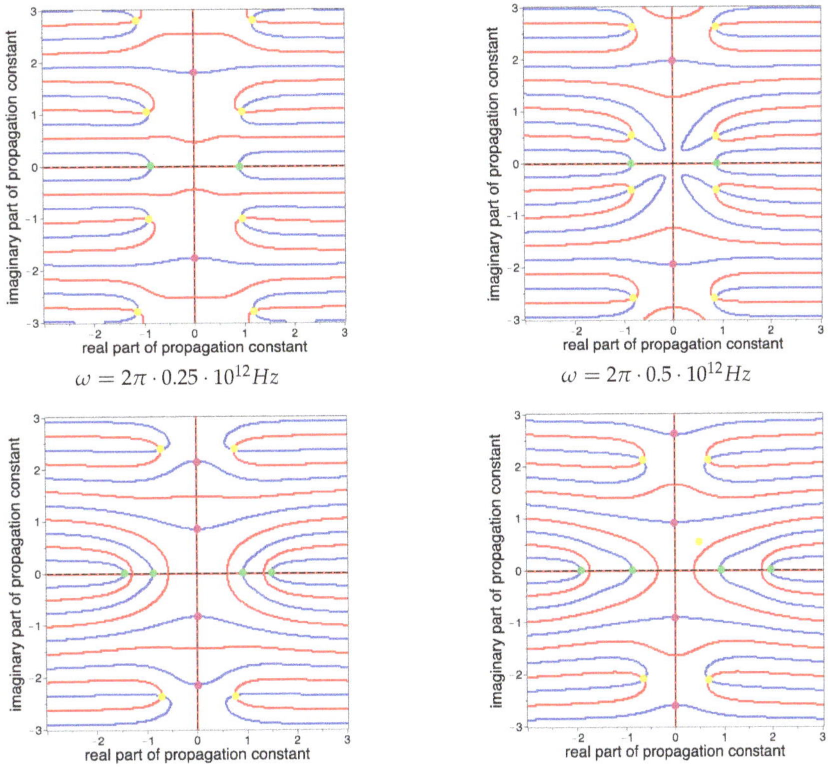

$\omega = 2\pi \cdot 0.25 \cdot 10^{12} Hz$ $\omega = 2\pi \cdot 0.5 \cdot 10^{12} Hz$

$\omega = 2\pi \cdot 0.75 \cdot 10^{12} Hz$ $\omega = 2\pi \cdot 1.0 \cdot 10^{12} Hz$

Figure 8. Leaky waves. Inhomogeneous dielectric with $\varepsilon(\rho) = 4 + \frac{\rho}{r - r_0}$**.** Numerical solution to the system (15): blue curves are a solution to the first equation of (15); red curves are a solution to the second equation of (15); green intersection points correspond to propagating leaky TE waves; purple intersection points correspond to evanescent leaky TE waves; yellow intersection points correspond to complex leaky TE waves.

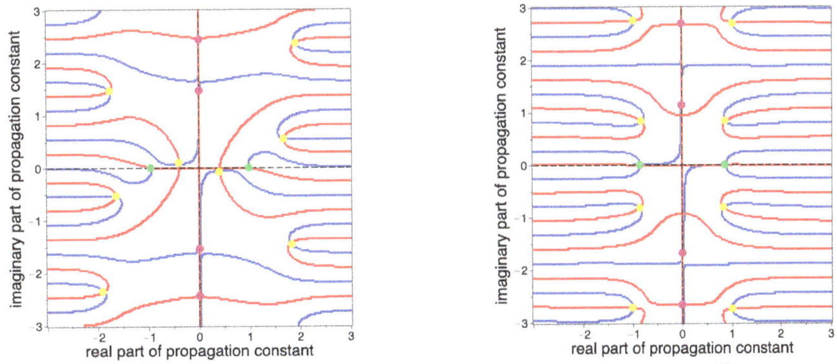

$\omega = 2\pi \cdot 0.25 \cdot 10^{12} Hz$ $\omega = 2\pi \cdot 0.5 \cdot 10^{12} Hz$

Figure 9. *Cont.*

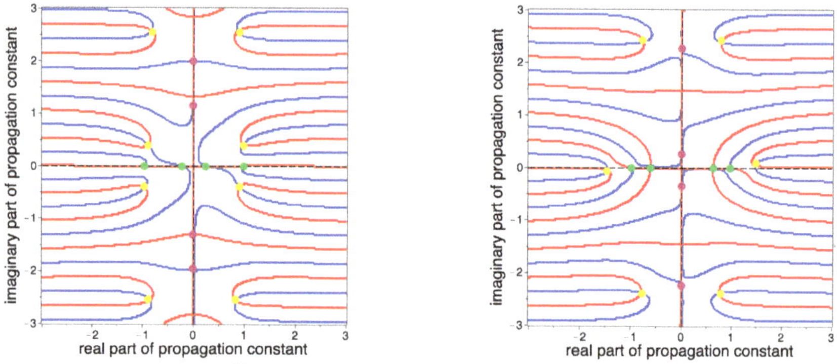

$\omega = 2\pi \cdot 0.75 \cdot 10^{12} Hz$ $\omega = 2\pi \cdot 1.0 \cdot 10^{12} Hz$

Figure 9. Leaky waves. Dielectric with losses with $\varepsilon = 4 + 0.1\omega i$. Numerical solution to the system (15): blue curves are a solution to the first equation of (15); red curves are a solution to the second equation of (15); green intersection points correspond to propagating leaky TE waves; purple intersection points correspond to evanescent leaky TE waves; yellow intersection points correspond to complex leaky TE waves.

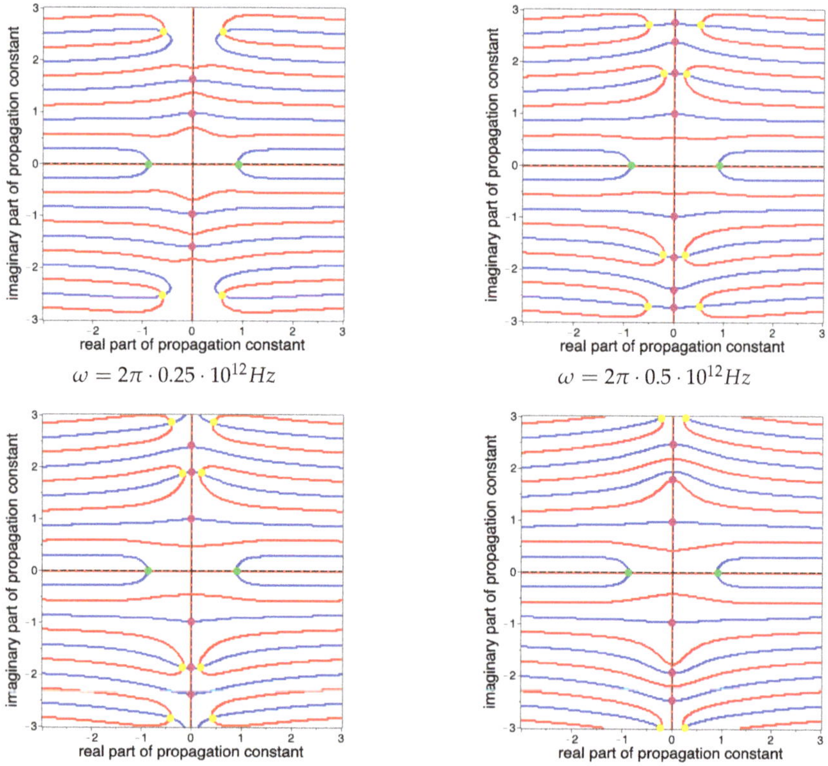

$\omega = 2\pi \cdot 0.75 \cdot 10^{12} Hz$ $\omega = 2\pi \cdot 1.0 \cdot 10^{12} Hz$

Figure 10. Leaky waves. Metamaterial with $\varepsilon = -4$. Numerical solution to the system (15): blue curves are a solution to the first equation of (15); red curves are a solution to the second equation of (15); green intersection points correspond to propagating leaky TE waves; purple intersection points correspond to evanescent leaky TE waves; yellow intersection points correspond to complex leaky TE waves.

In Figures 3–6, the solutions to the problem of surface TE waves in a Goubau line are presented. There are no complex surface TE waves (yellow points). In Figures 3 and 6, we have carried out numerical experiments for a dielectric with constant permittivity and a dielectric with losses. In these cases, propagating surface TE waves (green points) do not exist at all frequencies. If the waveguide is filled with an inhomogeneous dielectric (as in Figure 4), then with increasing frequency, the number of propagating surface TE waves increases; in addition, values of real propagating constants increase in the modulus. In the case of a metamaterial (see Figure 6), with increasing frequency, the absolute values of the propagation constants corresponding to the evanescent surface TE waves decrease, and the absolute values of the propagation constants corresponding to the propagating surface TE waves increase.

In Figures 7–10, solutions to the problem of leaky TE waves in a Goubau line are shown. In the case of a homogeneous dielectric with constant permittivity, all types of leaky TE waves exist at all chosen frequencies (see Figure 7). It can be seen, that with increasing frequency, the absolute values of the real propagation constants increase, the absolute values of the complex propagation constants decrease, and the number of the imaginary propagation constants increases. This trend continues in the case of homogeneous dielectric shown in Figure 8 and in the case of dielectric with losses shown in Figure 9. In the case of a metamaterial (see Figure 10), propagating leaky TE waves (green points) do not exist at all frequencies. In the case with increasing frequency, the absolute values of the imaginary part of the propagation constants corresponding to the complex surface TE waves (yellow points) increase, and the absolute values of the real part of the propagation constants tend to zero.

5. Conclusions

We have developed a numerical method of analysis of the surface TE wave propagation in an inhomogeneous metal–dielectric waveguide structure coated with graphene. The physical problem is reduced to a transmission eigenvalue problem for Maxwell's equations. Using the solutions to an auxiliary Cauchy problem with initial conditions on the boundary $\rho = r_0$ and transmission conditions on the boundary $\rho = r$, it is possible to obtain the dispersion equations for surface and leaky waves. The numerical determination of approximate solutions is performed using a version of the shooting method developed specifically for this class of problems.

It can be seen from the numerical results that propagating surface waves and propagating leaky waves exist only at certain frequencies. In addition, complex surface waves were not found.

6. Discussion

In this paper, the problem of propagation of an electromagnetic TE wave in an inhomogeneous Goubau line with a graphene layer was studied. More precisely, a circular concentric dielectric layer is covered with an infinitely thin layer of graphene. Unlike most similar works, graphene's nonlinearity is taken into account here, which appears in the terahertz and infrared frequency ranges. In this case, the nonlinearity is described using the Kerr-type law, that is, cubic nonlinearity is considered. Since the graphene layer has a thickness of the order of one atom, only tangential components of the electric field are taken into account in the nonlinearity. In addition, we assume that the nonlinearity is determined only by the amplitudes of the propagating waves and does not depend on the longitudinal coordinate.

We consider both homogeneous and inhomogeneous Goubau lines. Both surface and leaky waves are studied. In all cases, we obtain an implicit dispersion equation by applying a numerical method. With the help of the dispersion equation, the propagation constants corresponding to the eigenwave of the waveguide are determined. Note that the proposed method makes it possible to find not only real but also complex constants of the propagation of normal waves in the Goubau line.

It makes sense to take into account nonlinear effects in graphene when nonlinearity has a significant impact on the propagation of electromagnetic waves, that is, if the radiation is strong enough.

In [24], the similar nonlinear eigenvalue problem of electromagnetic wave propagation in a plane dielectric layer covered with graphene was considered. In that paper, the dispersion equation was obtained explicitly. Eigenwaves, as well as propagating constants of surface waves, were calculated. In this paper, we consider not only surface waves but also leaky and complex waves.

Note that we did not obtain an explicit dispersion equation. Our approach is very different. First, we reduce the boundary eigenvalue problem to a Cauchy problem using initial conditions. Then, we solve the Cauchy problem numerically via an appropriate code. Thus, we obtain dispersion equations at the external boundary. The dispersion equation is solved with respect to a spectral parameter, i.e., the propagating constant.

Usually, along with TE-polarized waves, TM-polarized waves in the waveguide structure are also considered. In this article, we have focused on the study of only TE-polarized waves, but TM waves can also be studied using the proposed method. But when studying TM waves, it is already necessary to change the formulation of the problem, so it is natural to consider this case separately. The case of hybrid waves is interesting when both electric and magnetic longitudinal components of an electromagnetic wave take place. In this case, the proposed research method is also applicable, but the system of differential equations becomes much more complicated.

In addition to the waveguide structure discussed in this article, it is interesting to study more complex configurations consisting of several dielectric concentric layers with graphene coatings on each of them. It is not difficult to see that the method will again be applicable to finding the propagation constants of surface, complex, and leaky waves. Such structures can be very useful in practice. It should also be noted that the problem investigated in the article is new from a mathematical point of view since nonlinear coupling conditions are very exotic in mathematical physics and, as far as the authors know, have not been considered before. Therefore, the development of new numerical methods for solving nonlinear eigenvalue problems of this type is an urgent problem in mathematical physics.

Author Contributions: Methodology, Y.S.; software, E.S.; investigation, E.S. and Y.S.; writing—original draft preparation, E.S.; writing—review and editing, Y.S.; visualization, E.S.; supervision, Y.S. All authors have read and agreed to the published version of the manuscript.

Funding: The research was funded by the Russian Science Foundation under the project 20-11-20087.

Institutional Review Board Statement: Not applicable.

Informed Consent Statement: Not applicable.

Data Availability Statement: The data presented in this study are available on request from the corresponding author.

Conflicts of Interest: The authors declare no conflict of interest.

References

1. Geim, A.K.; Novoselov, K.S. The rise of graphene. *Nat. Mater.* **2007**, *6*, 183–191. [CrossRef] [PubMed]
2. Heydari, M.B.; Samiei, M.H.V. Plasmonic Graphene Waveguides: A Literature Review. *arXiv* **2018**, arXiv: 1809.09937.
3. Huawei, L.; Shuangchen, R.; Min, Z.; Hong, S.; Ling, L.I. Graphene surface plasmon polaritons with opposite in-plane electron oscillations along its two surfaces. *Appl. Phys. Lett.* **2015**, *107*, 091602.
4. How, G.C.; Son, C.H.; Ping, L.E. Synthesis of highly confined surface plasmon modes with doped graphene sheets in the midinfrared and terahertz frequencies. *Phys. Rev. B* **2012**, *85*, 125431.
5. Zhu, B.; Ren, G.; Zheng, S.; Lin, Z.; Jian, S. Nanoscale dielectric-graphene-dielectric tunable infrared waveguide with ultrahigh refractive indices. *Opt. Express* **2013**, *21*, 17089–17096. [CrossRef]
6. Svintsov, D.; Vyurkov, V.; Ryzhii, V.; Otsuji, T. Voltage-controlled surface plasmon-polaritons in double graphene layer structures. *J. Appl. Phys.* **2013**, *113*, 053701. [CrossRef]
7. Belonenko, M.B.; Lebedev, N.G.; Yanyushkina, N.N. Solitons in a system of coupled graphene Waveguides. *Phys. Solid State* **2012**, *54*, 174–177. [CrossRef]

8. Buslaev, P.I.; Iorsh, I.V.; Shadrivov, I.V.; Belov, P.A.; Kivshar, Y.S. Plasmons in waveguide structures formed by two graphene layers. *JETP Lett.* **2013**, *97*, 535–539. [CrossRef]
9. Evseev, D.A.; Eliseeva, S.V.; Sementsov, D.I. Waves in a plane graphene - dielectric waveguide Structure. *Eur. Phys. J. Appl. Phys.* **2017**, *80*, 10501. [CrossRef]
10. Mikhailov, S.A. Non-linear electromagnetic response of graphene. *Europhys. Lett.* **2007**, *79*, 27002. [CrossRef]
11. Mikhailov, S.A.; Ziegler, K. Nonlinear electromagnetic response of graphene: Frequency multiplication and the self-consistent-field effects. *J. Phys. Condens. Matter* **2008**, *20*, 384204. [CrossRef]
12. Hendry, E.; Hale, P.J.; Moger, J.; Savchenko, A.K.; Mikhailov, S.A. Coherent nonlinear optical response of graphene. *Phys. Rev. Lett.* **2010**, *105*, 097401. [CrossRef]
13. Ooi, K.J.A.; Tan, D.T.H. Nonlinear graphene plasmonics. *Proc. R. Soc.* **2017**, *473*, 20170433. [CrossRef]
14. Goubau, G. Surface waves and their application to transmission lines. *J. Appl. Phys.* **1950**, *21*, 1119–1128. [CrossRef]
15. Harms, F. Elektromagnetische wellen an einem draht mit isolierender zylindrischer hülle. *Ann. Phys.* **1907**, *6*, 44–60. [CrossRef]
16. Sommerfeld, A. Über die Fortpflanzung elektrodynamischer Wellen langs eines Drahtes. *Ann. Phys.* **1899**, *67*, 233–290. [CrossRef]
17. Avouris, P. Graphene: Electronic and photonic properties and devices. *Nano Lett.* **2010**, *10*, 4285–4294. [CrossRef] [PubMed]
18. Geim, A.K. Graphene: Status and Prospects. *Science* **2009**, *324*, 1530–1534. [CrossRef]
19. Lee, C.; Wei, X.D.; Kysar, J.W.; Hone, J. Measurement of the elastic properties and intrinsic strength of monolayer grapheme. *Science* **2008**, *321*, 385–388. [CrossRef]
20. Yang, L.; Phua, S.L.; Toh, C.L.; Zhang, L.; Ling, H.; Chang, M.; Zhou, D.; Dong, Y.; Lu, X. Polydopamine-coated graphene as multifunctional nanofillers in polyurethane. *RSC Adv.* **2013**, *3*, 6377–6385. [CrossRef]
21. Naushad, M. *A New Generation Material Graphene: Applications in Water Technology;* Springer: Berlin/Heidelberg, Germany, 2019; Volume 7, pp. 187–208.
22. Hanson, G.W. Dyadic Green's functions and guided surface waves for a surface conductivity model of graphene. *J. Appl. Phys.* **2008**, *103*, 064302. [CrossRef]
23. Abramowitz, M.; Stegun, I.A. *Handbook of Mathematical Functions;* National Bureau of Standards: Washington, DC, USA, 1972.
24. Smirnov, Y.; Tikhov, S. The nonlinear eigenvalue problem of electromagnetic wave propagation in a dielectric layer covered with graphene. *Photonics* **2023**, *10*, 523. [CrossRef]

photonics

MDPI

Article

Tunnel Lining Crack Detection Method Based on Polarization 3D Imaging

Yue Zhang [1,*], Xuemin Zhang [1], Yun Su [1], Xuan Li [2], Shiwei Ma [3], Su Zhang [4], Weihe Ren [1] and Kang Li [1]

1 Beijing Institute of Space Mechanics & Electricity, Beijing 100094, China
2 School of Optoelectronic Engineering, Xidian University, Xi'an 710071, China
3 Tieke Chengdu Testing Technology Co., Ltd., Chengdu 610021, China
4 School of Optoelectronic Engineering, Changchun University of Science and Technology, Changchun 130012, China; susiezhang21@126.com
* Correspondence: yue3723302@126.com

Abstract: Non-contact and non-destructive polarization 3D imaging uses a passive, single-frame array image to calculate 3D information, making it possible to obtain high-precision 3D information about tunnel cracks, and offering outstanding technical advantages. Based on the introduction of the principle of crack detection with polarization 3D imaging, a tunnel lining crack detection plan was developed and a detection equipment was designed. The method and process of polarization 3D imaging for lining crack detection are described in detail. A model of the impact of the tunnel environment on 3D detection and a method for obtaining absolute information have been established to obtain high-precision 3D information about cracks. In a real tunnel environment, tests were conducted to detect wide cracks, narrow cracks, and artificial cracks. The crack detection accuracy with respect to the crack width was 0.2–0.3 mm, and with respect to crack length was 0.2–0.3 mm. At the same time, crack depth information could be obtained. The present research results can provide technical support for the application of polarization 3D imaging in tunnel crack detection.

Keywords: polarization 3D imaging; tunnel lining crack; crack detection; polarization camera

Citation: Zhang, Y.; Zhang, X.; Su, Y.; Li, X.; Ma, S.; Zhang, S.; Ren, W.; Li, K. Tunnel Lining Crack Detection Method Based on Polarization 3D Imaging. *Photonics* **2023**, *10*, 1085. https://doi.org/10.3390/photonics10101085

Received: 17 May 2023
Revised: 13 June 2023
Accepted: 21 July 2023
Published: 27 September 2023

1. Introduction

With the prosperity and development of China's economy and society, the number and length of tunnel constructions in China have rapidly increased in recent years, with railway tunnels becoming an important part of the transportation infrastructure. By the end of 2020, the operating mileage of railways in China reached 145,000 km, with a total of 16,798 railway tunnels in operation, spanning approximately 19,630 km, including 209 special-length railway tunnels with a total length of 2811 km. Tunnel safety has become particularly important, as the condition of tunnels directly affects railway safety and transport efficiency. Due to the different construction eras, materials, and structures, Itunnels have various problems. Tunnel lining cracks are among the most common tunnel problems; when such cracks extend to a certain degree, they inevitably change the stress state, increasing the probability of structural damage and potentially causing further problems, such as water leakages. Tunnel lining cracks and tunnel water leakages make up about 30% and 18%, respectively, of tunnel problems. Preventing and treating tunnel lining cracks, which requires the timely, accurate, and rapid detection of relevant crack data, is key to ensuring tunnel safety. However, tunnel cracks often have a similar color to the background structure, making them difficult to identify. Crack shapes are also varied and irregular, making recording and describing three-dimensional information about them complex.

Currently, non-contact and non-destructive detection techniques for tunnel lining cracks mainly include artificial detection, laser scanning-based detection, and camera-based detection [1–4]. The artificial detection method [1–4] has long detection times, a low efficiency, and high cost, while also yielding unreliable results. It also suffers from the inability to store the obtained data in digital archives and the inability to accurately express and record the morphology and nature of the observed cracks. The laser scanning-based detection method [1–4] scans the surface of the tunnel lining with a laser, calculates the three-dimensional coordinates of each point on the object surface based on the received signal, and obtains a tunnel surface image based on the positions of numerous points. This method requires a laser emission device and a receiving device, and the whole system is complex. Laser scanning requires the measurement of a large number of three-dimensional point coordinates and the reconstruction of the model based on point cloud data, resulting in the creation of large quantities of data and establishing a non-continuous three-dimensional surface. The camera-based detection method [1–4] mainly measures cracks using multiple linear sensors or array sensors. The image accuracy of linear sensors is influenced by the precision of the linear scanning motion; in comparison, array cameras offer better stability and intuitive imaging. This method has been applied in crack image acquisition systems for various inspection vehicles. To detect cracks in tunnel lining, this method mainly relies on two-dimensional image processing, using crack width, length, and area as the crack features, and obtaining feature values through image recognition. Due to the large amount of image data obtained with regard to tunnel lining surface and the complexity of tunnel environments, at present, the accuracy and efficiency of image recognition and crack identification fail to meet the practical requirements for high precision and rapid detection.

Currently, non-contact and non-destructive testing systems for tunnel lining cracks mainly include laser scanning detection systems, camera photography detection systems, and systems that contain laser scanning detection and camera photography detection [5]. The tunnel disease detection vehicle developed by the Korean Institute of Nondestructive Testing [6] uses CCD imaging technology for tunnel disease detection. It uses real-time photos and videos to generate static images, requires auxiliary lighting, and needs to identify cracks from a large number of images. The tunnel fast measurement system developed by China's Wuhan University Zhuoyue Technology Co., Ltd., Wuhan, China [6,7] consists of four subsystems: a crack subsystem, a main control subsystem, an infrared subsystem, and a deformation subsystem. The crack subsystem includes 34 LED lighting systems and 32 high-speed area array cameras, achieving a full-coverage detection of tunnel cross-sections. The deformation subsystem includes two laser scanning radar systems with a crack width detection accuracy of 0.2 mm. The method, combining laser scanning detection and camera photography detection, uses laser scanning detection to obtain the three-dimensional contours of the lining and camera photography to detect lining cracks. In March 2014, Mitsubishi Electric Corporation of Japan launched the MIMM-R highway tunnel inspection vehicle [7], which consists of twenty array cameras, three GPS systems, one IMU system, one odometer, three laser scanning radars, and three cameras. The inspection vehicle uses array cameras and LED lighting devices to capture tunnel lining problems and uses laser scanning radars to depict the 3D contour of the lining.

Although non-contact tunnel-inspection methods such as laser scanning technology and digital camera measurement and detection technology can basically meet the requirements of daily tunnel inspection and serve good auxiliary roles in tunnel maintenance work, they have the following drawbacks. (1) A complex data processing—the image quality is greatly affected by the tunnel environment, making this approach unsuitable for use in tunnels during the construction period. Additionally, the massive quantity of generated data cannot be quickly and efficiently processed, resulting in a low crack-recognition accuracy and efficiency. (2) The detection system is complex, and crack recognition and tunnel 3D profile acquisition require the separate implementation of different pieces of equipment.

A tunnel lining crack detection method based on polarization 3D imaging was proposed in [8]. This approach uses a passive array polarization camera to obtain tunnel lining images and uses the polarization information of a single-frame image to reconstruct 3D information about the tunnel wall. Polarization 3D imaging is a passive, polarized imaging method. It utilizes the Fresnel equation to establish the function between the polarization characteristics of reflected light and the normal direction of the target surface to achieve 3D reconstructions. With this approach, 3D inversion can be achieved with a single image. As it does not require active illumination and is not limited by baseline length, it can better meet the application requirements of high-precision 3D information acquisition for crack detection in tunnel safety inspections.

Through elevation information calculation, cracks in tunnels that have different depths from the background structure can be quickly extracted, and crack width, length, and other information can be output simultaneously. Polarization 3D imaging is particularly suitable for identifying cracks with similar colors to the background structure; however, it is also affected by complex environmental factors, such as humidity and dust in the tunnel during the construction period. Therefore, it is necessary to consider tunnel environmental factors when using existing polarization 3D imaging technology to improve detection accuracy. Compared with existing detection methods, polarization 3D imaging technology has the advantages of a low power consumption, simple and small equipment, simple data processing, a high crack-detection accuracy, and the potential to build 3D tunnel profiles and generate 3D data for the identification of problems.

Polarization 3D imaging can also detect tunnel lining leakages caused by cracks, lining peeling, and lining corrosion. Among tunnel diseases, tunnel lining cracks and tunnel leakages account for approximately 50%. Monitoring lining cracks is more difficult than monitoring other tunnel diseases, so as long as tunnel cracks can be detected, other tunnel diseases can also be detected.

2. Detection Principle of Tunnel Lining Cracks Based on Polarization 3D Imaging

Polarization 3D imaging utilizes the relationship between the normal of the surface of the object to be detected and the degree of polarization of the reflected light on the surface. According to the reflection and refraction laws of light waves, the propagation direction of light wave on the surface of an object is determined by the direction of incidence of light wave and the shape of the object's surface. The Fresnel formula and definition of the degree of polarization show the relationship between the polarization of the reflected light on the surface of the object and the direction of incidence of light wave. Therefore, the information about the direction of incident light can be obtained from the polarization of the reflected light on the surface of the object. As shown in Figure 1 [9], the light beam is incident on the surface of the object to be detected and is reflected after incidence. According to the law of light reflection, the incident angle of the incident light is equal to the reflection angle, which is equal to the angle θ in Figure 1. The propagation direction of the reflected light is the z-axis, and the plane where the detector is located is the xoy plane. The projection of the normal of the surface of the object $\vec{n}$ at the point of the incident light on the xoy plane is $\vec{n}'$, and the angle between it and the x-axis is φ. θ_{pol} is the linear polarization direction of the polarizer, and φ is the azimuth angle of the normal of the exit surface with respect to the x-axis.

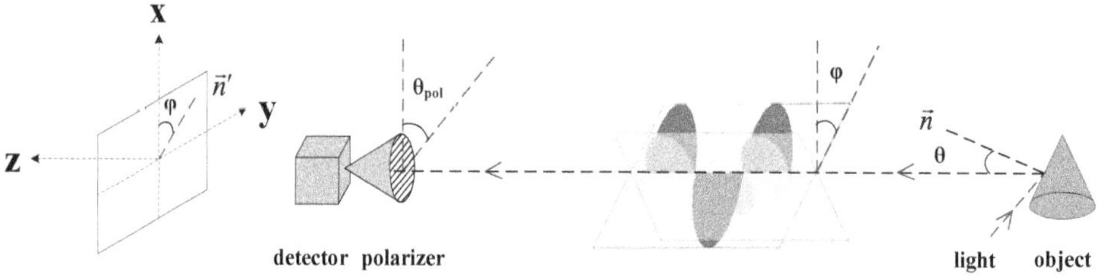

Figure 1. Schematic diagram of normal measurement on the surface of object.

To better present the relationship between the surface normal $\vec{n}$, incident angle θ, and incident azimuth angle of the object φ, the normal $\vec{n}$ in Figure 1 is represented in polar coordinates as shown in Figure 2. The incident angle θ is the zenith angle of the normal $\vec{n}$, and the incident azimuth angle φ is the azimuth angle of the normal $\vec{n}$.

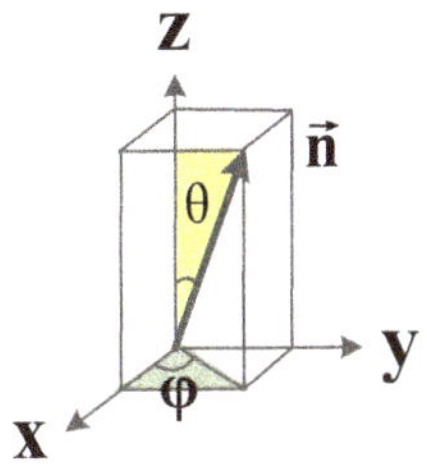

Figure 2. Polar coordinates of the normal $\vec{n}$ on the surface of object.

From Figure 2, it can be seen that the zenith angle θ and azimuth angle φ determine the surface normal $\vec{n}$ of the object, which is expressed as:

$$\vec{n} = -\frac{\partial}{\partial x}z(x,y)\hat{\mathbf{x}} - \frac{\partial}{\partial y}z(x,y)\hat{\mathbf{y}} + \hat{\mathbf{z}} \tag{1}$$

$$\vec{n}_x = -\frac{\partial}{\partial x}z(x,y) = \tan\theta\cos\varphi \ \vec{n}_y = -\frac{\partial}{\partial y}z(x,y) = \tan\theta\sin\varphi \tag{2}$$

where $\theta \in [0°, 90°]$ and $\varphi \in [0°, 360°]$. Therefore, the target surface normal can be determined by the zenith angle and azimuth angle of the normal. $z(x,y)$ is the relationship between the surface height of the object and the Cartesian coordinates (x,y). $z(x,y)$ is obtained by calculating the normal of the microfacets based on the polarization characteristics of the diffused reflection of the object surface in natural scenes [10].

In practical applications, the Stokes vector, composed of Stokes parameters, is often used to represent the polarization state of the light wave. When the incident angle θ_{pol} is 0°, 45°, 90°, and 135°, respectively, the cameras can obtain the intensity I_0, I_{45}, I_{90}, and I_{135} of the object to be detected in four polarization directions. I_0 represents the intensity of the horizontal polarization component. I_{90} represents the intensity of the vertical polarization component. I_{45} represents the intensity of the polarization component in the 45° direction. I_{135} represents the intensity of the polarization component in the 135° direction. I_R and I_L are the intensities of the right-handed and left-handed circular polarization components, respectively. The Stokes parameters S_0, S_1, S_2, and S_3 are defined as:

$$S_0 = I_0 + I_{90} \ S_1 = I_0 - I_{90} \ S_2 = I_{45} - I_{135} = 2I_{45} - (I_0 + I_{90}) \ S_3 = I_R - I_L \tag{3}$$

According to the definition of the degree of polarization, the degree of polarization P is given by:

$$\mathrm{P} = \frac{\sqrt{\mathrm{S_1}^2 + \mathrm{S_2}^2 + \mathrm{S_3}^2}}{\mathrm{S_0}} \tag{4}$$

where S_3 is related to circular polarization and can be ignored when considering mainly linear polarized light as complete polarized light. Therefore, the degree of polarization can be represented by Stokes parameters.

$$\mathrm{P} = \frac{\sqrt{\mathrm{S_1}^2 + \mathrm{S_2}^2}}{\mathrm{S_0}} = \frac{\sqrt{(I_0 - I_{90})^2 + (2I_{45} - (I_0 + I_{90}))^2}}{I_0 + I_{90}} \tag{5}$$

The zenith angle of polarized light from the object is:

$$\theta = \arccos\sqrt{\frac{n^4(1-P^2)+2n^2(2P^2+P-1)+P^2+2P-4n^3P\sqrt{1-P^2}+1}{(P+1)^2(n^4+1)+2n^2(3P^2+2P-1)}} \quad (6)$$

and the azimuth angle is:

$$\varphi = \frac{1}{2}\begin{cases} \arctan(S_2/S_1)+90^\circ, & S_1 \le 0 \\ \arctan(S_2/S_1)+180^\circ, & S_1 > 0 \;\; S_2 < S_1 \\ \arctan(S_2/S_1)+0^\circ, & S_1 > 0 \;\; S_2 \ge S_1 \end{cases} \quad (7)$$

By substituting the zenith θ and azimuth angles φ of the polarized light into Equations (1) and (2), respectively, the normal $\vec{n}$ can be obtained, and the surface function of the object $z(x,y)$ can be reconstructed.

3. Tunnel Lining Crack Detection Scheme

3.1. Detection Scheme

Based on polarization 3D imaging, the tunnel lining crack detection scheme is shown in Figure 3a. The detection system consists of three area-array polarization cameras, two lighting sources and one data processing system. Three area-array polarization cameras are used to simultaneously obtain images of approximately 50% of the entire tunnel section. As shown in Figure 3b, six area-array polarization cameras are required to obtain images of the entire tunnel section simultaneously [11,12]. During the experimental stage, the detection system used three area-array polarization cameras to verify the technical ability. Based on the result of crack detection for approximately 50% of the entire tunnel section, the six area-array polarization cameras can be used in the future to expand the detection to the entire tunnel section at the same time.

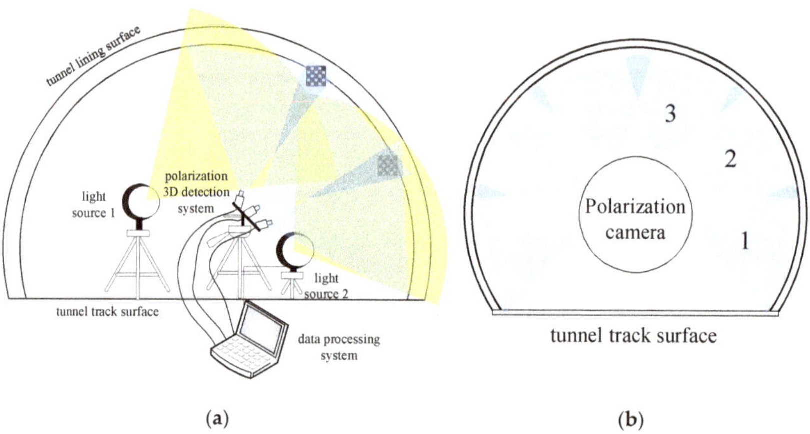

Figure 3. (**a**) Detection scheme of tunnel lining crack with plane array polarization cameras. (**b**) Placement of polarization cameras for detecting the entire tunnel section.

Two white light sources were used during the tunnel crack detection test, which was carried out during the construction period, as lighting conditions were poor in the tunnel. Light sources will not be necessary in the tunnel during an operation period with good internal lighting conditions. The lights from two white light sources were non-polarized, which had a negligible impact on the accuracy of polarization 3D inversion. A data processing system was used to process the polarized images obtained by the three area-array polarization cameras in real-time. The system calculated the 3D information of the tunnel lining cracks and output information such as the width, length, and depth of the cracks.

A test chart was arranged at the overlap of the fields of view from three area-array polarization cameras, which was used to calibrate the stitching parameters of the three cameras' fields of view and improve the stitching precision. When the crack test in the tunnel during the operation period, the ground track, detection system, and tunnel wall are all at a constant relative distance, only one stitching calibration is needed. Using the stitching calibration results, the images obtained by three area-array polarization cameras can be polarization 3D inverted, and the 3D images can be stitched together to obtain the 3D contour information and 3D crack information of the entire tunnel section.

3.2. Instrumentation and Equipment

Three area-array polarization cameras are shown in Figure 4. Three area-array polarization cameras, with a pixel resolution of 2.4 K × 2 K, were arranged at a certain angle in the same horizontal plane to ensure a 5–10% overlap of the fields of view. They were used to simultaneously capture images of 50% of the tunnel section. As technology advances, a larger pixel resolution of the polarization cameras can effectively reduce the number of cameras to obtain images of the entire tunnel section. Three area-array polarization cameras were fixed on a connecting plate, which was attached to the instrument bracket. The instrument bracket can be adjusted to change the imaging angle of the cameras. The laser radar system can obtain a 3D point cloud image of the tunnel wall for the comparison with the 3D imaging results obtained by the polarization cameras. Additionally, a caliper was used to accurately measure the width of cracks in the tunnel lining at a certain position, which was used to compare with the width obtained by the polarization cameras at the same position.

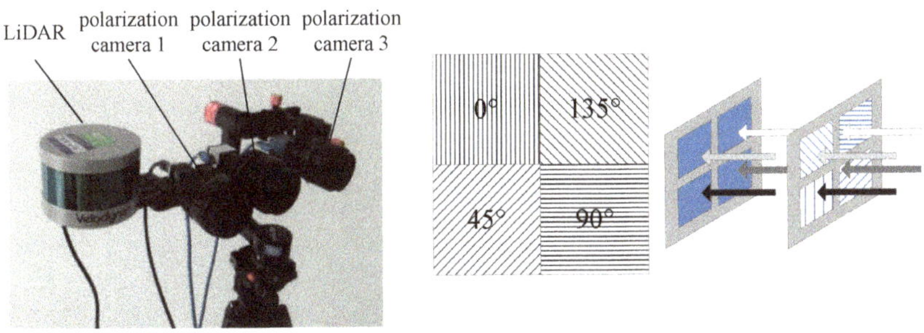

Figure 4. (**a**) Polarization 3D imaging system used in tunnel lining crack detection. (**b**) Schematic of focal plane array in polarization camera.

The polarization camera is a focal plane polarization camera, and micro-polarized optical elements are attached directly to the focal plane array. There are four polarization directions on four adjacent pixels, 0°, 45°, 90°, and 135°. Three Stokes vectors can be obtained simultaneously, and the polarization information of incident light can be obtained in real-time.

3.3. Crack Detection Method and Process

There are many studies on crack detection based on image processing. The general method is as follows: (1) image pre-processing to enhance the contrast and clarity of the image [13,14]; (2) image enhancement [15], edge detection [16], highlighting the target area; (3) identifying cracks based on image features such as gray level, texture, and morphology [17,18]; and (4) providing feature information of the cracks. Crack recognition algorithms based on threshold segmentation, morphology, and graph search methods have been developed [19–21]. These methods are used to highlight crack information in two-dimensional images through processing for identification, but which cannot obtain the depth information of the cracks.

In our work, the polarization information in the tunnel lining images was used to calculate the three-dimensional information of the tunnel lining, including the depth information of the cracks. The background wall of the tunnel lining was a continuous plane, its elevation was continuous. By setting the elevation threshold, the cracks can be separated from the background wall, and the crack information can be obtained quickly and directly. The edge extraction method can be used to directly obtain feature parameters such as crack length, width, depth, position, direction, and distribution. At the same time, by using the three-dimensional image of the entire tunnel lining from multiple polarization cameras, the three-dimensional profile of the tunnel can be obtained. Compared with the design profile of the tunnel, the quantitative information of the over-excavation and under-excavation during the construction period can be obtained. This part of the work will be described in detail in subsequent papers.

The method and process for detecting cracks in tunnel lining based on polarization 3D imaging are shown in Figure 5. Firstly, polarization cameras were used to obtain the polarization image of the tunnel lining, and the polarization characteristics of the tunnel lining were solved by the Stokes formula. Combined with the analysis results of the polarization light affecting the observed surface in the tunnel environment, the normal of the micro-surface elements was calculated using the method described in the first section. Secondly, the surface integral algorithm was used to obtain the 3D image of the tunnel lining and the 3D shape of the cracks. The elevation threshold was set to separate the background wall from the cracks, and the edge detection algorithm was used to extract the crack edges and obtain information such as crack width, depth, length, and position. By using scale factor calibration, the relative information was converted into absolute information, and the accurate 3D information of the cracks was output. The relevant details will be explained in detail in the following sections.

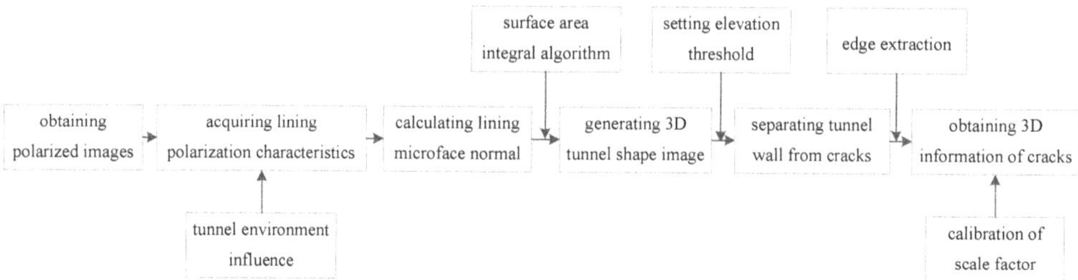

Figure 5. Method and process of tunnel lining cracks based on polarization 3D imaging.

4. Tunnel Lining Crack Detection Scheme

4.1. Impact Analysis of Tunnel Environment on Polarization 3D Imaging

During the construction period, the tunnel environment was unfavorable, such as high humidity, dust, and poor lighting conditions. The light from the tunnel wall passed through the tunnel environment atmosphere and entered the polarization 3D cameras. The obtained polarization information was affected by the environment humidity and dust, and reduced the monitoring accuracy. Monte Carlo's method [22,23] was used to analyze the influence of humidity and dust on polarization information to obtain the accurate polarization information of the tunnel wall and a high level of precision detection.

The polarization characteristics of the diffused reflected light from the tunnel lining are represented as S in Figure 6. The reflected light from the tunnel lining transmitted through the tunnel environment, and the polarization characteristics became S′ when the reflected light arrived at the entrance of the polarization camera. The method described below can be used to obtain the model of environmental transmission impact. At a certain location, S could be obtained through a close observation of the tunnel lining, as well as S′. S and S′ could be used to validate the environmental transmission impact model. Since the internal environment of the tunnel would not change in a short time, when observing other parts of the tunnel lining, this model could also be used to remove the influence of environmental transmission and obtain accurate polarization characteristics of the tunnel lining.

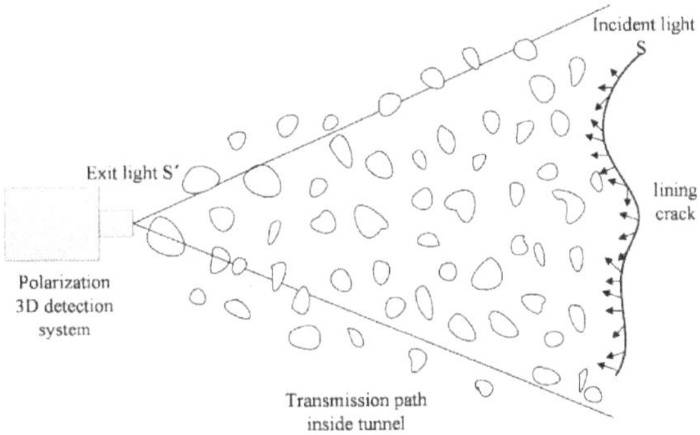

Figure 6. Schematic diagram of crack detection in a tunnel environment.

The Monte Carlo method can simulate the scattering and absorption process of each photon in atmospheric transmission, and infer their radiation intensity distribution through statistical methods. The following relationship exists throughout the scattering process:

$$S' = R(-\gamma_n) \cdot M(\theta_n) \cdot R(\phi_n) \cdot (-\gamma_{n-1}) \cdot M(\theta_{n-1}) \cdot R(\phi_{n-1}) \cdot \ldots \cdot R(-\gamma_1) \cdot M(\theta_1) \cdot R(\phi_1) S \quad (8)$$

where M is the Muller matrix, R is the rotation matrix, θ_s is the scattering angle, and ϕ and Y are the angles from the incident Stokes vector to the scattering plane, and back to the reference plane after scattering, respectively. The rotation angle Y can be expressed as:

$$\cos\gamma = \frac{-\mu_{zn} + \mu_{z(n+1)} \cos\theta_{sn}}{\pm\sqrt{(1-\cos^2\theta_{sn})\left(1-\mu^2_{z(n+1)}\right)}} \quad (9)$$

where μ_{zn} and $\mu_{z(n+1)}$ are the direction cosines of the photon before scattering and after scattering, respectively.

The rotation matrices $R(-\phi)$ and $R(Y)$ can be expressed as:

$$R(\phi) = \begin{bmatrix} 1 & 0 & 0 & 0 \\ 0 & \cos(2\phi) & \sin(2]\phi) & 0 \\ 0 & -\sin(2\phi) & \cos(2\phi) & 0 \\ 0 & 0 & 0 & 1 \end{bmatrix} \quad (10)$$

The Mueller matrix, scattering angles θ_s, and azimuth angles ϕ can be determined by the work of S. Zhang [24].

The crack detection test in a tunnel was carried out in summer, and the relative humidity in the tunnel was high. Relative humidity had a significant impact on the refractive index of aerosol particles.

Relative humidity, f, and the aerosol particle radius, r_h, are related to the dry aerosol particle radius, r_0, as shown in the following equation [24,25]:

$$r_h = (1 - f)^{-1/u} r_0 \tag{11}$$

where the u is constant, which represents the hygroscopicity of aerosol particles and varies with different types of aerosol particles.

Various particle radii also lead to change in the complex refractive index of the particles.

$$m_{re} = m_{rw} + (m_{r0} - m_{rw})[r_h / r_0] \tag{12}$$

$$m_{ie} = \frac{m_{iw}}{m_{rw}^2} + \left(\frac{m_{i0}}{m_{r0}^2 + 2} - \frac{m_{iw}}{m_{rw}^2} \right) \left[\frac{r_h}{r_0} \right]^{-3} \cdot \left(m_{re}^2 + 2 \right) \tag{13}$$

where m_{re} is the real part and m_{ie} is the imaginary part of the refractive index after moisture absorption. The refractive index after moisture absorption is $m_e = m_{re} + m_{ie}$.

During the construction period, the particles in the tunnel were of the dust type, with a dry aerosol particle radius of 0.6 μm and a refractive index of $m_0 = 1.53 + 0.008i$. The constant u was 4.8 [24].

Based on the formula, the influence of humidity and dust on the degree of polarization and azimuth angle under different linear polarization directions was analyzed, assuming perfectly polarized light in an ideal state, as shown in Figure 7. The analysis shows that the changes in the degree of polarization and azimuth angle due to humidity in the tunnel environment were similar for different linear polarization directions. When the humidity was less than 80%, the changes were relatively small, ranging from 0.67 to 0.75 and from 0.21 to 0.25, respectively. However, when the humidity was higher than 80%, the degree of polarization and azimuth angle increased significantly, ranging from 0.75 to 0.88 and from 0.24 to 0.29, respectively. High humidity had a significant impact on the precision of polarization 3D imaging.

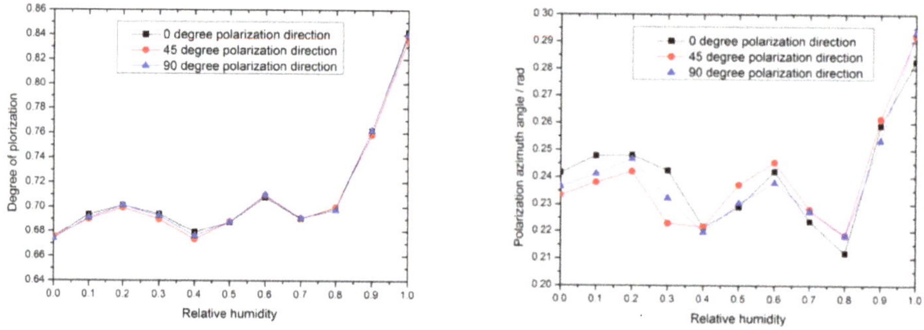

(a) (b)

Figure 7. The influence of relative humidity on the degree of polarization (**a**) and polarization azimuth angle (**b**) under different polarization directions.

By imaging the tunnel wall in close, the initial Stokes vector was obtained and substituted into Equation (8) to analyze the influence of the tunnel environment on the transmission of polarized light from the tunnel wall, as shown in Figure 8. The measured humidity in the tunnel was 79%. The degree of polarization becomes 0.42 after transmission through the tunnel environment, as shown in Figure 8a. The azimuth angle became 0.22 after transmission through the tunnel environment, as shown in Figure 8b. To improve the precision of polarization 3D imaging to realize high-precision detection, the influence of the tunnel environment transmission was incorporated into the polarized characteristics of the lining obtained by the polarization cameras, and the influence of the tunnel environment transmission was removed.

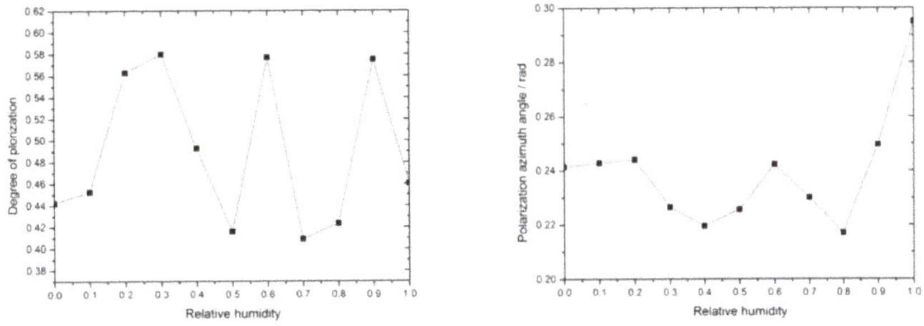

(**a**) (**b**)

Figure 8. Through tunnel environment transmission, the degree of polarization entering the polarization camera (**a**) and azimuth angle entering the polarization camera (**b**).

4.2. Absolute Scale Factor Calibration Method

The integrated gradient information based on Equations (1) and (2) provided the relative depth in the pixel coordinate system of the tunnel lining surface [26]. Although it reflected the 3D shape of the tunnel lining surface, it could not accurately obtain the absolute depth information of the cracks. To obtain accurate absolute depth information, ultra-light clay similar in color and material to the tunnel wall was coated onto the surface of the tunnel lining, and a ping-pong ball was embedded in the clay to form a known 3D shape, with a known absolute depth and a known absolute width of a half-spherical groove, as shown in Figure 9. Then, the polarization 3D camera was used to obtain a polarization imaging of the half-spherical groove, and the 3D shape of the half-spherical groove was calculated. The calculated result was compared with the real shape of the ping-pong ball to obtain the corresponding relationship between relative depth and width, and absolute depth and width. This relationship was used to obtain accurate information with absolute depth and width.

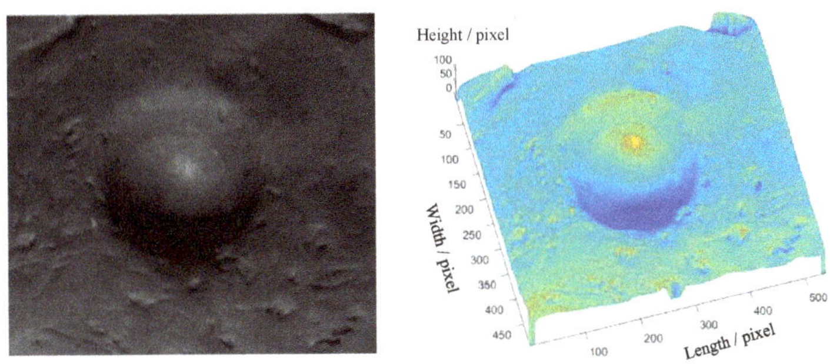

(**a**) (**b**)

Figure 9. (**a**) A hemispherical groove formed by embedding a ping-pong ball into ultra-light clay. (**b**) Polarization 3D inversion result of hemispherical grooves.

During the operation period, cracks were observed in the tunnel, and the relative distances between the ground track, the detection system composed of polarization cameras, and the tunnel wall were constant. Therefore, a one-time absolute distance calibration was sufficient.

5. Tunnel Lining Crack Detection Results and Analysis

5.1. Field Testing for Tunnel Lining Crack Detection

The tunnel lining crack test was conducted in a tunnel during a construction period in Sichuan, as shown in Figure 10. Artificial cracks were made to simulate the detection of shallow new cracks that may exist in the tunnel during the operation period. There were relatively few cracks in the construction-period portion of the tunnel. The environment in the construction-period tunnel was poor, with high humidity, dust, and poor lighting conditions. Based on the polarization 3D imaging, the tunnel lining crack detection system could effectively detect cracks in the construction-period tunnel under poor conditions, and the effectiveness of the detection system is expected to be even better during the operation period due to better conditions.

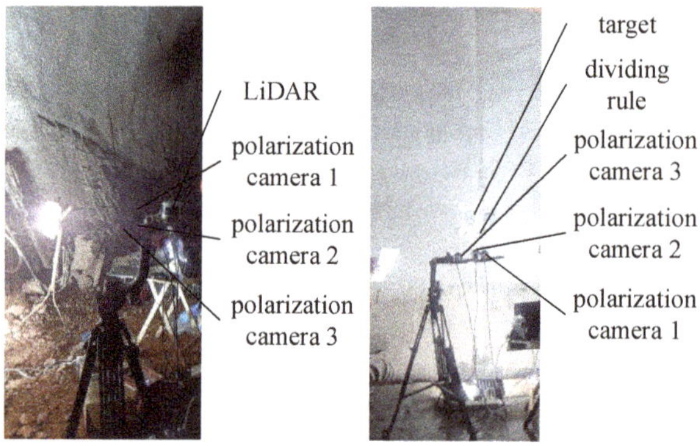

Figure 10. Tunnel lining crack detection system in tunnel.

During the tunnel lining crack detection test, test charts were used to calibrate the stitching parameters of three cameras to improve their accuracy, and a calibration scale was used to obtain the imaging resolution of the polarization 3D detection system.

5.2. Tunnel Lining Crack Test Results

5.2.1. Test Results of Wide Crack Detection in Tunnel

As shown in Figure 11a, the actual measurement of wide cracks in the tunnel was obtained. To determine the actual width value of the cracks, a scale was placed at various positions in the vertical crack direction. A caliper was also used to accurately measure the width of cracks in the tunnel lining at different locations and mark the locations, such as the red line in Figure 11a. According to the results of the scale and the caliper, the actual measured value of the crack at the position indicated by the red line was 3.8 mm.

The polarization 3D inversion result in Figure 11b showed that the width of the crack at the location of the red line was 4.0 mm, with a width error of 0.2 mm. The crack length in Figure 11a was 30.6 mm, whereas the crack length obtained by polarization 3D inversion in Figure 11b was 30.8 mm, with a length error of 0.2 mm. The polarization 3D inversion result in Figure 11b showed that the crack depth at the location of the red line was 9.2 mm. Based on the principle of polarization 3D imaging, it was known that the accuracy of the inverted crack depth was consistent with the accuracy of crack width and length. Compared to other detection methods based on camera photography, polarization 3D imaging had outstanding advantages, and it could obtain not only the width and length information of the crack, but also the depth information of the crack.

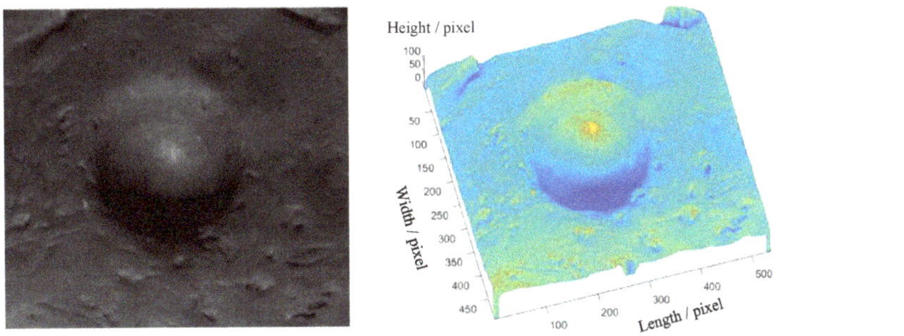

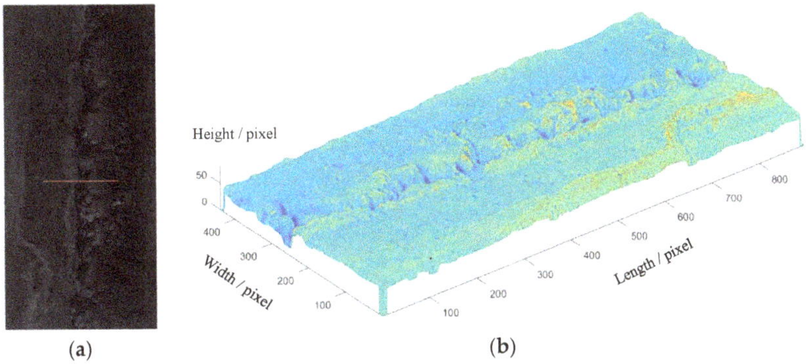

Figure 11. (**a**) Actual measurement images of a wide crack in tunnel lining. (**b**) Polarization 3D inversion result of wide crack.

5.2.2. Test Results of Narrow Crack Detection in Tunnel

As shown in Figure 12a, the actual measurement of narrow cracks in the tunnel was obtained. To determine the actual width value of the cracks, the scale and the caliper were also placed at various positions in the vertical crack direction. According to the results of the scale and the caliper, the actual measured value of the crack at the position indicated by the red line was 1.2 mm. The polarization 3D inversion result in Figure 12b showed that the width of the crack at the location of the red line was 0.9 mm, with a width error of 0.3 mm. The crack length in Figure 12a was 25.3 mm, and the crack length obtained by polarization 3D inversion in Figure 12b was 25.1 mm, with a length error of 0.2 mm. The polarization 3D inversion result in Figure 12b showed that the crack depth at the location of the red line was 4.8 mm. If light could not reach the deepest parts of the crack, there would be no reflected light from that position, and it would be impossible to invert the depth information without reflected light. Two white light sources were used during the tunnel crack detection test, so the light can enter the crack of 0.9 mm width. But we were unable to determine whether the light reached the deepest portions of the crack or not. The 4.8 mm depth was the result obtained by polarization 3D imaging. The 4.8 mm indicates the depth that light could reach. The 4.8 mm might be or might not be the deepest position.

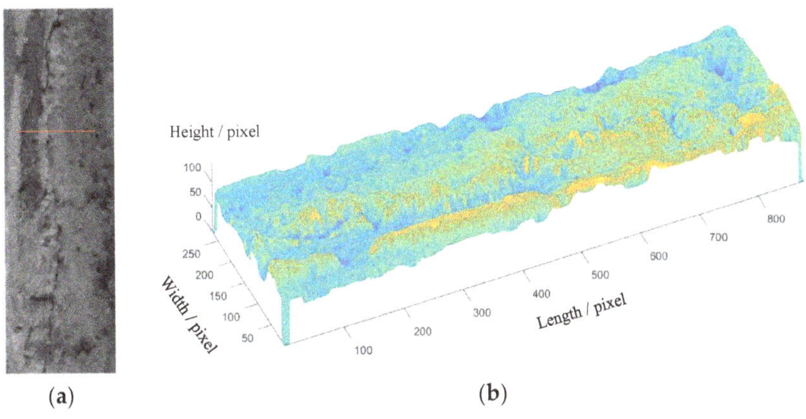

Figure 12. (**a**) Actual measurement images of a narrow crack in tunnel lining. (**b**) Polarization 3D inversion result of narrow crack.

5.2.3. Results of Artificial Crack Detection in Tunnel

After wide crack detection and narrow crack detection, artificial crack detection was conducted to simulate the detection of new shallow cracks in the tunnel during the operation period. The actual measured image of the artificial crack in the tunnel is shown in Figure 13a. To determine the actual width value of the artificial crack, the scale and the caliper were also placed at multiple positions in the direction of the vertical crack. The actual measurement value of the artificial crack at the position indicated by the red line was 1.5 mm. The results of the polarization 3D inversion shown in Figure 13b revealed that the width of the crack at the position of the red line was 1.8 mm, with a width error of 0.3 mm. The crack length in Figure 13a was 32.7 mm, while the length obtained by polarization 3D inversion in Figure 13b was 33.0 mm, with a length error of 0.3 mm. The crack depth at the position of the red line obtained by polarization 3D inversion in Figure 13b was 0.2 mm. If light could not reach the deepest parts of the crack, there would be no reflected light from that position, and it would be impossible to invert the depth information without reflected light.

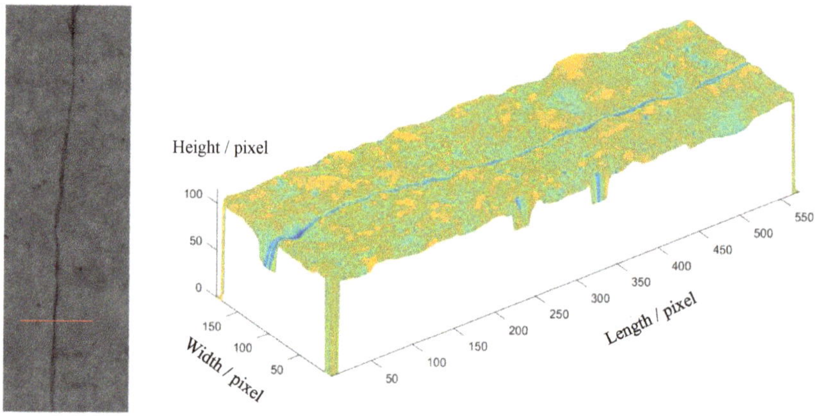

(a) (b)

Figure 13. (**a**) Actual measurement images of an artificial scratch in tunnel lining. (**b**) Polarization 3D inversion result of artificial crack.

6. Conclusions

Based on the polarization 3D imaging technology for the detection of cracks in tunnel lining, passive area-array polarization cameras were used to obtain tunnel lining images, and the polarization information was used to reconstruct the 3D structure of the tunnel wall. This allowed for the quick extraction of cracks with different depths from the background structure of the tunnel, and for obtaining 3D information about the cracks. It can effectively compensate for the drawbacks of the laser scanning method, which has a high power consumption and produces discontinuous 3D point cloud images. It also effectively compensates for the drawbacks of the camera photography detection method, which has a complex crack detection process. Based on the principle of the polarization 3D detection of cracks in tunnel lining, our work designed a scheme to detect the tunnel cross-section using three array polarization cameras and developed a polarization 3D detection system. We established a crack detection method for tunnel lining based on the real environment conditions of the tunnel. A model for the influence of the tunnel environment on polarized light transmission was established to address the impact of polarization imaging in complex environments such as strong local light, high humidity, and dust during the construction period of tunnels. A quantitative analysis of the impact of a tunnel's environment on polarization 3D imaging was conducted to obtain accurate information on tunnel lining crack detection and achieve high-precision detection. Tests were carried out in the tunnel during the construction period to detect wide cracks, narrow cracks, and artificial cracks. The crack width detection errors were 0.2 mm, 0.3 mm, and 0.3 mm, respectively. The crack length detection errors were 0.2 mm, 0.2 mm, and 0.3 mm, respectively, and the crack depth detection errors were the same as the width and length detection errors, which met the requirements of tunnel crack detection. Based on the principle of polarization 3D imaging, it can be inferred that the precision of the inverted crack depth is consistent with the precision of the crack width and length. In subsequent research, the precision of obtaining 3D data of tunnel cracks by polarization 3D imaging will be improved, and a 3D profile information on the tunnel section will be constructed.

Author Contributions: Conceptualization Y.Z. and Y.S.; methodology Y.Z. and X.Z.; software, X.L.; validation Y.Z., X.Z. and Y.S.; formal analysis Y.Z.; investigation Y.Z., X.Z. and Y.S.; resources, Y.Z.; data curation S.Z., W.R. and K.L.; writing—original draft preparation, Y.Z.; writing—review and-editing Y.Z.; visualization Y.Z.; supervision Y.Z.; project administration X.Z.; funding acquisition S.M. All authors have read and agreed to the published version of the manuscript.

Funding: This research received no external funding.

Institutional Review Board Statement: Not applicable.

Informed Consent Statement: Not applicable.

Data Availability Statement: The data that support the findings of this study are available from the corresponding author, Y.Z., upon reasonable request.

Conflicts of Interest: The authors declare no conflict of interest.

References

1. Sheerin, S.N.; Kavitha, S.; Raghuraman, G. Review and analysis of crack detection and classification techniques based on crack types. *Int. J. Appl. Eng. Res.* **2021**, *13*, 6056–6062.
2. Chen, X.S.; Xu, Z.H.; Bao, X.H. Overview of research on tunnel defects monitoring and detection technology. *Hazard Control Tunn. Undergr. Eng.* **2020**, *2*, 1–12.
3. Wang, S.L.; Gao, Y.; Qi, F.L.; Ke, Z.T.; Li, H.Y.; Lei, Y.; Peng, Z. Review on inspection technology of railway operation tunnels. *J. Traffic Transp. Eng.* **2020**, *20*, 40–57.
4. Gong, Y.F.; Xiao, M.Q.; Wang, S.F.; Tang, Z. Review and developing trend of railway tunnel detection technology. *Railw. Stand. Des.* **2019**, *63*, 93–98.
5. Xue, C.M. Research statuf and technical analysis of tunnel intelligent detection technolgoy based on machine vision. *Shanxi Sci. Technol. Commun.* **2019**, *6*, 66–68.
6. Liu, X.R. *Tunnel Crack Imaging Acquisition and Detection Technology*; Beijing Jiaotong University: Beijing, China, 2019.
7. Yang, J.; Liu, X.; Liu, X.; Zhang, P.; Peng, F. Review of Rapid Test Vehicles for Highway Tunnel Structure. *J. East China Jiaotong Univ.* **2018**, *4*, 30–38.
8. Atkinson, G.A.; Hancock, E.R. Recovery of surface orientation from diffuse polarization. *IEEE Trans. Image Process.* **2006**, *15*, 1653–1664. [CrossRef]
9. Wolff, L.B. Surface orientation from polarization images. In *SPIE Optics, Illumination, and Image Sensing for Machine Vision II*; SPIE: Bellingham, WA, USA, 1987; Volume 850, pp. 110–121.
10. Zhang, Y.; Zhang, X.M.; Liu, Y.; Li, X.; Zhang, S.; Yu, Y.; Jiao, J.C.; Su, Y. Calibration method of relative linear deviation angle of polarization three-dimensional imager. *J. Infrared Millim. Waves* **2022**, *41*, 1062–1071.
11. Li, J.C.; Zhang, C.B.; Chai, X.S.; Xue, F. Reseach on crack detection system of tunnel lining based on image recognition technology. *J. Zhejiang Univ. (Eng. Sci.)* **2018**, *58*, 20–24.
12. Song, Q.; Wu, Y.Q.; Xin, X.S.; Yang, L.; Yang, M.; Chen, H.M.; Liu, C.; Hu, M.J.; Chai, X.S.; Li, J.C. Real-time tunnel crack analysis system via deep learning. *IEEE Access* **2019**, *7*, 64186–64197. [CrossRef]
13. Wang, J.F.; Qiu, Y.; Liu, S.Z. Tunnel lining crack detection method based on image compensation. *J. Zhejiang Univ. (Eng. Sci.)* **2022**, *56*, 1404–1415.
14. Qiu, Y. *Research on Error Compensation Technology of Tunnel Inspection Vehicle Image Inspection System*; Chang'an University: Xi'an, China, 2021.
15. Zhang, W.; Zhang, Z.; Qi, D. Automatic crack detection and classification method of subway tunnel safety monitoring. *Sensors* **2014**, *14*, 19307–19328. [CrossRef] [PubMed]
16. Li, C.; Xu, P.; Niu, L.; Sheng, Y.; Liu, L.S.; Cun, M. Tunnel crack detection using coarse-to-fine region localization and edge detection. *WIREs Data Min. Knowl. Discov.* **2019**, *9*, e1308. [CrossRef]

17. Zhu, L.Q.; Bai, B.; Wang, Y.D.; Yu, Z.J.; Guo, B.Q. Subway tunnel crack identification algorithm based on feature analysis. *Chin. J. Sci. Instrum.* **2015**, *37*, 64–70.
18. Yin, G.S.; Gao, J.G.; Shi, M.H.; Jin, M.Z.; Tuo, H.L.; Li, C.; Zhang, B. Tunnel crack recognition method under image block. *J. Traffic Transp. Eng.* **2022**, *22*, 148–159.
19. Dang, L.M.; Wang, H.X.; Li, Y.F.; Park, Y.; Oh, C.M.; Nguyen, T.N.; Moon, H. Automatic tunnel lining crack evaluation and measurement using deep learning. *Tunn. Undergr. Space Technol.* **2022**, *124*, 104472. [CrossRef]
20. Zhou, Z.; Zhang, J.J.; Gong, C.J.; Wu, W. Automatic tunnel lining crack detection vis deep learning with generative adversarial network-based data augmentation. *Undergr. Space* **2021**, *9*, 140–154. [CrossRef]
21. Li, G.; Ma, B.; He, S.H.; Ren, X.L.; Liu, Q.W. Automation tunnel crack detection based on U-net and a convolutional neural network with alternately updated clique. *Sensors* **2020**, *20*, 717. [CrossRef]
22. Min, X. Electric Monte Carlo simulation of polarized light propagation in turbid media. *Opt. Express* **2004**, *12*, 6350–6539.
23. Jessica, C.R.; Scott, A.P.; Steve, L.J. Three Monte Carlo programs of polarized light transpaort into scattering media: Part I. *Opt. Express* **2005**, *13*, 4420–4438.
24. Zhang, S.; Fu, Q.; Duan, J.; Zhan, J.T. Research of influence of humidity on transmission characteristics of ploarized light. *Infrared Laser Eng.* **2016**, *45*, 0511001. [CrossRef]
25. Cai, J.; Gao, J.; Fan, Z.G.; Feng, S.; Fang, J. The polarization characteristics research of aerosol particles under the humidity influence. *Chin. J. Lumin.* **2013**, *34*, 639–644.
26. Ping, X.X.; Liu, Y.; Dong, X.M.; Zhao, Y.Q.; Zhang, Y. 3D reconstruction of textrueless and high reflective target by polarization and binocular stereo vision. *J. Infrared Millim. Waves* **2017**, *36*, 432–438.

photonics

MDPI

Article

Development and Calibration of a Vertical High-Speed Mueller Matrix Ellipsometer

Jiamin Liu [1,†], Song Zhang [1,†], Bowen Deng [1], Lei Li [1], Honggang Gu [1], Jinlong Zhu [1], Hao Jiang [1,*] and Shiyuan Liu [1,2,*]

[1] State Key Laboratory of Intelligent Manufacturing Equipment and Technology, Huazhong University of Science and Technology, Wuhan 430074, China; jiaminliu@hust.edu.cn (J.L.); songzhang@hust.edu.cn (S.Z.); bowendeng@hust.edu.cn (B.D.); lei_li@hust.edu.cn (L.L.); hongganggu@hust.edu.cn (H.G.); jinlongzhu03@hust.edu.cn (J.Z.)

[2] Optics Valley Laboratory, Wuhan 430074, China

* Correspondence: hjiang@hust.edu.cn (H.J.); shyliu@hust.edu.cn (S.L.)

† These authors contributed equally to this work.

Abstract: In order to meet the requirements of dynamic monitoring from a bird's eye view for typical rapidly changing processes such as mechanical rotation and photoresist exposure reaction, we propose a vertical high-speed Mueller matrix ellipsometer that consists of a polarization state generator (PSG) based on the time-domain polarization modulation and a polarization state analyzer (PSA) based on division-of-amplitude polarization demodulation. The PSG is realized using two cascaded photoelastic modulators, while the PSA is realized using a six-channel Stokes polarimeter. On this basis, the polarization effect introduced by switching the optical-path layout of the instrument from the horizontal transmission to the vertical transmission is fully considered, which is caused by changing the incidence plane. An in situ calibration method based on the correct definition of the polarization modulation and demodulation reference plane has been proposed, enabling the precise calibration of the instrument by combining it with a time-domain light intensity fitting algorithm. The measurement experiments of SiO_2 films and an air medium prove the accuracy and feasibility of the proposed calibration method. After the precise calibration, the instrument can exhibit excellent measurement performance in the range of incident angles from 45° to 90°, in which the measurement time resolution is maintained at the order of 10 μs, the measurement accuracy of Mueller matrix elements is better than 0.007, and the measurement precision is better than 0.005.

Keywords: vertical optical layout; polarization effect calibration; polarization modulation and demodulation reference plane; incidence plane switching; high-speed Mueller matrix ellipsometer

Citation: Liu, J.; Zhang, S.; Deng, B.; Li, L.; Gu, H.; Zhu, J.; Jiang, H.; Liu, S. Development and Calibration of a Vertical High-Speed Mueller Matrix Ellipsometer. *Photonics* **2023**, *10*, 1064. https://doi.org/10.3390/photonics10091064

Received: 5 September 2023
Revised: 19 September 2023
Accepted: 20 September 2023
Published: 21 September 2023

1. Introduction

High-speed dynamic processes, such as liquid-interface reactions [1], high-temperature loading [2–4], fast two-phase coupling reactions [5–7], directed self-assembly [8,9], mechanical rotation [10,11], directional displacement [12,13], etc., often involve rich physical and chemical properties related to temporal resolution, which have a significant impact on human production and life. It is of great significance to accurately and effectively characterize and monitor such dynamic processes. Since the above dynamic processes usually have the characteristics of short duration [14], orientation dependence [15], sample morphology diversity, and susceptibility to interference from external factors [16], extremely high requirements are being placed on the measurement instruments regarding temporal resolution, wide sample adaptability, and non-destructive measurement.

Currently, the measurement methods that can characterize the modification of samples by external loading conditions mainly include in situ scanning electron microscopy [17], thermos–gravimetric analysis [18], and in situ X-ray diffraction [19], which are widely used in their respective fields. However, these methods make it challenging to obtain the sample's transient and dynamic optical properties and morphological parameters in a non-invasive manner. The Mueller matrix ellipsometer (MME) can measure the Mueller matrix of the sample in a non-invasive way and then extract its optical characteristics and morphological parameters via an inversion reconstruction algorithm [20,21]. However, the polarization modulation and demodulation based on dual rotating compensators usually utilized in the instrument will limit the measurement temporal resolution to the order of seconds in principle [22], which makes it challenging to meet the real-time monitoring requirements of various high-speed dynamic processes. With the birth of high-frequency polarization, phase modulation devices such as the photoelastic modulator (PEM) [23], liquid–crystal phase variable retarder (LCVR) [24], and spatial light modulators [25], the measurement time resolution of Muller matrix ellipsometers has been improved accordingly. Zhang et al. proposed a high-speed MME with a horizontal light-path layout [26], which enables the measurement of the Mueller matrix with an 11 μs temporal resolution. The instrument was used to realize the precise measurement of the dynamic phase retardation of a nematic LCVR and the transient attitude angle of a birefringent waveplate [26,27], which has produced crucial academic influence. However, its horizontal optical-path layout leads to the vertical arrangement of the stage, which leads to the vertical clamping of samples, which significantly limits the types of samples to be tested and the application scenarios. In particular, this vertical clamping of samples cannot be compatible with currently interesting samples such as micro-domain two-dimensional materials, photoresists, liquids, etc.

In this work, we propose a vertical high-speed MME, in which the polarization modulation and demodulation of the probe light refer to the scheme adopted by the original horizontal MME. The polarization effect introduced by the vertical optical-path layout is fully considered and accurately corrected by the proposed in situ calibration method based on the definition of a reference plane for polarization modulation and demodulation. Then, the measurement experiments of SiO_2 films and an air medium prove the accuracy and feasibility of the proposed calibration method.

2. Instrument Prototype

Still following the traditional double-rotating compensator ellipsometer architecture [22], the vertical instrument uses a polarizer and two cascaded PEMs with different modulation frequencies as the PSG, and a six-channel Stokes polarimeter based on division-of-amplitude (DOA) as the PSA [27,28], which enables the avoidance of the mechanical rotation modulation of the compensator. The corresponding principle optical path is shown in Figure 1a. The whole-system settings of the instrument in order of light propagation are L-PSG-S-PSA-D, where L, S, and D stand for the laser source, the sample, and six photomultiplier tubes, respectively. By using a 5 mW He-Ne Laser (HNL050LB, Thorlabs, NJ, USA) with a central wavelength of 632.8 nm, a pre-polarized beam is projected into the PSG module, in which the fixed-azimuth linear polarizer (LPVIS100-MP2, Thorlabs, NJ, USA) and two PEMs (II/FS47 and II/FS50LR, Hinds, OR, USA) with phase modulation frequencies of 47.112 kHz and 50.006 kHz can jointly generate probe light with time-varying polarization. Using the period division method proposed by Zhang et al. [26], the probe light whose polarization state changes with time at a period of 11 μs can be generated by the PSG module, which is the key to realizing the high measurement time resolution. It should be noted that, with the longitudinal mode spacing of the He-Ne laser less than 435 MHz, the bandwidth of the probe beam is less than 0.0004 nm, which ensures the monochromaticity of the instrument's measurement results.

Then, the PSA module collects the reflection or transmission light from the sample and splits it into three branches equally using two non-polarizing beam splitters (BS019 and BS013, Thorlabs, NJ, USA) with splitting ratios of 70:30 and 50:50, respectively. With three branches for polarization demodulation, the three Stokes parameters of the reflection or transmission light can be determined according to the DOA principle. Each polarization–demodulation branch consisted of a polarizing beam splitter (CCM1-PBS25-633/M, Thorlabs, NJ, USA) and two photomultiplier tubes (H10721 Series, Hamamatsu, Shizuoka Pref.,

Japan). Notably, a half-wave plate (WPMH05M-633, Thorlabs, NJ, USA) with a fast-axis azimuthal angle of $-45°$, and a quarter-wave plate (WPMQ05M-633, Thorlabs, NJ, USA) with a fast-axis azimuthal angle of 22.5° are used in the second and third polarization–demodulation branches, respectively. Since the photomultiplier tubes with a response time of 0.57 ns and two oscilloscopes (WaveSurfer-3000, Teledyne Lecroy, Chestnut Ridge, NY, USA) with bandwidth of 200 MHz are used as the signal detection module, it is possible to capture all the Stokes parameters of the reflection beam simultaneously in several nanoseconds. Thus, the Mueller matrix measurement with a period of 11 μs can be achieved. In order to ensure the synchronization of the signals collected by the six detection channels, a synchronous triggering and acquisition method based on the phase reference signal output by the PEM controller was utilized [26].

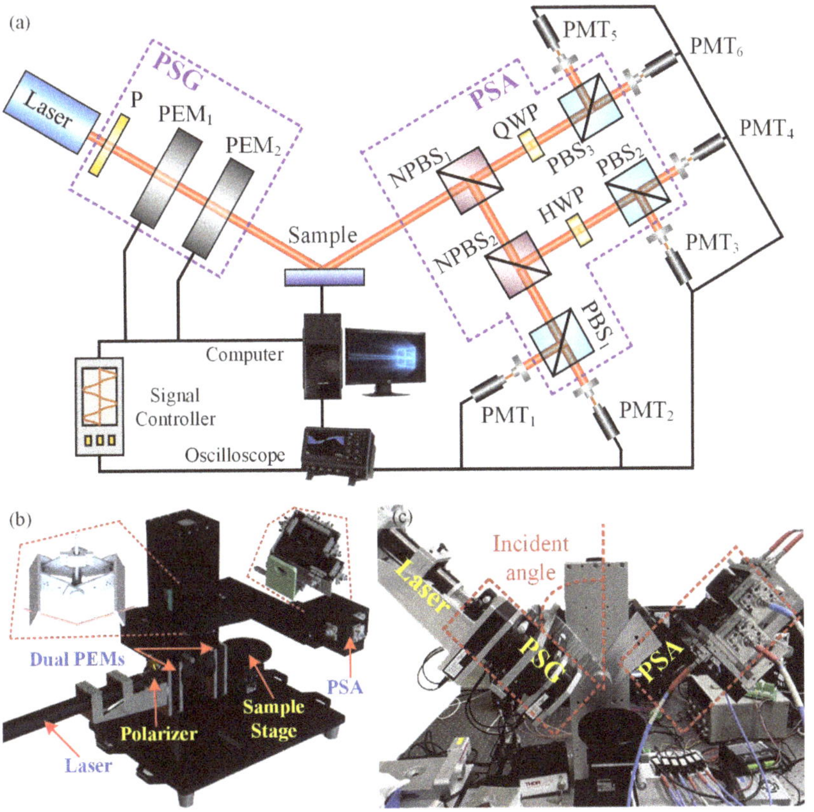

Figure 1. Vertical high-speed Mueller matrix ellipsometer. (**a**) Principle-optical-path schematic diagram; (**b**) 3D-modeling schematic diagram; and (**c**) self-developed MME prototype. P, polarizer; PEM_1 and PEM_2, photoelastic modulators; $NPBS_1$ and $NPBS_2$, non-polarization beam splitters with splitting ratios of 70:30 and 50:50, respectively; PBS_1, PBS_2, and PBS_3, polarization beam splitters with the same specifications; QWP, quart-wave plate; HWP, half-wave plate; PMT1, PMT2, PMT3, PMT4, PMT5, and PMT6, photomultiplier tubes with the same specifications; PSG, polarization state generator; PSA: polarization state analyzer. The left inset in (**b**) shows the varying angle mechanism based on the ball screw assembly, while the right inset in (**b**) exhibits the actual PSA module.

In the vertical optical path layout shown in Figure 1b, the ball screw assembly was used to alter the incident angle, which can realize an variable angle range of 45°~90° and an angular resolution of 0.1°. As shown in the left inset in Figure 1b, the slider driven by the motor makes a reciprocating linear motion on the guide rail, causing the rotation of the rotating arms via the four-bar linkage, which finally realizes the smooth adjustment of the incidence angle. Figure 1b shows the 3D structural model for the vertical high-speed MME, and the right inset presents the arrangement of the PSA module in detail. Correspondingly, Figure 1c exhibits the prototype of vertical high-speed MME.

3. Calibration Method

Compared with the previous self-developed horizontal high-speed MME [26], the incidence plane of the vertical high-speed MME is in the vertical plane under the configuration of oblique incidence, which means that the reference plane for defining the polarization state has rotated 90°, as shown in Figure 2. Meanwhile, the incidence plane for the straight-through measurement mode of the vertical instrument has uncertainty and multi-solution, which is attributed to the parallel relationship between the probe light's wave vector and the normal vector of the sample surface in this measurement mode. That is why the systematic model of the horizontal instrument can be applied to the system calibration of the vertical instrument in the direct-through measurement mode rather than in the oblique measurement mode. Based on the above considerations, an updated systematic model was proposed, especially for the vertical instrument. Supposing the light source's Stokes vector is $\mathbf{S}_{\text{in}} = [1, 1, 0, 0]^{\text{T}}$ and the PEM can be considered as a retarder with time-varying retardance [29], the Stokes vector $\mathbf{S}_{\text{PSG}}$ of the probe beam can be expressed as the following formula:

$$\begin{aligned}\mathbf{S}_{\text{PSG}} = \mathbf{R}(-\theta_{\text{PEM2}})\mathbf{M}_{\text{RET}}(\delta_{\text{PEM2}})\mathbf{R}(\theta_{\text{PEM2}})\mathbf{R}(-\theta_{\text{PEM1}})\mathbf{M}_{\text{RET}}(\delta_{\text{PEM1}})\mathbf{R}(\theta_{\text{PEM1}}) \\ \cdot\mathbf{R}(-90°)\mathbf{R}(-\theta_{\text{P}})\mathbf{M}_{\text{P}}\mathbf{R}(\theta_{\text{P}})\mathbf{S}_{\text{in}}\end{aligned}, \quad (1)$$

where $\mathbf{M}_{\text{P}}$ and $\mathbf{M}_{\text{RET}}$ are the Mueller matrices of the polarizer and the retarder, respectively. The detail expression of $\mathbf{M}_{\text{P}}$ and $\mathbf{M}_{\text{RET}}$ can be found in the literature [26,30]. $\mathbf{R}$ represents the rotation matrix. The angles θ_{P}, θ_{PEM1}, and θ_{PEM2} are the azimuthal angles of the polarizer, and the first and second PEMs, respectively. The parameters δ_{PEM1} and δ_{PEM2} symbolize the time-varying phase retardance of the first and second PEMs, respectively, which can be expressed as the following formula:

$$\delta_i = \delta_{\text{peak},\,i} \cdot \sin(2\pi f_i t + \varphi_i) + \delta_{\text{static},\,i} i = \text{PEM1, PEM2}, \quad (2)$$

where $\delta_{\text{peak},\,i}$ is the peak retardance of the ith PEM driven by a certain voltage. The parameter $\delta_{\text{static},\,i}$ symbolizes the static retardance of the ith PEM, related to the birefringence properties of crystal in the PEM. The parameters f_i and φ_i are the modulation frequency and the initial phase of the ith PEM, respectively, in which the nominal values of f_1 and f_2 are 42 kHz and 59 kHz, respectively. And t is the time variable.

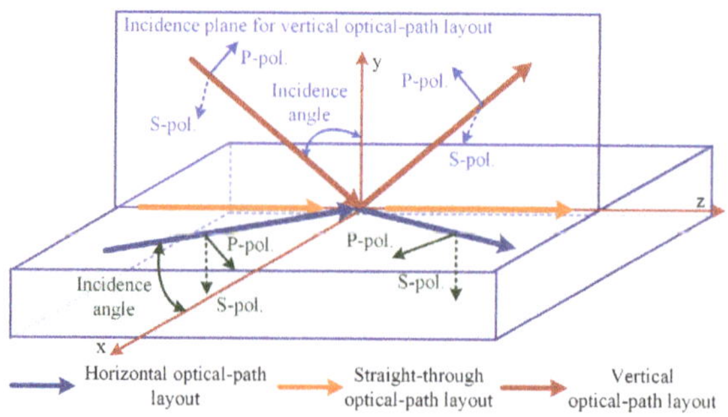

Figure 2. Reference planes for defining the polarization state basis vectors in the vertical and horizontal instrument.

In the six-channel PSA module, each detection channel can be represented by an ordered cascade of Mueller matrices for each optical element used in the channel. Since each channel only captures the light intensity, only the first row in the ordered cascade of Mueller matrices is required to represent polarization demodulation and sensing for each channel, as shown in the following formula:

$$\mathbf{a}_1 = k_1\begin{bmatrix}1 & 0 & 0 & 0\end{bmatrix}\cdot\left[\mathbf{R}(-90^\circ)\mathbf{M}^{\mathrm{R}}_{\mathrm{PBS}}\mathbf{R}(90^\circ)\right]\cdot\mathbf{M}^{\mathrm{T}}_{\mathrm{NPBS55}}\mathbf{M}^{\mathrm{R}}_{\mathrm{NPBS73}}, \tag{3}$$

$$\mathbf{a}_2 = k_2\begin{bmatrix}1 & 0 & 0 & 0\end{bmatrix}\cdot\left[\mathbf{R}(-90^\circ)\mathbf{M}^{\mathrm{T}}_{\mathrm{PBS}}\mathbf{R}(90^\circ)\right]\cdot\mathbf{M}^{\mathrm{T}}_{\mathrm{NPBS55}}\mathbf{M}^{\mathrm{R}}_{\mathrm{NPBS73}}, \tag{4}$$

$$\begin{array}{l}\mathbf{a}_3 = k_3\begin{bmatrix}1 & 0 & 0 & 0\end{bmatrix}\cdot\left[\mathbf{R}(-90^\circ)\mathbf{M}^{\mathrm{R}}_{\mathrm{PBS}}\mathbf{R}(90^\circ)\right]\\ \cdot\left[\mathbf{R}(-\theta_{\mathrm{HWP}})\mathbf{M}_{\mathrm{HWP}}\mathbf{R}(\theta_{\mathrm{HWP}})\right]\cdot\mathbf{M}^{\mathrm{R}}_{\mathrm{NPBS55}}\mathbf{M}^{\mathrm{R}}_{\mathrm{NPBS73}}\end{array}, \tag{5}$$

$$\begin{array}{l}\mathbf{a}_4 = k_4\begin{bmatrix}1 & 0 & 0 & 0\end{bmatrix}\cdot\left[\mathbf{R}(-90^\circ)\mathbf{M}^{\mathrm{T}}_{\mathrm{PBS}}\mathbf{R}(90^\circ)\right]\\ \cdot\left[\mathbf{R}(-\theta_{\mathrm{HWP}})\mathbf{M}_{\mathrm{HWP}}\mathbf{R}(\theta_{\mathrm{HWP}})\right]\cdot\mathbf{M}^{\mathrm{R}}_{\mathrm{NPBS55}}\mathbf{M}^{\mathrm{R}}_{\mathrm{NPBS73}}\end{array}, \tag{6}$$

$$\mathbf{a}_5 = k_5\begin{bmatrix}1 & 0 & 0 & 0\end{bmatrix}\cdot\left[\mathbf{R}(-90^\circ)\mathbf{M}^{\mathrm{R}}_{\mathrm{PBS}}\mathbf{R}(90^\circ)\right]\cdot\left[\mathbf{R}(-\theta_{\mathrm{QWP}})\mathbf{M}_{\mathrm{QWP}}\mathbf{R}(\theta_{\mathrm{QWP}})\right]\cdot\mathbf{M}^{\mathrm{T}}_{\mathrm{NPBS73}}, \tag{7}$$

$$\mathbf{a}_6 = k_6\begin{bmatrix}1 & 0 & 0 & 0\end{bmatrix}\cdot\left[\mathbf{R}(-90^\circ)\mathbf{M}^{\mathrm{T}}_{\mathrm{PBS}}\mathbf{R}(90^\circ)\right]\cdot\left[\mathbf{R}(-\theta_{\mathrm{QWP}})\mathbf{M}_{\mathrm{QWP}}\mathbf{R}(\theta_{\mathrm{QWP}})\right]\cdot\mathbf{M}^{\mathrm{T}}_{\mathrm{NPBS73}}, \tag{8}$$

where the coefficient k_i with i = 1~6 represents the gain factor of each intensity detection channel in the PSA module. $\mathbf{M}^{\mathrm{R}}_{\mathrm{PBS}}$ and $\mathbf{M}^{\mathrm{T}}_{\mathrm{PBS}}$ are the Mueller matrices of the polarizing beam splitter in the reflection and transmission mode, respectively. $\mathbf{M}^{\mathrm{R}}_{\mathrm{NPBS55}}$ and $\mathbf{M}^{\mathrm{T}}_{\mathrm{NPBS55}}$ symbolize the Mueller matrices of the non-polarizing beam splitter, with a splitting ratio of 50:50 in the reflection and transmission mode, respectively. Similarly, $\mathbf{M}^{\mathrm{R}}_{\mathrm{NPBS73}}$ and $\mathbf{M}^{\mathrm{T}}_{\mathrm{NPBS73}}$ are the Mueller matrices of the non-polarizing beam splitter, with a splitting ratio of 70:30 in the reflection and transmission mode, respectively. $\mathbf{M}_{\mathrm{HWP}}$ and $\mathbf{M}_{\mathrm{QWP}}$ are the Mueller matrices of the half-wave plate and the quarter-wave plate, respectively. Parameters θ_{HWP} and θ_{QWP} are the azimuthal angles of the half-wave plate and the quarter-wave plate, respectively. The detailed expressions of these above Mueller matrices can be found in the literature [28,30,31]. Compared with the horizontal system reported in ref. [26], the essential difference is reflected in the spatial topological relationship between the reference plane for the amplitude division realized by the PSA module and the incidence plane of the measured sample. As for the horizontal optical-path layout design, the reference plane for the amplitude division is parallel with the incidence plane of the measured sample, while the orthogonal relationship between the two planes can be found in the current vertical optical-path layout design. Correspondingly, the reference planes for the polarization definition of the two optical-path layouts are completely different, so the measurement models of the two measurement systems are entirely different. That is to say, the Formulas (1) and (3)–(8) in the manuscript are different from Formulas (13) and (15) in ref. [26].

By combining these six vectors in order, an instrument matrix characterizing the PSA module can be generated, as shown in the following formula:

$$\mathbf{A} = \begin{bmatrix}\mathbf{a}_1 & \mathbf{a}_2 & \mathbf{a}_3 & \mathbf{a}_4 & \mathbf{a}_5 & \mathbf{a}_6\end{bmatrix}^{\mathrm{T}}, \tag{9}$$

Then, the systematic model for the vertical high-speed MME can be established as the following formula:

$$\mathbf{B} = \mathbf{A}\mathbf{M}_{\mathrm{S}}\mathbf{W} = \mathbf{A}\mathbf{M}_{\mathrm{S}}\cdot\begin{bmatrix}\mathbf{S}_{\mathrm{PSG}}(t_1) & \mathbf{S}_{\mathrm{PSG}}(t_2) & \cdots & \mathbf{S}_{\mathrm{PSG}}(t_j) & \cdots & \mathbf{S}_{\mathrm{PSG}}(t_{\mathrm{N}})\end{bmatrix}, \tag{10}$$

where $\mathbf{S}_{\mathrm{PSG}}(t_j)$ is the Stokes vector of the probe beam generated by the PSG module at the moment t_j. $\mathbf{M}_{\mathrm{s}}$ represents the Mueller matrix of the sample to be tested. $\mathbf{B} = [\mathbf{I}_1, \mathbf{I}_2, \cdots,$

$\mathbf{I}_j, \cdots, \mathbf{I}_N]$ is the intensity matrix, in which $\mathbf{I}_j$ symbols the intensity vector consisted of six intensity values detected by the vertical high-speed MME at the moment t_j.

The systematic model shown in Equation (10) can be used not only for calibrating the instrument's system parameters, but also for extracting the samples' measurands. By turning off the dual PEMs in the vertical high-speed MME under the straight-through measurement mode, the instrument matrix or the systematic parameters of the PSA module can be obtained by utilizing the air as the sample and fitting the light intensity matrix collected by the six detection channels. Then, with the first and second PEMs turning on, respectively, the systematic parameters of the two PEMs and the polarizer can be obtained by measuring the air again under the straight-through mode. The above calibration process will be shown as a flow chart in Figure 3. Correspondingly, a χ^2 function was introduced to estimate the goodness-of-fit for the nonlinear least-squares regression analysis adopted in the calibration process.

$$\chi^2(\mathbf{p}) = [\mathbf{B}_\mathrm{m} - \mathbf{B}_\mathrm{c}(\mathbf{p}, t)]^\mathrm{T} \Sigma_\mathbf{B} [\mathbf{B}_\mathrm{m} - \mathbf{B}_\mathrm{c}(\mathbf{p}, t)], \tag{11}$$

where $\mathbf{p}$ is the systematic parameter set of the PSG or PSA module. $\mathbf{B}_\mathrm{m}$ and $\mathbf{B}_\mathrm{c}$ are the measured and calculated intensity matrices, respectively, which have been normalized via the gain factors of the six detection channels. $\mathbf{\Sigma}_\mathrm{B}$ represents the Moore–Penrose inverse of the covariance matrix of the measured intensity matrix $\mathbf{B}_\mathrm{m}$. The initial values of parameter set $\mathbf{p}$ used for the regression iteration come from the offline calibration of corresponding optical devices using the commercial MME (RC2, J. A. Woollam Inc., Lincoln, NE, USA). It is worth emphasizing that the in situ calibration results of the PSA module will be displayed in terms of the instrument matrix rather than the system parameters of the optical components involved, while the in situ calibration results of the PSA module will be displayed in terms of the system parameter values of the PEMs. It should be noted that the in situ calibration of the PSA module means the determination of 31 systematic parameters via the regression analysis, which includes all elements of the 6 × 4 instrument matrix, the gain factors of the six detection channels, and the azimuth angle of the polarizer. Regression analysis to achieve this goal requires offline calibration of each component to provide fitting initial values of the instrument matrix. Regression analysis to achieve this goal requires the offline calibration of each component to provide iteration initial values of the instrument matrix in the fitting process.

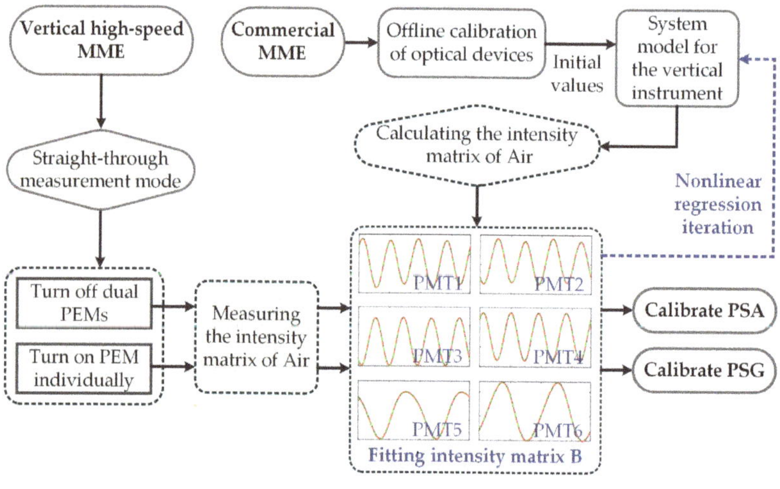

Figure 3. Flow chart of the calibration process for the vertical instrument.

4. Results and Discussion

4.1. Calibration of PSA and PSG Module

Figure 4 shows the air's Muller matrices, which are reported by the instrument without dual PEMs in the straight-through measurement mode, and reported by the theoretical calculation, respectively. According to the comparison results shown in Figure 4, it can be seen that the measured Mueller matrix generated by the calibration process of the PSA module is very close to the theoretical Mueller matrix of air, in which the absolute deviation of each element is about 0.005. Correspondingly, the instrument matrix characterizing the PSA module can be extracted from the calibration process, as shown in the following equation:

$$\mathbf{A} = \begin{bmatrix} 0.287 & -0.268 & -0.035 & 0.006 \\ 0.320 & 0.294 & 0.040 & -0.006 \\ 0.284 & -0.120 & -0.087 & 0.225 \\ 0.233 & 0.096 & 0.074 & -0.184 \\ 0.276 & -0.119 & 0.148 & 0.179 \\ 0.264 & 0.104 & -0.147 & -0.179 \end{bmatrix}, \quad (12)$$

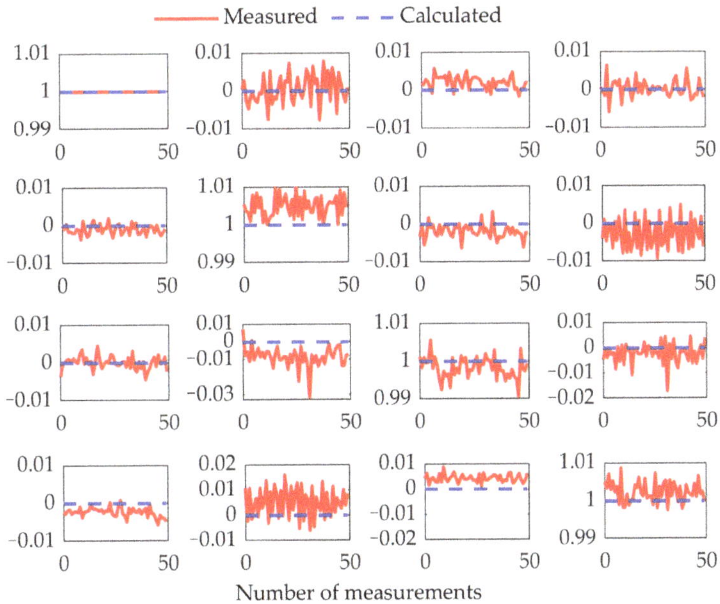

Figure 4. Mueller matrices of the air reported by the vertical instrument without the PEMs working under the straight-through mode. The results were obtained from 50 repeated-measurement experiments.

Subsequently, each PEM driven by different voltages was considered the sample to be tested, which was measured by the vertical instrument under the straight-through measurement mode. It is highly convenient for the in situ calibration of the PSG module by setting the azimuth angles of the polarizer, PEM1, and PEM2 to 45°, 0°, and 45°, respectively. Four systematic parameters can be determined from the calibration process, in which the analysis method has been previously reported [26]. Taking a PEM with a retardance modulation frequency of 42 kHz as an example, with the driving voltage changing from 0 V to 4.8 V, the PEM in the entire operating range can be in situ calibrated by repeating the above calibration process. Figure 5a,b shows the peak and static-phase retardance of the PEM under different driving voltages, respectively. Through performing linear fitting on the peak-phase retardance δ_{peak} under different driving voltages U_{driv}, it can be found that the peak retardance δ_{peak} had a strong linear dependence on the driving voltage, in which the linear factor and the truncation value were 218.34°/V and 17.00°, respectively. As the driving voltage gradually increased, the static phase retardance δ_{static} fluctuated around the average value of 0.043°, indicating no significant correlation between the two. Meanwhile, the static-phase retardance δ_{static} was close to 0, consistent with the PEM manufacturer's setting.

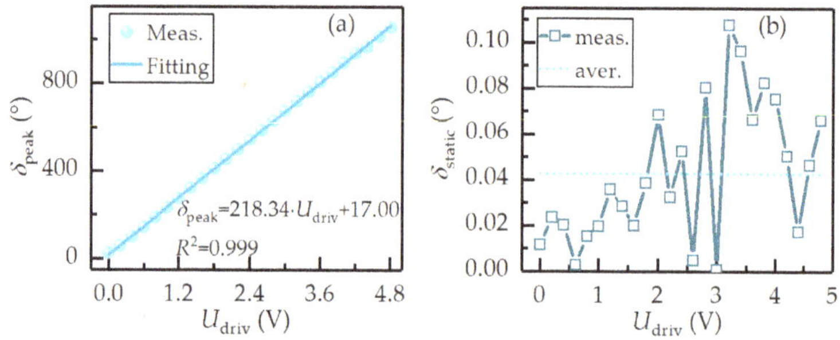

Figure 5. Peak- (**a**) and static (**b**)-phase retardance of the PEM with retardance modulation frequency of 42 kHz under different driving voltages.

The detailed parameters of dual PEMs used in the vertical high-speed MME, reported from the calibration process, are shown in Table 1. The retardance modulation frequencies of the two PEMs were very close to the nominal frequency, with both relative deviations less than 0.6%. The static retardance δ_{static} of the first and second PEMs were 0.041° and 0.036°, respectively, which are in accordance with the theoretical static retardance of 0°. The azimuthal angles of −0.34° and 44.82° for the PEMs are consistent with the pre-set values. The consistency observed above proves the accuracy and effectiveness of the constructed calibration method.

Table 1. Systematic parameters of the dual PEMs obtained from the in situ calibration process.

Specification	δ_{peak} (°)	f (kHz)	φ (°)	δ_{static} (°)	θ_{PEM} (°)
II/FS42LR	1056.7	42.05	260.9	0.043	−0.34
II/FS60LR	1078.8	59.64	129.9	0.036	44.82

4.2. Measurement Performance of the Vertical Instrument

After implementing the above calibration process in the transmission mode, a standard SiO_2 film with a nominal thickness of 20 nm was used as the sample under the incident angle of 45°, which allows for validating the above instrument matrix. The maximum deviation for all the elements in these two instrument matrices was less than 0.045, which indicates the high consistency between the results reported by the two calibration modes. Then, other standard SiO_2 films with thicknesses of 1.7, 18, 25, and 31 nm were characterized by the vertical high-speed MME at the incident angle of 45°. The corresponding measured results were compared with those reported by the commercial MME to judge the measurement accuracy of the built instrument. In fact, the exceptionally high measurement accuracy of the RC2 Mueller matrix ellipsometer in the thickness measurement of standard SiO_2 films has been widely demonstrated [32,33], which inspired us to use the results measured by the commercial ellipsometer as the reference values. Taking the 18nm thick SiO_2 film as an example, the Mueller matrices reported by five repeated measurement experiments are shown in Figure 6, in which the results measured by a commercial ellipsometer are used as a reference to facilitate comparative analysis.

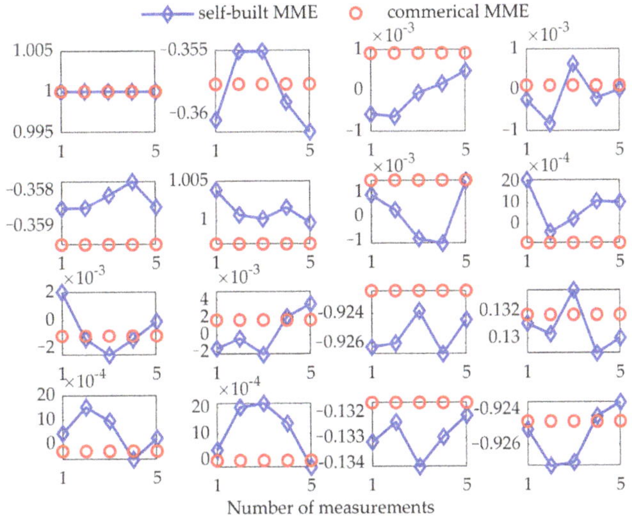

Figure 6. Mueller matrices of 18 nm thick SiO_2 film reported by the self-built MME and the commercial MME.

It can be easily noticed that the maximum deviation of most elements in the main diagonal block of the measured Mueller matrix from that reported by commercial MME was less than 0.004, and the maximum deviation of all elements in the off-diagonal block from that reported by commercial MME was less than 0.003. Only m_{22} and m_{33} in the main diagonal block of the measured Mueller matrix deviated from that reported by the commercial MME, with the deviations reaching 0.007 and 0.004, respectively, which might be attributed to the cumulative depolarization effect of each optical component in the instrument. Although there was a certain amount of deviation between individual elements and the reference value, the consistency of the above comparison results still fully illustrates the validity of the above calibration method and the measurement accuracy of the instrument. Correspondingly, the comparison between the thicknesses of the SiO_2 films measured by the instrument and that reported by commercial MME is shown in Table 2. It should be noted that the measurement and reference values in Table 2 are the average thickness of five repeated film-thickness metrology experiments, while the standard deviation of the five thickness measurement results was used to evaluate the uncertainty of the measured thickness. As for the SiO_2 films with thicknesses larger than 18.00nm, the measured thicknesses were very close to the reference values reported by the commercial MME, with relative deviations of less than 2.3%, which reflects the reliability and accuracy of measurement results. Although the relative thickness deviation for the 1.70 nm thick SiO_2 film from the reference result is about 8.15%, their absolute deviation of less than 0.15nm still indicates the measurement accuracy.

Table 2. Comparison between the SiO_2 films' thicknesses determined by the two instruments.

Nominal Value (nm)	Measured Value (nm)	Reference Value (nm)	Deviation (nm)	Relative Deviation
1.70	1.84 ± 0.62	1.69 ± 0.01	0.15	8.15%
18.00	17.86 ± 0.23	18.10 ± 0.02	−0.24	1.34%
25.00	25.13 ± 0.13	25.31 ± 0.02	−0.18	0.72%
31.00	31.42 ± 0.32	30.72 ± 0.04	0.72	2.29%

Furthermore, the measurement temporal resolution of the instrument can be quantitatively evaluated by carrying out 50 repeated measurements of the air under the straight-through measurement mode in tandem with setting the sampling rate of the oscilloscope at 2 GHz. In each temporal-resolution test experiment, the instrument's measurement configuration can be set as the optimal configuration under each corresponding measurement period, which can be obtained using an optimization algorithm satisfying specific constraints, such as the multi-objective genetic algorithm optimizing the Pareto optimal frontier [34]. The light-intensity signal sampling rate for each preset period is as optimal as possible. During the temporal-resolution test, the instrument performance was evaluated using each element's average deviation and standard deviation for the 50 Mueller matrices of the air. The corresponding analysis results are shown in Figure 7. With a temporal resolution less than or equal to 5 μs, the average deviation for most Mueller matrix elements is on the order of 0.01, indicating that achieving the high temporal resolution in the measurement experiments often sacrifices the measurement precision. When the temporal resolution increased to 10 μs and above, the average deviation and standard deviation of each Mueller matrix element improved to less than 0.005, indicating that the instrument's reliable temporal resolution in the measurement was about 10 μs.

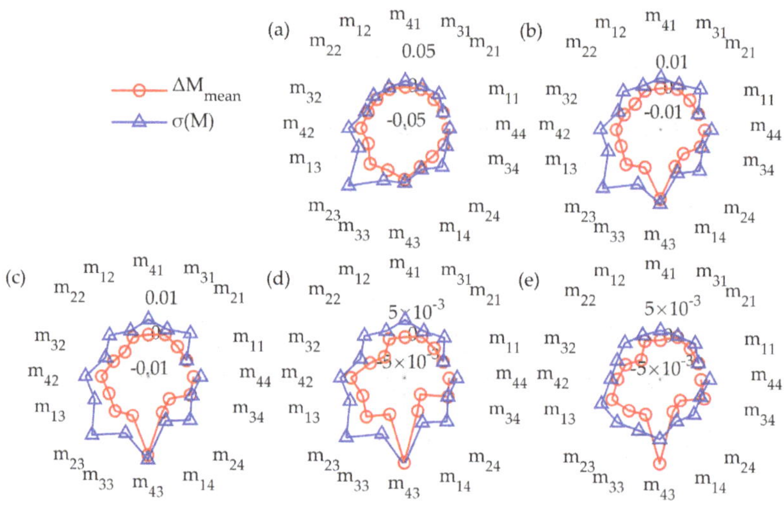

Figure 7. Mueller matrix elements' average deviation and standard deviation generated by 50 repeated measurements of the air under different temporal resolutions. From (**a**) to (**e**), the results correspond to the preset temporal resolutions of 2 μs, 4 μs, 5 μs, 10 μs, and 50 μs, respectively.

5. Conclusions

In this work, we proposed a vertical high-speed MME that consists of a PSG based on the time-domain polarization modulation and a PSA based on division-of-amplitude polarization demodulation. The PSG was realized using a polarizer and two cascaded PEMs, while the PSA was realized using a six-channel Stokes polarimeter. On this basis, an in situ dual-step calibration method based on the correct definition of the polarization modulation and demodulation reference plane was proposed, enabling the precise calibration of the instrument by combining it with a time-domain light-intensity fitting algorithm. The measurement experiments of SiO_2 films and an air medium prove the accuracy and feasibility of the proposed calibration method. After the precise calibration, the instrument can exhibit excellent measurement performance in the range of incident angles from 45° to 90°, in which the measurement time resolution is maintained at the order of 10 μs, the measurement accuracy of Mueller matrix elements is better than 0.007, and the measurement precision is better than 0.005. The above analysis fully demonstrates the potential application value of the proposed instrument.

Author Contributions: Conceptualization, J.L., S.Z., H.J. and S.L.; methodology, J.L., H.J. and S.Z.; software, J.L., S.Z., B.D., H.G. and J.Z.; validation, J.L., S.Z., H.J. and B.D.; formal analysis, H.G., J.Z., L.L. and S.L.; investigation, J.L., S.Z. and B.D.; resources, J.L., S.Z., B.D., H.G. and J.Z.; data curation, J.L. and B.D.; writing—original draft preparation, J.L., S.Z., H.J. and S.L.; writing—review and editing, J.L., S.Z., H.G., J.Z., H.J. and S.L.; visualization, J.L., S.Z. and L.L.; supervision, H.J. and S.L.; project administration, H.J. and S.L.; funding acquisition, H.J. and S.L. All authors have read and agreed to the published version of the manuscript.

Funding: This research was funded by National Natural Science Foundation of China (Grant Nos 52130504, 51975232, 52205592, 52305577), the Key Research and Development Plan of Hubei Province (Grant No. 2022BAA013), and the 2021 Postdoctoral Innovation Research Positions in Hubei Province (Grant No. 0106100226).

Institutional Review Board Statement: Not applicable.

Informed Consent Statement: No applicable.

Data Availability Statement: Data will be made available on request.

Acknowledgments: The authors would like to thank the technical support from the Experiment Center for Advanced Manufacturing and Technology in School of Mechanical Science & Engineering of HUST.

Conflicts of Interest: The authors declare no conflict of interest.

References

1. Benjamin, I. Reaction dynamics at liquid interfaces. *Annu. Rev. Phys. Chem.* **2015**, *66*, 165–188. [CrossRef] [PubMed]
2. Ferrera, M.; Magnozzi, M.; Bisio, F.; Canepa, M. Temperature-dependent permittivity of silver and implications for thermoplasmonics. *Phys. Rev. Mater.* **2019**, *3*, 105201. [CrossRef]
3. Magnozzi, M.; Ferrera, M.; Mattera, L.; Canepa, M.; Bisio, F. Plasmonics of Au nanoparticles in a hot thermodynamic bath. *Nanoscale* **2019**, *11*, 1140–1146. [CrossRef] [PubMed]
4. Xu, M.; Yang, J.; Zhang, S.; Liu, L. Role of electron-phonon coupling in finite-temperature dielectric functions of Au, Ag, Cu. *Phys. Rev. B* **2017**, *96*, 115154. [CrossRef]
5. Manukyan, K.V.; Avetisyan, A.G.; Shuck, C.E.; Chatilyan, H.A.; Rouvimov, S.; Kharatyan, S.L.; Mukasyan, A.S. Nickel oxide reduction by hydrogen: Kinetics and structural transformations. *J. Phys. Chem. C* **2015**, *119*, 16131–16138. [CrossRef]
6. Vendelbo, S.B.; Elkjær, C.F.; Falsig, H.; Puspitasari, I.; Dona, P.; Mele, L.; Morana, B.; Nelissen, B.J.; van Rijn, R.; Creemeer, J.F.; et al. Visualization of oscillatory behaviour of Pt nanoparticles catalysing CO oxidation. *Nat. Mater.* **2014**, *13*, 884–890. [CrossRef]
7. Yu, J.; Yuan, W.; Yang, H.; Xu, Q.; Wang, Y.; Zhang, Z. Fast gas-solid reaction kinetics of nanoparticles unveiled by millisecond in situ electron diffraction at ambient pressure. *Angew. Chem. Int. Ed.* **2018**, *57*, 11344–11348. [CrossRef]
8. Raeburn, J.; Cardoso, A.Z.; Adams, D. The importance of the self-assembly process to control mechanical properties of low molecular weight hydrogels. *Chem. Soc. Rev.* **2013**, *42*, 5143–5156. [CrossRef]
9. Jiang, H.; Peng, H.; Chen, G.; Gu, H.; Chen, X.; Liao, Y.; Liu, S.; Xie, X. Nondestructive investigation on the nanocomposite ordering upon holography using Mueller matrix ellipsometry. *Eur. Polym. J.* **2019**, *110*, 123–129. [CrossRef]
10. Gillmer, S.R.; Yu, X.; Wang, C.; Ellis, J.D. Robust high-dynamic-range optical roll sensing. *Opt. Lett.* **2015**, *40*, 2497–2500. [CrossRef]
11. Saito, Y.; Arai, Y.; Gao, W. Detection of three-axis angles by an optical sensor. *Sens. Actuator. A Phys.* **2009**, *150*, 175–183. [CrossRef]
12. Long, D.A.; Reschovsky, B.J.; LeBrun, T.W.; Gorman, J.J.; Hodges, J.T.; Plusquellic, D.F.; Stroud, J.R. High dynamic range electro-optic dual-comb interrogation of optomechanical sensors. *Opt. Lett.* **2022**, *47*, 4223–4326. [CrossRef] [PubMed]
13. Bag, A.; Neugebauer, M.; Mick, U.; Christiansen, S.; Schulz, S.A.; Banzer, P. Towards fully integrated photonic displacement sensors. *Nat. Commun.* **2020**, *11*, 2915. [CrossRef]
14. Petkovšek, R.; Petelin, J.; Možina, J.; Bammer, F. Fast ellipsometric measurements based on a single crystal photoelastic modulator. *Opt. Express* **2010**, *18*, 21410–21418. [CrossRef] [PubMed]
15. Zhang, S.; Wang, L.; Yi, A.; Gu, H.; Chen, X.; Jiang, H.; Liu, S. Dynamic modulation performance of ferroelectric liquid crystal polarization rotators and Mueller matrix polarimeter optimization. *Front. Mech. Eng.* **2020**, *15*, 256–264. [CrossRef]
16. Zhang, L.; Liu, J.; Zhu, J.; Jiang, H.; Liu, S. Femtosecond laser induced damaging inside fused silica observed by a single-pulse ultrafast measurement system. *Opt. Express* **2022**, *30*, 26111–26119. [CrossRef]
17. Barroo, C.; Wang, Z.; Schlögl, R.; Willinger, M. Imaging the dynamics of catalysed surface reactions by in situ scanning electron microscopy. *Nat. Catal.* **2020**, *3*, 30–39. [CrossRef]

18. Zhou, L.; Jiang, C.; Lin, Q. Entropy analysis and grey cluster analysis of multiple indexes of 5 kinds of genuine medicinal materials. *Sci. Rep.* **2022**, *12*, 6618. [CrossRef]
19. Wehrenberg, C.E.; McGonegle, D.; Bolme, C.; Higginbotham, A.; Lazicki, A.; Lee, H.J.; Nagler, B.; Park, H.-S.; Remington, B.A.; Rudd, R.E.; et al. In situ X-ray diffraction measurement of shock-wave-driven twinning and lattice dynamics. *Nature* **2017**, *550*, 496–499. [CrossRef]
20. Chen, X.; Gu, H.; Liu, J.; Chen, C.; Liu, S. Advanced Mueller matrix ellipsometry: Instrumentation and emerging applications. *Sci. China Technol. Sc.* **2022**, *65*, 2007–2030. [CrossRef]
21. Bian, S.; Cui, C.; Arteaga, O. Mueller matrix ellipsometer based on discrete-angle rotating Fresnel rhomb compensators. *Appl. Opt.* **2021**, *60*, 4964–4971. [CrossRef] [PubMed]
22. Liu, S.; Chen, X.; Zhang, C. Development of a broadband Mueller matrix ellipsometer as a powerful tool for nanostructure metrology. *Thin Solid Films* **2015**, *584*, 176–185. [CrossRef]
23. Atalar, O.; Van Laer, R.; Safavi-Naeini, A.H.; Arbabian, A. Longitudinal piezoelectric resonant photoelastic modulator for efficient intensity modulation at megahertz frequencies. *Nat. Commun.* **2022**, *13*, 1526. [CrossRef] [PubMed]
24. López-Téllez, J.M.; Bruce, N.C. Mueller-matrix polarimeter using analysis of the nonlinear voltage–retardance relationship for liquid-crystal variable retarders. *Appl. Opt.* **2014**, *53*, 5359–5366. [CrossRef] [PubMed]
25. Marco, D.; López-Morales, G.; Sanchez-López, M.M.; Lizana, Á.; Moreno, I.; Campos, J. Customized depolarization spatial patterns with dynamic retardance functions. *Sci. Rep.* **2021**, *11*, 9415. [CrossRef] [PubMed]
26. Zhang, S.; Jiang, H.; Gu, H.; Chen, X.; Liu, S. High-speed Mueller matrix ellipsometer with microsecond temporal resolution. *Opt. Express* **2020**, *28*, 10873–10887. [CrossRef]
27. Zhang, S.; Jiang, H.; Gu, H.; Chen, X.; Liu, S. Attitude metrology based on the field-of-view effect of birefringence using high-speed polarimetry. *Opt. Lett.* **2020**, *45*, 2074–2077. [CrossRef]
28. Liu, J.; Zhang, C.; Zhong, Z.; Gu, H.; Chen, X.; Jiang, H.; Liu, S. Measurement configuration optimization for dynamic metrology using Stokes polarimetry. *Meas. Sci. Technol.* **2018**, *29*, 054010. [CrossRef]
29. Arteaga, O.; Freudenthal, J.; Wang, B.; Kahr, B. Mueller matrix polarimetry with four photoelastic modulators: Theory and calibration. *Appl. Opt.* **2012**, *51*, 6805–6817. [CrossRef]
30. Fujiwara, H. *Spectroscopic Ellipsometry Principles and Applications*; John Wiley & Sons, Ltd.: Chichester, UK, 2007; pp. 52–79.
31. Zhang, S.; Gu, H.; Liu, J.; Jiang, H.; Chen, X.; Zhang, C.; Liu, S. Characterization of beam splitters in the calibration of a six-channel Stokes polarimeter. *J. Opt.* **2018**, *20*, 125606. [CrossRef]
32. Ehrstein, J.; Richter, C.; Chandler-Horowitz, D.; Vogel, E.; Young, C.; Shah, S.; Maher, D.; Foran, B.; Hung, P.Y.; Diebold, A. A comparison of thickness values for very thin SiO_2 films by using ellipsometric, capacitance-voltage, and HRTEM measurements. *J. Electrochem. Soc.* **2006**, *153*, F12–F19. [CrossRef]
33. Herzinger, C.M.; Johs, B.; McGahan, W.A.; Woollam, J.A. Ellipsometric determination of optical constants for silicon and thermally grown silicon dioxide via a multi-sample, multi-wavelength, multi-angle investigation. *J. Appl. Phys.* **1998**, *83*, 3323–3336. [CrossRef]
34. Ramirez-Atencia, C.; Bello-Orgaz, G.; R-Moreno, M.D.; Camacho, D. Solving complex multi-UAV mission planning problems using multi-objective genetic algorithms. *Soft Comput.* **2017**, *21*, 4883–4900. [CrossRef]

photonics

MDPI

Article

Polarization-Sensitive Structural Colors Based on Anisotropic Silicon Metasurfaces

Xiao Shang [1,2], Jiebin Niu [1,2], He Li [1,2], Longjie Li [1,2], Huakui Hu [1,2], Cheng Lu [1] and Lina Shi [1,*]

[1] State Key Lab of Fabrication Technologies for Integrated Circuits, Institute of Microelectronics of the Chinese Academy of Sciences, No. 3 West Road, Beitucheng, Beijing 100029, China; shangxiao@ime.ac.cn (X.S.)
[2] University of Chinese Academy of Sciences, 19 Yuquan Road, Shijingshan District, Beijing 100049, China
* Correspondence: shilina@ime.ac.cn

Abstract: Structural colors based on all-dielectric metasurfaces hold great promise for a wide range of applications, including high-density optical storage, ultra-high-resolution 3D displays, imaging security certification, and so on. However, achieving dynamic tunable structural color with a compact and simple Si platform remains a great challenge. Here, we propose a dynamic tunable structural coloration with polarization-sensitive metasurfaces consisting of arrays of Si elliptical nanopillars, enabling full-colored images to be displayed and switched through the control of the polarization of incident light. A distinct feature of our design is that the color phase is independent of the viewing angle, which is fundamental for real applications. Moreover, we demonstrated that dual and multiple colors can be obtained by varying the angle of either the polarizer or the analyzer. Our scheme provides a simple yet general approach for potential applications in the fields of virtual reality, ultra-high-resolution 3D displays, and high-density information storage.

Keywords: structural color; metasurface; Si nanostructures; polarization

Citation: Shang, X.; Niu, J.; Li, H.; Li, L.; Hu, H.; Lu, C.; Shi, L. Polarization-Sensitive Structural Colors Based on Anisotropic Silicon Metasurfaces. *Photonics* **2023**, *10*, 448. https://doi.org/10.3390/photonics10040448

Received: 17 March 2023
Revised: 3 April 2023
Accepted: 10 April 2023
Published: 13 April 2023

1. Introduction

Color, the main visual vehicle for human perception of the real world, is a visual perception of light generated by the eye, brain, and life experience [1,2]. Visible light is reflected or absorbed by objects, which can produce different colors and bring us a colorful world [3,4]. Although chemical pigments have many applications in industrial products, it has obvious drawbacks: they cannot endure durative high-temperature or high-intensity illumination, and they threaten the environment and human health [5]. Additionally, these pigments have spots that are around 25 nm, which leads to a low resolution of less than 1000 dpi. It is difficult to meet the permanence of coloring and, most importantly, the low resolution of the colors produced by these conventional dyes and pigments limits their practical application in advanced color displays and high-resolution images. By contrast, structural colors offer an alternative coloring mechanism that is physically derived from the controlled scattering, diffraction, and interference of light through nanostructures or microstructures [6–13]. It is known that the Mie resonances appear when the incident electromagnetic wave is comparable to the physical dimension of the resonator for high-index dielectric nanostructures [14,15]. Thus, vibrant colors can be obtained by simply tuning the structural parameters rather than relying on their chemical nature, holding potential for diverse applications, such as high-resolution color printings [16–33], high-resolution digital displays [10,34–37], focusing hyperlens or holographic hyperlens [38–40], and information storage [41,42].

Although some typical configurations, such as nanoparticles, nanogratings, and multilayer thin films have been utilized to generate highly saturated and pure structural colors, the practical applications of structural color in data storage, optical security, and optical steganography are still restricted because most systems can only generate static colors. In order to overcome these limitations, more scholars began to work on polarization-tunable anisotropic nanostructures and actively tunable metasurface [43–62]. For instance, Olson et al. modulated the brightness of color by incorporating aluminum nanorod array pixels into a liquid crystal display with an applied voltage [36], Xu et al. demonstrated fast monochromatic and full-color electrochromic switching using electrochromic polymer nanoslit arrays [63]. However, the incomplete full-color spectral response and complicated electrical device structures significantly hinder their practical application. Hence, challenges remain in simultaneously achieving dynamic control and polarization dependence structural color with a compact, single-layer structure.

In this work, we present a strategy to realize polarization-sensitive color generation by exploring the orientation properties of anisotropic elliptic Si nanopillars directly deposited on a Si substrate. Owning to the geometrical anisotropy, the incident white light with different polarizations was backscattered into distinct colors by anisotropic scatters. We realized the dramatic contrast of colors under two polarization states of incident light by tuning the diameters of the nanopillars along the x- and y-axes, and we experimentally obtained basic colors, such as red, green, blue, and yellow, across the visible spectrum by carefully designing the structural parameters. The calculated results showed that the color phase had only a slight variation in intensity as the viewing angle changed owning to the Mie resonance origin. Furthermore, the method suggested here can be utilized to change the output color by simple rotation of the polarizer and analyzer with efficiently and smoothly controlled structural color across the entire visible spectral range. Thus, our designed full-color anisotropic Si metasurface with the controllable and flexible generation of arbitrary polarization distributions for light paves the way toward many practical applications, such as virtual reality, ultrahigh-resolution 3D displays, and high-density information storage.

2. Methods and Materials

2.1. Sample Fabrication

The Si metasurfaces on Si wafers (100 mm diameter, prime grade, double-sided polished) were fabricated with the electron beam lithography (EBL) technique followed by inductively coupled plasma (ICP) etching. First, we cleaned the Si wafer in an ultrasound bath in acetone and anhydrous ethanol for 10 min. Second, a 150 nm PMMA (A3) film was spin-coated onto the cleaned Si wafer and baked at 180 °C for 5 min. After that, the PMMA resist was exposed to the electron beam to form the PMMA nanostructures. Then, the sample was transferred into an electron beam evaporator and directly coated with 30 nm Cr films, and then acetone was used to finally lift off the Cr mask. Then, the silicon was etched away with inductively coupled plasma (ICP) using C_3F_8 and SF_6 gases. Finally, by immersing the sample into the chromium etchant to remove the Cr, the Si metasurfaces were obtained.

2.2. Optical Characterization

The samples were assessed using an optical microscope (OLYMPUS BX51), a charge-coupled device (CCD) camera (Canon EOS 750D), and a fiber-coupled spectrometer (Ocean Optics HR4000). The structures were illuminated by a white halogen lamp (100W) and a homemade optical setup to control the polarization and incident angle. The reflection spectra were recorded by the spectrometer with a spectrometer slit size of 0.2 mm, and bright-field microscopy images were taken by the CCD camera. All reflection spectra were normalized to the reflection from an aluminum mirror.

3. Results and Discussion

The proposed structures were composed of asymmetric elliptic Si nanopillars on a Si substrate with different diameters along the x- and y-directions, as illustrated in Figure 1a. In the case of the same structure, x-polarization and y-polarization incident white light was scattered back to distinct colors due to the asymmetry of the nanopillars. The major axes and minor axes of the structure were D_x and D_y along the x- and y-directions, respectively. The lattice dimension P was fixed at 300 nm. The schematic of the single-pixel and two top-view SEM images of typical elliptic Si nanopillar arrays are shown in the right half of Figure 1a. Figure 1b,c present a bright field optical microscope image of the fabricated elliptic Si nanopillar arrays under white light illumination through an objective of 20X (NA: 0.45). We changed the diameter of nanopillars along the x- and y-directions from 96 to 244 nm to cover the full color palette. Each area of 15 × 15 μm^2 presented an individual color corresponding to an array of nanopillars with a specific D_x and D_y. The height of the elliptic Si nanopillar was 162 nm measured by SEM. As shown in Figure 1b,c, distinct vivid color pixels of the color palette exhibited red, yellow, green, and blue. The color changed from red to blue with the increasing diameter of the nanopillar, indicating that the optical response is mainly determined by the diameter of the nanopillar. For each nanostructure group, a sharp color contrast could be obtained under the two polarized incident lights. Figure 1b shows that the color changed from red to yellow to green to blue with an increasing D_x under x-polarization, where D_y was fixed at 96 nm. For the y-polarization incidence, all pixels exhibited a distinct saturation of red. It was also possible to achieve hue- and saturation-tuned colors by varying D_x and D_y for y-polarization, as shown in Figure 1c. Particularly, we obtained a vivid blue color by simply tuning the geometric parameters of the ellipse nanopillars, which is very difficult to realize with Si nanostructures due to the higher loss of Si in the blue and violet bands.

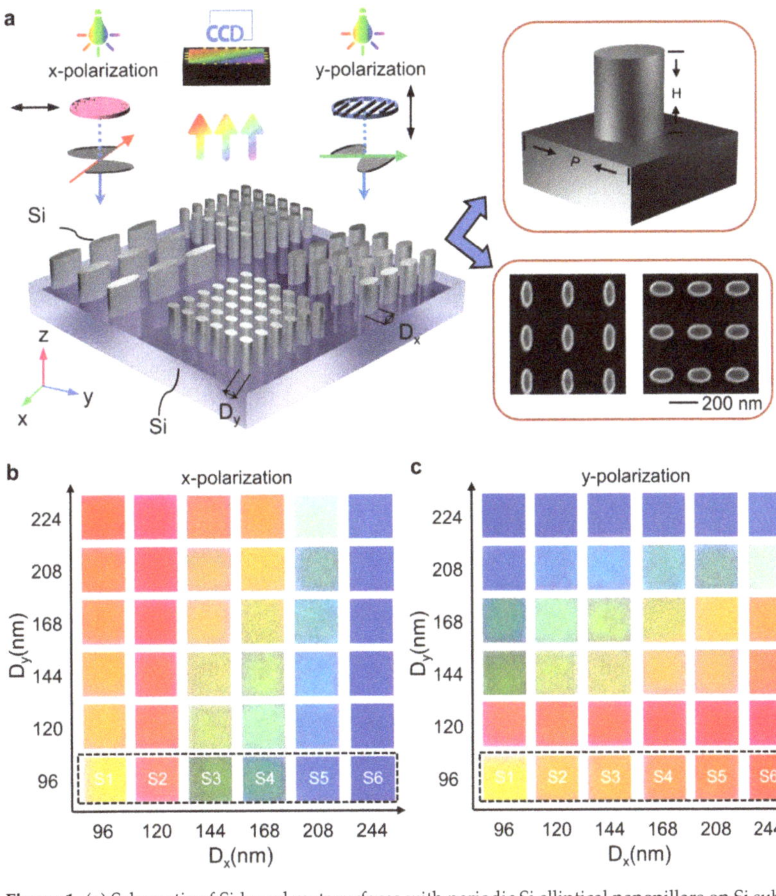

Figure 1. (**a**) Schematic of Si-based metasurfaces with periodic Si elliptical nanopillars on Si substrate. Two top-view SEM images of a typical elliptic Si nanopillar array. Optical bright microscopy image of the color palette under (**b**) x-polarization and (**c**) y-polarization light.

To analyze the reflection characteristics of the nanopillars under the x- and y-polarization states, we measured the reflection spectrums of six nanopillars under x- and y-polarization incident light, as shown in Figure 2, which correspond to the geometry parameters in the dashed box in Figure 1. For the incident light polarized along the major axis of the nanopillar, as the diameter increased, the reflection dip shifted toward longer wavelengths, and a new dip appeared at the short wavelength with a further increase in the diameter, which was determined by the period in accordance with the Rayleigh anomaly. When the nanopillars varied from S1 to S6, the main dip wavelengths of the reflected spectra were 415, 501, 539, 572, 650, and 663 nm, respectively. As a result, by simply tailoring the nanopillar size, perfect absorption of light at specific wavelengths across the visible spectrum was achieved. For the incident light polarized along the minor axis of the nanopillar, the dips in the calculated reflection spectra in Figure 2b had no apparent shift as D_x increased, which also demonstrates the modification of the saturation with a nearly unchanged hue. Given that the periods of the proposed structure in the subwavelength regime were much smaller, the slight red shift in the resonance wavelength might have been due to the near-field coupling between the adjacent nanopillars. However, the efficiency and bandwidth of the reflected spectrum were slightly different under y-polarization. We calculated the chromaticity of the colors for the x- and y- polarization incidence on the CIE 1931 color map in Figure 2c,d. The black dots in Figure 2c show that the chromaticity diagram has significantly different hues due to the presence of various main dip wavelengths, and a wide color variety can be achieved by increasing the major axis of the Si nanopillar under x- polarization incidence. By contrast, the center wavelength only moved around 10 nm; the saturation of the colors shows a slightly rising trend from S1 to S6 in Figure 2d. Therefore, we simultaneously realized the saturation- and hue-tuning structure color in two orthogonal polarization states, which is desirable for practical applications, such as anticounterfeiting, 3D display, data storage, and so on.

To illustrate the underlying mechanism of the above polarization-sensitive structural colors, we next simulated the light–matter interaction by the finite difference time domain method with periodic boundary conditions in the x- and y-directions to model infinite arrays and perfect match layers in the z-direction to mimic a free space. The refractive indices of Si were taken from the book of Edward D. Palik. It is well known that high-refractive dielectric scatters exhibit resonance when the wavelength of the incident light is similar to the physical size of the scatters and that the magnetic and electrical responses of these scatters have similar intensities. In general, the dipole mode resonance wavelength of nanostructures in air is $\lambda/n \sim d$, and when the diameter is large enough, the nanostructures also support higher-order modes in the visible spectrum. We present the magnetic field distributions of S3 at the resonant wavelength of A at 432 nm and B at 539 nm under x- and y-polarization states in Figure 3. Figure 3a,b show the magnetic field distributions $|H|$ in the x–z plane at wavelengths of 432 nm under two polarization states. Figure 3c,d show $|H|$ at wavelengths of 539 nm under two polarization states. All the magnetic field distributions at resonant wavelength were the strongest at the center of the nanopillar in view of the observed field profiles. In general, magnetic dipole resonance with high absorption and minimal scattering compared with an electric dipole at resonances with longer wavelengths, we found that some of the energy leaked into the substrate, and they could decrease the resonant reflection and form a deep resonant dip. The magnetic field distribution was weak and primarily located in the interior of the Si nanopillar for the non-resonant wavelength, as shown in Figure 3a,d. The above phenomenon indicates that the resonance mechanism of the two polarization states is identical. It depends on the inherent properties of Si; the periodic arrangement of Si nanopillars can support dipole resonance and form channels that can introduce light energy to the Si substrate, deepening the dip of the reflection spectrum.

We also investigated the angle independence of the anisotropic metasurface designed on the basis of Si elliptic nanopillars. In order to investigate how the angle of view affects the structural color through reflectance, we calculated the periodic structural reflection spectra of D_x = 120 nm and D_y = 96 nm under x- and y-polarization states, respectively, in Figure 4a,c. The reflection efficiency was a function of the incidence angle and wavelength. For x- and y-polarizations, the central resonant positions were 490 nm and 435 nm, respectively, and they remained nearly constant, with only a slight variation in intensity from 0 to 30° as the angle of incidence changed. As a result, color hardly changed with increasing incidence angles, and color phase remained almost constant on the CIE 1931 diagrams when the incident angle was less than 30°, as shown in Figure 4b,d. However, when the incidence angle of the light was greater than 30°, the view angle change of the structural color was slightly different for the x- and y-polarization. For the x-polarization, when the incident angle was greater than 30°, there was a specific resonance around the main resonant dip, which gradually grew. The reflected light became less monochromatic, resulting in a decrease in saturation, so the color changed from magenta to deep orange to brownish yellow. At the same time, under the y-polarization, with an increase in the incident angle, the position of the resonance dip remained almost unchanged. When the incidence angle was greater than 50°, the reflection increased precipitously to nearly 600 nm, so the color changed from orange to yellow.

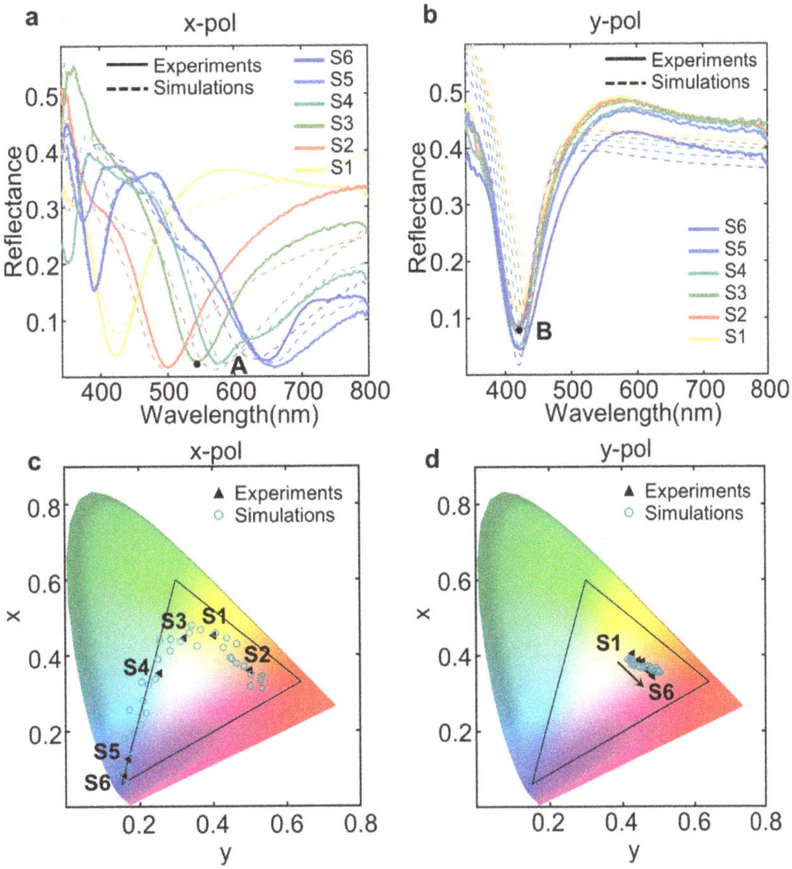

Figure 2. Simulated and experimental reflection spectra under (**a**) x- and (**b**) y-polarization light. Corresponding CIE 1931 chromaticity coordinates for (**c**) x- and (**d**) y-polarization, respectively.

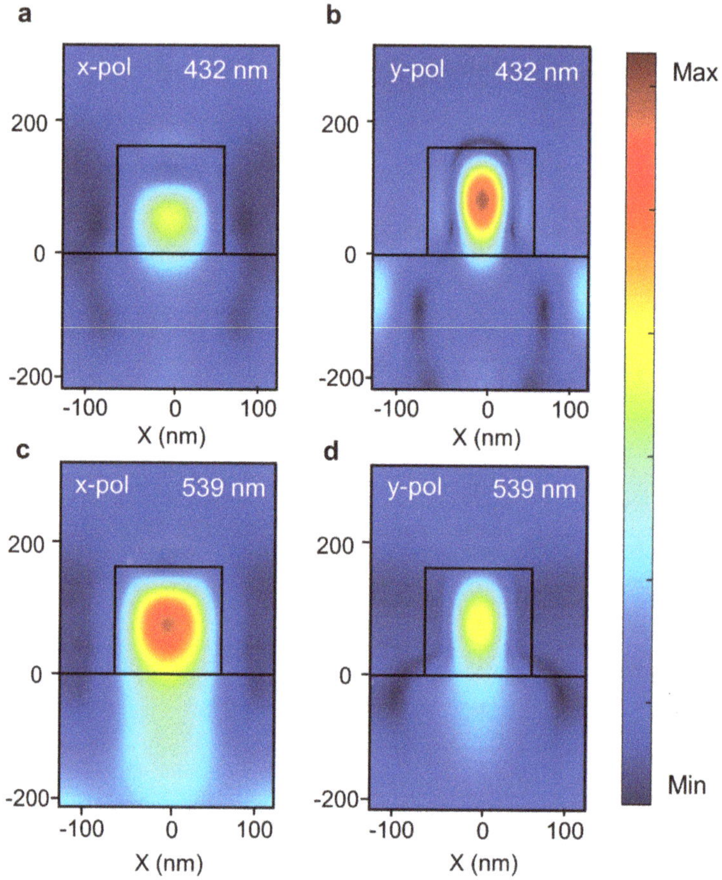

Figure 3. Simulated magnetic field distributions |H| in x–z plane at resonant wavelength of 432 nm under (**a**) x-polarization and (**b**) y-polarization, respectively, and at resonant wavelength of 539 nm under (**c**) x-polarization and (**d**) y-polarization, respectively.

Figure 4e shows a simulated color image with an equivalent hue at a 10° interval of oblique incident light. The color under two polarization states remained basically constant from 0° to 30°, which was consistent with the results in Figure 4a,c. In accordance with the visual field of a single human eye, the symbol recognition angle was about 30°, so our Si nanostructure color perspective independence is suitable for small-angle display applications. In addition, it can be used for wide-angle detectors and angle-sensitive microscopes due to the great change in color saturation with a varying incident angle when the incident angle is greater than 30°.

The designed elliptical Si nanostructure possesses strong anisotropy so that we can establish a polarization-controlled structural color generation mechanism by changing its polarization state. The proposed elliptical Si metasurfaces have different colors under different observation conditions. To investigate the basic working process of manipulating the polarization, various combinations of the input polarizer and output analyzer angle differences (φ_p—φ_a) were utilized. According to Malus's law, the output intensity of reflection through a polarizer and analyzer depends on the relative angle between the polarization axis of the polarizer and the electric field orientation of the reflected light. Figure 5 shows the measured reflection spectra for different settings of the input polarization angle (φ_p = 0°, 45°, 90°, and 135°) as a function of the output analyzer angle

($0^\circ < \varphi_a < 360^\circ$, in steps of 1°). For the $\varphi_p = 0^\circ$ and 90° states, the reflection was blocked if the analyzing polarizer had incident polarization. Thus, if $\varphi_p = \varphi_a$, the reflection was at a maximum, and if φ_p was at a right angle to the φ_a, the reflection was zero, as shown in Figure 5b,f. When the input polarization angles were 45° and 135°, the situation was more intricate. Each elliptical nanopillar can generate two independent phase shifts of reflected light in two directions. Very different colors were thus obtained for different settings of the analyzing polarizer (φ_a). Figure 5d,h show that the reflection spectra sensitively depend on the setting of φ_a. The maxima shifted as φ_a was altered, but the overall intensities were preserved, which clearly indicated a phase shift. In the diagonal case ($\varphi_p = 45^\circ$ and 135°), the phase shift was responsible for the observation of four separate output colors, and only a single color was observed for each x-polarization and y-polarization ($\varphi_p = 0^\circ$ and 90°).

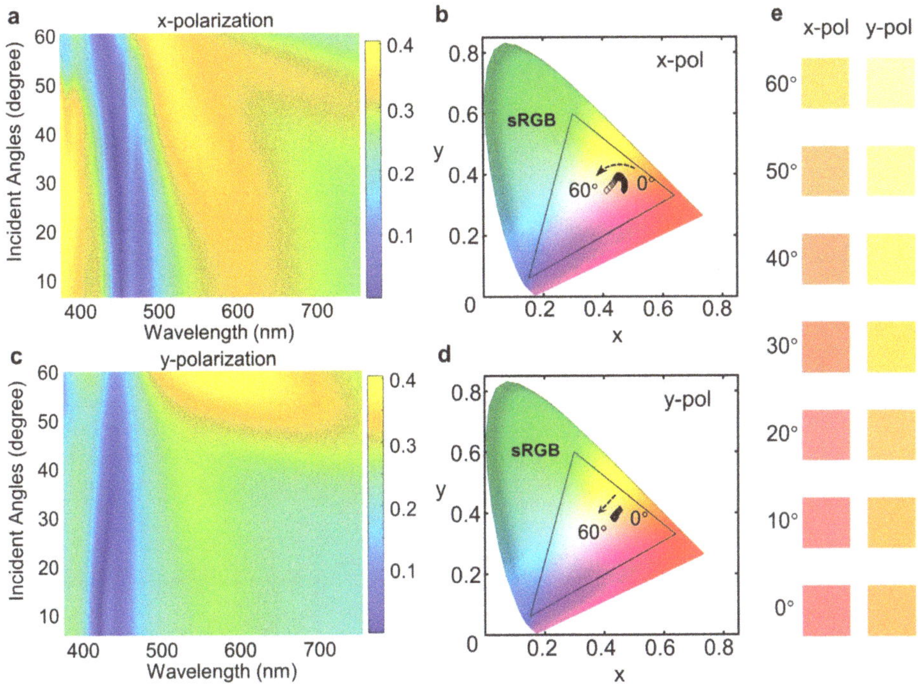

Figure 4. Simulated reflection spectra of S2 (D_x = 120 nm, D_y = 96 nm) under (**a**) x- and (**c**) y-polarization. Corresponding CIE 1931 chromaticity coordinates under (**b**) x- and (**d**) y-polarization. (**e**) Corresponding colors for varying incident angles from 0° to 90° with 10° interval.

As a result, the input linear polarization light (45° with respect to the long axis of the elliptical nanopillar) was converted into elliptical polarized states. This color dependence on the polarization angle is particularly useful, leading to the dynamic control of the color by changing the polarization angle of the incident light and output light, which can be used in 3D displaying and high-density storage and is essential for encryption and anticounterfeiting.

Corresponding CIE 1931 chromaticity coordinates of the reflection spectra in Figure 5b,d are shown in Figure 6. When the polarizer was orientated at 0°, which corresponded to the x-polarization incident light, the corresponding color hue was unchanged, and only the brightness changed with a varying analyzer angle. When the input polarization angle was 45°, the colors showed a very strong contrast with varying analyzer angles. These appear in Figure 5c as purple, dark blue, brown, and dark green.

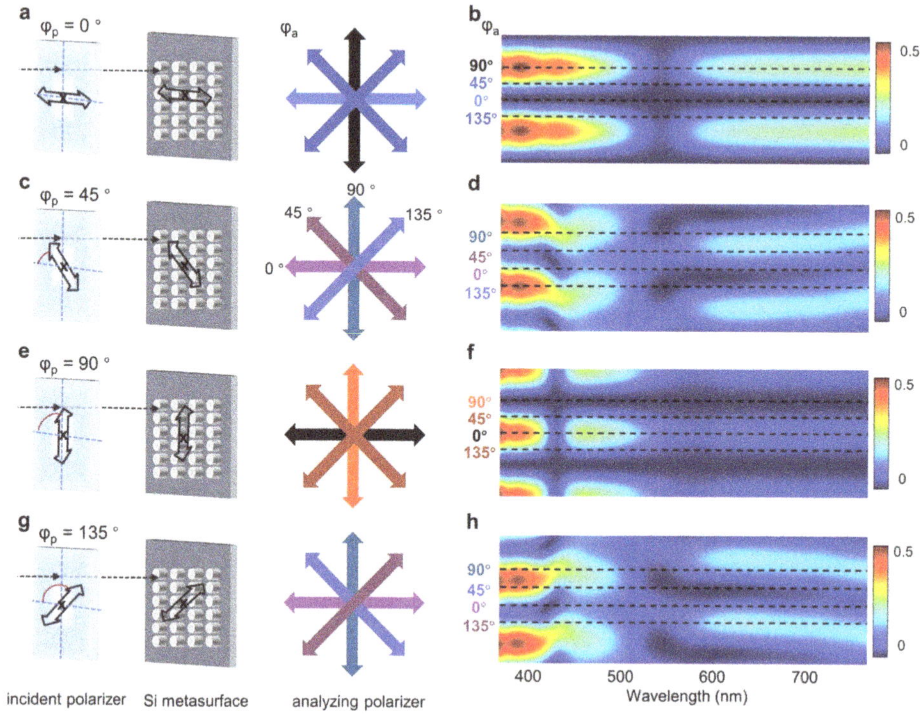

Figure 5. Reflection spectra as a function of polarization angles. (**a**,**c**,**e**,**g**) Measurement schemes for different angles of the incident polarizer (φ_p) and the analyzing polarizer (φ_a); each output arrow is colored in the observed color. (**b**,**d**,**f**,**h**) Corresponding calculated reflection plots for rotation of the analyzing polarizer ($0° < \theta < 360°$, 1° steps).

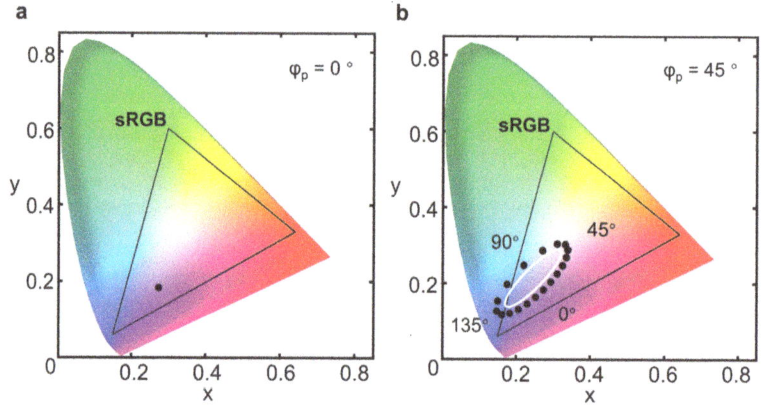

Figure 6. Corresponding CIE 1931 chromaticity coordinates of the reflection spectra in (**a**) Figure 5d and (**b**) Figure 5f.

4. Conclusions

In conclusion, we demonstrated polarization-sensitive structural colors based on anisotropic metasurfaces consisting of elliptical Si nanopillar arrays directly on a Si substrate, which generate colors using Mie resonances confined in Si nanopillars. Regular colors, such as red, green, blue, and yellow, across the visible spectrum were generated by simply tuning the axis of elliptical nanopillars. A distinct feature of our design is that the pixel preserves the color phase independently of the viewing angle, which is fundamental for real applications. More importantly, multiple colors can be obtained by changing the polarizer and analyzer angles. Our proposed metasurface has a lower cost and is more compatible with standard CMOS technology, holding great potential for many practical applications, such as virtual reality, ultrahigh-resolution 3D displays, and high-density information storage.

Author Contributions: Conceptualization, X.S.; methodology, X.S. and L.S.; software, X.S., H.L., L.L. and H.H.; validation, X.S., H.H., C.L. and J.N.; formal analysis, X.S. and L.L.; resources, H.H. and C.L.; writing—original draft preparation, X.S.; writing—review and editing, X.S. and L.S.; supervision, X.S. and H.L.; funding acquisition, L.S. All authors have read and agreed to the published version of the manuscript.

Funding: This research was funded by the National Key R&D Program of China 2019YFB2205002, the National Natural Science Foundation of China (Grant No. 61875225, 61888102). The APC was funded by the National Natural Science Foundation of China (Grant No. 61875225).

Institutional Review Board Statement: Not applicable.

Informed Consent Statement: Not applicable.

Data Availability Statement: Not applicable.

Conflicts of Interest: The authors declare no conflict of interest.

References

1. Vukusic, P.; Sambles, J.R. Photonic structures in biology. *Nature* **2003**, *424*, 852–855. [CrossRef] [PubMed]
2. Rouw, R.; Scholte, H.S. Increased structural connectivity in grapheme-color synesthesia. *Nat. Neurosci.* **2007**, *10*, 792–797. [CrossRef] [PubMed]
3. Bohren, C.F.; Huffman, D.R. *Absorption and Scattering of Light by Small Particles*; John Wiley & Sons: Hoboken, NJ, USA, 2008.
4. Kerker, M. *The Scattering of Light and Other Electromagnetic Radiation*; Elsevier: Amsterdam, The Netherlands, 2016.
5. Brown, M.A.; De Vito, S.C. Predicting azo dye toxicity. *Crit. Rev. Environ. Sci. Technol.* **1993**, *23*, 249–324. [CrossRef]
6. Baek, K.; Kim, Y.; Mohd-Noor, S.; Hyun, J.K. Mie Resonant Structural Colors. *ACS Appl. Mater. Interfaces* **2020**, *12*, 5300–5318. [CrossRef] [PubMed]
7. Kinoshita, S.; Yoshioka, S.; Miyazaki, J. Physics of structural colors. *Rep. Prog. Phys.* **2008**, *71*, 076401. [CrossRef]
8. Shen, Y.; Rinnerbauer, V.; Wang, I.; Stelmakh, V.; Joannopoulos, J.D.; Soljačić, M. Structural Colors from Fano Resonances. *ACS Photonics* **2015**, *2*, 27–32. [CrossRef]
9. Song, M.; Wang, D.; Peana, S.; Choudhury, S.; Nyga, P.; Kudyshev, Z.A.; Yu, H.; Boltasseva, A.; Shalaev, V.M.; Kildishev, A.V. Colors with plasmonic nanostructures: A full-spectrum review. *Appl. Phys. Rev.* **2019**, *6*, 041308. [CrossRef]
10. Tittl, A. Tunable structural colors on display. *Light Sci. Appl.* **2022**, *11*, 155. [CrossRef]
11. Yang, B.; Cheng, H.; Chen, S.; Tian, J. Structural colors in metasurfaces: Principle, design and applications. *Mater. Chem. Front.* **2019**, *3*, 750–761. [CrossRef]
12. Yang, W.; Xiao, S.; Song, Q.; Liu, Y.; Wu, Y.; Wang, S.; Yu, J.; Han, J.; Tsai, D.P. All-dielectric metasurface for high-performance structural color. *Nat. Commun.* **2020**, *11*, 1864. [CrossRef]
13. Zhao, Y.; Xie, Z.; Gu, H.; Zhu, C.; Gu, Z.J.C.S.R. Bio-inspired variable structural color materials. *Chem. Soc. Rev.* **2012**, *41*, 3297–3317. [CrossRef] [PubMed]
14. Van de Groep, J.; Polman, A. Designing dielectric resonators on substrates: Combining magnetic and electric resonances. *Opt. Express* **2013**, *21*, 26285–26302. [CrossRef] [PubMed]
15. Hentschel, M.; Koshelev, K.; Sterl, F.; Both, S.; Karst, J.; Shamsafar, L.; Weiss, T.; Kivshar, Y.; Giessen, H. Dielectric Mie voids: Confining light in air. *Light Sci. Appl.* **2023**, *12*, 3. [CrossRef] [PubMed]
16. Xu, T.; Wu, Y.-K.; Luo, X.; Guo, L.J. Plasmonic nanoresonators for high-resolution colour filtering and spectral imaging. *Nat. Commun.* **2010**, *1*, 59. [CrossRef]
17. Cheng, F.; Gao, J.; Luk, T.S.; Yang, X. Structural color printing based on plasmonic metasurfaces of perfect light absorption. *Sci. Rep.* **2015**, *5*, 11045. [CrossRef]
18. Cheng, F.; Gao, J.; Stan, L.; Rosenmann, D.; Czaplewski, D.; Yang, X. Aluminum plasmonic metamaterials for structural color printing. *Opt. Express* **2015**, *23*, 14552–14560. [CrossRef]
19. Dai, Q.; Zhu, G.; Zhang, W.; Li, J.; Li, Z.; Cui, H.; Wei, K.; He, Z.; Guan, Z.; Zheng, G. Dual-channel anticounterfeiting color-nanoprinting with a single-size nanostructured metasurface. *Opt. Express* **2022**, *30*, 33574–33587. [CrossRef]
20. Dong, Z.; Ho, J.; Yu, Y.F.; Fu, Y.H.; Paniagua-Dominguez, R.; Wang, S.; Kuznetsov, A.I.; Yang, J.K.W. Printing Beyond sRGB Color Gamut by Mimicking Silicon Nanostructures in Free-Space. *Nano Lett.* **2017**, *17*, 7620–7628. [CrossRef]

21. Kumar, K.; Duan, H.; Hegde, R.S.; Koh, S.C.W.; Wei, J.N.; Yang, J.K.W. Printing colour at the optical diffraction limit. *Nat. Nanotechnol.* **2012**, *7*, 557–561. [CrossRef]
22. Liu, X.; Huang, Z.; Zang, J. All-Dielectric Silicon Nanoring Metasurface for Full-Color Printing. *Nano Lett.* **2020**, *20*, 8739–8744. [CrossRef]
23. Miyata, M.; Hatada, H.; Takahara, J. Full-Color Subwavelength Printing with Gap-Plasmonic Optical Antennas. *Nano Lett.* **2016**, *16*, 3166–3172. [CrossRef] [PubMed]
24. Nagasaki, Y.; Hotta, I.; Suzuki, M.; Takahara, J. Metal-Masked Mie-Resonant Full-Color Printing for Achieving Free-Space Resolution Limit. *ACS Photonics* **2018**, *5*, 3849–3855. [CrossRef]
25. Nagasaki, Y.; Suzuki, M.; Hotta, I.; Takahara, J. Control of Si-Based All-Dielectric Printing Color through Oxidation. *ACS Photonics* **2018**, *5*, 1460–1466. [CrossRef]
26. Sun, S.; Zhou, Z.; Zhang, C.; Gao, Y.; Duan, Z.; Xiao, S.; Song, Q. All-dielectric full-color printing with TiO_2 metasurfaces. *ACS Nano* **2017**, *11*, 4445–4452. [CrossRef]
27. Tan, S.J.; Zhang, L.; Zhu, D.; Goh, X.M.; Wang, Y.M.; Kumar, K.; Qiu, C.-W.; Yang, J.K.W. Plasmonic Color Palettes for Photorealistic Printing with Aluminum Nanostructures. *Nano Lett.* **2014**, *14*, 4023–4029. [CrossRef] [PubMed]
28. Zhang, Y.; Zhang, Q.; Ouyang, X.; Lei, D.Y.; Zhang, A.P.; Tam, H.-Y. Ultrafast Light-Controlled Growth of Silver Nanoparticles for Direct Plasmonic Color Printing. *ACS Nano* **2018**, *12*, 9913–9921. [CrossRef]
29. Shang, X.; Niu, J.; Wang, C.; Li, L.; Lu, C.; Zhang, Y.; Shi, L. Mie Resonances Enabled Subtractive Structural Colors with Low-Index-Contrast Silicon Metasurfaces. *ACS Appl. Mater. Interfaces* **2022**, *14*, 55933–55943. [CrossRef]
30. Li, L.; Niu, J.; Shang, X.; Chen, S.; Lu, C.; Zhang, Y.; Shi, L. Bright Field Structural Colors in Silicon-on-Insulator Nanostructures. *ACS Appl. Mater. Interfaces* **2021**, *13*, 4364–4373. [CrossRef]
31. Shi, L.; Niu, J.; Li, L.; Wang, C.; Shang, X.; Zhang, P.; Liu, Y.; Zhang, Y. Deep Subwavelength Wide-Angle Structural Colors at the Single Pixel Level. *Adv. Opt. Mater.* **2022**, *10*, 2200552. [CrossRef]
32. Shi, S.; Lu, H.; Li, Y.; Bo, S.; Li, D.; Zhao, J. Asymmetric nanocavities with wide reflection color gamut for color printing. *Nanotechnology* **2022**, *34*, 025201. [CrossRef]
33. Liu, H.; Wang, H.; Wang, H.; Deng, J.; Ruan, Q.; Zhang, W.; Abdelraouf, O.A.; Ang, N.S.S.; Dong, Z.; Yang, J.K. High-Order Photonic Cavity Modes Enabled 3D Structural Colors. *ACS Nano* **2022**, *16*, 8244–8252. [CrossRef] [PubMed]
34. Hu, T.; Tseng, C.K.; Fu, Y.H.; Xu, Z.; Dong, Y.; Wang, S.; Lai, K.H.; Bliznetsov, V.; Zhu, S.; Lin, Q.; et al. Demonstration of color display metasurfaces via immersion lithography on a 12-inch silicon wafer. *Opt. Express* **2018**, *26*, 19548–19554. [CrossRef] [PubMed]
35. Joo, W.J.; Kyoung, J.; Esfandyarpour, M.; Lee, S.H.; Koo, H.; Song, S.; Kwon, Y.N.; Song, S.H.; Bae, J.C.; Jo, A.; et al. Metasurface-driven OLED displays beyond 10,000 pixels per inch. *Science* **2020**, *370*, 459–463. [CrossRef] [PubMed]
36. Olson, J.; Manjavacas, A.; Basu, T.; Huang, D.; Schlather, A.E.; Zheng, B.; Halas, N.J.; Nordlander, P.; Link, S. High Chromaticity Aluminum Plasmonic Pixels for Active Liquid Crystal Displays. *ACS Nano* **2016**, *10*, 1108–1117. [CrossRef]
37. Song, M.; Li, X.; Pu, M.; Guo, Y.; Liu, K.; Yu, H.; Ma, X.; Luo, X. Color display and encryption with a plasmonic polarizing metamirror. *Nanophotonics* **2018**, *7*, 323–331. [CrossRef]
38. Xu, G.; Kang, Q.; Fan, X.; Yang, G.; Guo, K.; Guo, Z. Influencing Effects of Fabrication Errors on Performances of the Dielectric Metalens. *Micromachines* **2022**, *13*, 2098. [CrossRef] [PubMed]
39. Wang, W.; Zhao, R.; Chang, S.; Li, J.; Shi, Y.; Liu, X.; Sun, J.; Kang, Q.; Guo, K.; Guo, Z. High-efficiency spin-related vortex metalenses. *Nanomaterials* **2021**, *11*, 1485. [CrossRef]
40. Wang, W.; Yang, Q.; He, S.; Shi, Y.; Liu, X.; Sun, J.; Guo, K.; Wang, L.; Guo, Z. Multiplexed multi-focal and multi-dimensional SHE (spin Hall effect) metalens. *Opt. Express* **2021**, *29*, 43270–43279. [CrossRef]
41. Song, M.; Wang, D.; Kudyshev, Z.A.; Xuan, Y.; Wang, Z.; Boltasseva, A.; Shalaev, V.M.; Kildishev, A.V. Enabling Optical Steganography, Data Storage, and Encryption with Plasmonic Colors. *Laser Photonics Rev.* **2021**, *15*, 2000343. [CrossRef]
42. Heydari, E.; Sperling, J.R.; Neale, S.L.; Clark, A.W. Plasmonic Color Filters as Dual-State Nanopixels for High-Density Microimage Encoding. *Adv. Funct. Mater.* **2017**, *27*, 1701866. [CrossRef]
43. Shrestha, V.R.; Lee, S.S.; Kim, E.S.; Choi, D.Y. Aluminum plasmonics based highly transmissive polarization-independent subtractive color filters exploiting a nanopatch array. *Nano Lett.* **2014**, *14*, 6672–6678. [CrossRef] [PubMed]
44. Raj Shrestha, V.; Lee, S.-S.; Kim, E.-S.; Choi, D.-Y. Polarization-tuned dynamic color filters incorporating a dielectric-loaded aluminum nanowire array. *Sci. Rep.* **2015**, *5*, 12450. [CrossRef] [PubMed]
45. Balaur, E.; Sadatnajafi, C.; Kou, S.S.; Lin, J.; Abbey, B. Continuously Tunable, Polarization Controlled, Colour Palette Produced from Nanoscale Plasmonic Pixels. *Sci. Rep.* **2016**, *6*, 28062. [CrossRef] [PubMed]
46. Li, Z.; Clark, A.W.; Cooper, J.M. Dual color plasmonic pixels create a polarization controlled nano color palette. *ACS Nano* **2016**, *10*, 492–498. [CrossRef] [PubMed]
47. Vashistha, V.; Vaidya, G.; Gruszecki, P.; Serebryannikov, A.E.; Krawczyk, M. Polarization tunable all-dielectric color filters based on cross-shaped Si nanoantennas. *Sci. Rep.* **2017**, *7*, 8092. [CrossRef]
48. Yang, B.; Liu, W.; Li, Z.; Cheng, H.; Chen, S.; Tian, J. Polarization-Sensitive Structural Colors with Hue-and-Saturation Tuning Based on All-Dielectric Nanopixels. *Adv. Opt. Mater.* **2018**, *6*, 1701009. [CrossRef]
49. Zang, X.; Dong, F.; Yue, F.; Zhang, C.; Xu, L.; Song, Z.; Chen, M.; Chen, P.Y.; Buller, G.S.; Zhu, Y.; et al. Polarization Encoded Color Image Embedded in a Dielectric Metasurface. *Adv. Mater.* **2018**, *30*, e1707499. [CrossRef]

50. Gao, S.; Park, C.S.; Lee, S.S.; Choi, D.Y. All-dielectric metasurfaces for simultaneously realizing polarization rotation and wavefront shaping of visible light. *Nanoscale* **2019**, *11*, 4083–4090. [CrossRef]
51. Fan, J.; Cheng, Y. Broadband high-efficiency cross-polarization conversion and multi-functional wavefront manipulation based on chiral structure metasurface for terahertz wave. *J. Phys. D Appl. Phys.* **2020**, *53*, 025109. [CrossRef]
52. Jung, C.; Yang, Y.; Jang, J.; Badloe, T.; Lee, T.; Mun, J.; Moon, S.-W.; Rho, J. Near-zero reflection of all-dielectric structural coloration enabling polarization-sensitive optical encryption with enhanced switchability. *Nanophotonics* **2020**, *10*, 919–926. [CrossRef]
53. Yue, W.; Gao, S.; Li, Y.; Zhang, C.; Fu, X.; Choi, D.-Y. Polarization-encrypted high-resolution full-color images exploiting hydrogenated amorphous silicon nanogratings. *Nanophotonics* **2020**, *9*, 875–884. [CrossRef]
54. Zhang, C.; Jing, J.; Wu, Y.; Fan, Y.; Yang, W.; Wang, S.; Song, Q.; Xiao, S. Stretchable All-Dielectric Metasurfaces with Polarization-Insensitive and Full-Spectrum Response. *ACS Nano* **2020**, *14*, 1418–1426. [CrossRef] [PubMed]
55. Zhao, J.; Yu, X.; Zhou, K.; Zhang, W.; Yuan, W.; Yu, Y. Polarization-sensitive subtractive structural color used for information encoding and dynamic display. *Opt. Lasers Eng.* **2021**, *138*, 106421. [CrossRef]
56. Guo, X.; Zhong, J.; Li, B.; Qi, S.; Li, Y.; Li, P.; Wen, D.; Liu, S.; Wei, B.; Zhao, J. Full-Color Holographic Display and Encryption with Full-Polarization Degree of Freedom. *Adv. Mater.* **2022**, *34*, e2103192. [CrossRef] [PubMed]
57. Guo, K.; Li, X.; Ai, H.; Ding, X.; Wang, L.; Wang, W.; Guo, Z. Tunable oriented mid-infrared wave based on metasurface with phase change material of GST. *Results Phys.* **2022**, *34*, 105269. [CrossRef]
58. Kang, Q.; Li, D.; Wang, W.; Guo, K.; Guo, Z. Multiband tunable thermal camouflage compatible with laser camouflage based on GST plasmonic metamaterial. *J. Phys. D Appl. Phys.* **2021**, *55*, 065103. [CrossRef]
59. Ai, H.; Kang, Q.; Wang, W.; Guo, K.; Guo, Z. Multi-beam steering for 6G communications based on graphene metasurfaces. *Sensors* **2021**, *21*, 4784. [CrossRef]
60. Kang, Q.; Li, D.; Guo, K.; Gao, J.; Guo, Z. Tunable thermal camouflage based on GST plasmonic metamaterial. *Nanomaterials* **2021**, *11*, 260. [CrossRef]
61. Ding, X.; Kang, Q.; Guo, K.; Guo, Z. Tunable GST metasurfaces for chromatic aberration compensation in the mid-infrared. *Opt. Mater.* **2020**, *109*, 110284. [CrossRef]
62. Moon, J.-B.; Lee, S.-J.; Kim, S.-H.; Chang, Y.; Hwang, H.; Yi, G.-R. Mie-Resonant Rutile TiO_2 Nanospheres for Unclonable Structural Colors. *Chem. Mater.* **2022**, *34*, 7897–7905. [CrossRef]
63. Xu, T.; Walter, E.C.; Agrawal, A.; Bohn, C.; Velmurugan, J.; Zhu, W.; Lezec, H.J.; Talin, A.A. High-contrast and fast electrochromic switching enabled by plasmonics. *Nat. Commun.* **2016**, *7*, 10479. [CrossRef] [PubMed]

photonics

MDPI

Communication

Full-Color and Anti-Counterfeit Printings with All-Dielectric Chiral Metasurfaces

Longjie Li [1,2,3], He Li [1,2,3], Huakui Hu [1,2,3], Xiao Shang [1,2,3], Huiwen Xue [1,2,3], Jinyu Hu [1,2,3], Cheng Lu [1,2], Shengjie Zhao [1,2], Jiebin Niu [1,2,*] and Lina Shi [1,2,*]

1 State Key Lab of Fabrication Technologies for Integrated Circuits, Institute of Microelectronics of the Chinese Academy of Sciences, No. 3 West Road, Beitucheng, Beijing 100029, China
2 Laboratory of Microelectronic Devices and Integrated Technology, Institute of Microelectronics of the Chinese Academy of Sciences, No. 3 West Road, Beitucheng, Beijing 100029, China
3 University of Chinese Academy of Sciences, 19 Yuquan Road, Shijingshan District, Beijing 100049, China
* Correspondence: niujiebin@ime.ac.cn (J.N.); shilina@ime.ac.cn (L.S.)

Simple Summary: Structural color is a vibrant color produced by optical effects such as interference, diffraction, or scattering of incident light interacting with the nanostructure, which has the advantages of environmental protection, high saturation, and high resolution. High saturation and dynamic tunability are the current research hotspots for structural color. We achieve full-color wide gamut structural colors and anti-counterfeit functions by an all-dielectric chiral metasurface consisting of half-gammadion-shaped resonators embedded in PMMA and a top TiO_2 layer on quartz coated with an ITO layer. This research has significant implications in micro-display, nano-printing, anti-counterfeiting, and information encryption.

Abstract: Structural color is anticipated to replace pigmented or chemical color due to its superior saturation, resolution, environmental friendliness, and longevity. We achieve a full-color gamut of structural colors and anti-counterfeit functions by an all-dielectric chiral metasurface consisting of half-gammadion-shaped resonators embedded in PMMA and a top TiO_2 layer on quartz coated with an ITO layer. The prominent resonance peaks of this embedded chiral metasurface under the cross-polarization condition, which are provided by the polarization conversion features of the chiral structure, lead to extremely saturated structural colors. The color phase mainly depends on the structure's period, opening up a full-color range well beyond sRGB. Especially, we demonstrate a star with hidden information of the letter "A" and the Chinese word " 李龙杰" by taking advantage of the difference in the color phase change under different polarization conditions to decode the information, thus realizing the anti-counterfeiting function. Our proposed embedded chiral metasurface provides dual-function structural colors and is highly promising for micro-display, nanoprinting, anti-counterfeiting, data storage, and information encryption.

Keywords: structural color; metasurface; chiral; anti-counterfeit

Citation: Li, L.; Li, H.; Hu, H.; Shang, X.; Xue, H.; Hu, J.; Lu, C.; Zhao, S.; Niu, J.; Shi, L. Full-Color and Anti-Counterfeit Printings with All-Dielectric Chiral Metasurfaces. *Photonics* **2023**, *10*, 401. https://doi.org/10.3390/photonics10040401

Received: 6 March 2023
Revised: 21 March 2023
Accepted: 24 March 2023
Published: 3 April 2023

1. Introduction

When compared to pigments and bioluminescence, structural color is a gift from nature that may be used to create a colorful and sustainable world due to the advantages of non-polluting, high resolution, high saturation, and durability, which are generated through the interference and scattering of light from the nanostructure [1]. An improvement in the utilization of plasmonic and all-dielectric metasurfaces to creative artificial structural colors has been attributed to recent developments in nanofabrication techniques [2]. Optical responses for plasmonic structural colors represented by Au [3–5], Ag [6–10], and Al [11–15] are mainly driven by plasmonic resonances based on a variety of metallic nanostructures, but this is constrained by the inherent ohmic loss of the metal, leading to a subpar structural color performance. Yet the electric and magnetic dipole resonances generated from high-index dielectric materials simultaneously adapt to govern the spectrum and achieve high-saturation colors [16], as shown by the dielectric structural colors represented by Si [17–27], TiO_2 [28–37], and Si_3N_4 [38,39].

Static structured color has great potential for applications in the full-color display, Meta-OLED, etc. [40]. However, it is usually difficult to alter the color scheme of a structure once it has been built. As a result, the question of how to control structural color dynamically is rising in importance. The use of phase-change materials, voltage-controlled coupling, and thermal stimulation to tune the metasurface are just a few examples of the methods that have been employed to achieve the dynamic tuning of structural color. Another method that promises to be faster to tune and at the same time allow the multiplexing of structures is the combination of anisotropic structures and polarized light [41]. Brilliant structural colors under a cross-polarization condition can be achieved by the anisotropic structure's high polarization conversion [5,7,8,14,33,37,42]. Dynamically tuned electromagnetic metasurfaces using different materials properties can enable applications, such as information processing, anti-counterfeiting, and information encryption [43–48]. Inspired by the dynamic modulation of electromagnetic metasurfaces, the dynamic adjustment of structural color in the visible range by a chiral metasurface is feasible because of the synergistic effects of the chiral metasurface and polarized light [49–51].

In this work, we propose an all-dielectric chiral metasurface consisting of half-gammadion-shaped resonators embedded in PMMA and a top TiO_2 layer on quartz coated with an indium tin oxide (ITO) layer to realize the full-color gamut of the structural colors and anti-counterfeit functions. TiO_2 was selected as the material for the embedded chiral half-gammadion-shaped resonators in our design due to its near-transparency in the visible spectrum and moderate refractive index (2.4–2.7). We show a full-color gamut of the structural colors that extends well beyond sRGB by properly tuning the duty cycle (f_L) and period (P). The resonance peaks of the cross-polarization reflectance spectra depend on the period, thus being noticeable in color. Color mode and gray mode can be switched under the different polarization conditions. The anti-counterfeiting function is experimentally implemented by concealing the letter "A" and the Chinese word " 李龙杰" inside a star pattern.

2. Results and Discussions

The proposed embedded chiral metasurface is schematically depicted in Figure 1a, which unfolds each layer by section. It is a square lattice of TiO_2 chiral half-gammadion-shaped resonators embedded in PMMA and a top TiO_2 layer on quartz coated with an ITO layer. The spatial inversion symmetry in the two-dimensional plane is broken by the asymmetrically protruding arms of the chiral half-gammadion-shaped resonators on the left and right sides of the central rectangular bar. The square lattice has period P; the length and width of the chiral half-gammadion-shaped resonators are, respectively, L and W; the thickness of the top layer TiO_2 layer is $H1$; the height of the embedded TiO_2 chiral half-gammadion-shaped resonators and PMMA is $H2$; and the thickness of the ITO layer is $H3$, as shown in Figure 1b,c. We introduce two new factors: f_W and f_L, defined as the ratios of W over the $L/2$, and L over the P, respectively, to simplify the design and description. Unless otherwise stated, f_W, $H1$, $H2$, and $H3$ are, respectively, 0.6, 40 nm, 100 nm, and 50 nm. The embedded chiral metasurface was fabricated using a standard electron-beam lithography (EBL) and atomic layer deposition (ALD) technique followed by a selectable inductively coupled plasma (ICP)-etching process. To begin, the quartz substrate was cleaned in the ultrasound bath with acetone and absolute ethyl alcohol for 10 min, respectively, then rinsed well with deionized water. Then, a 50 nm thick ITO was deposited on the substrate with a magnetron sputtering process, improving the efficiency and saturation and facilitating the scanning electron microscope (SEM) inspection (see Figure S1). Following this step, a PMMA film was spin-coated onto the substrate and baked at 180 °C for 2 min, and the pattern layout was transferred to the PMMA resist by EBL. Next, TiO_2 was filled by the conformal ALD process, where the total required film thickness is larger than the half of the maximum width of the nanostructures. At last, the top TiO_2 layer with different thicknesses was etched by the ICP process. Our proposed embedded chiral metasurface is beneficial for the preservation and the long-term use of the metasurface. Figure 1d presents the scanning electron microscope (SEM) images of the typical arrays of the TiO_2 chiral half-gammadion-shaped resonators with the ratio f_L of 0.7 and periods of 220, 300, and 380 nm. It looks a little blurry because the PMMA and top TiO_2 layer are not conducive. The chiral metasurface is well-fabricated except for the slightly rounded corners.

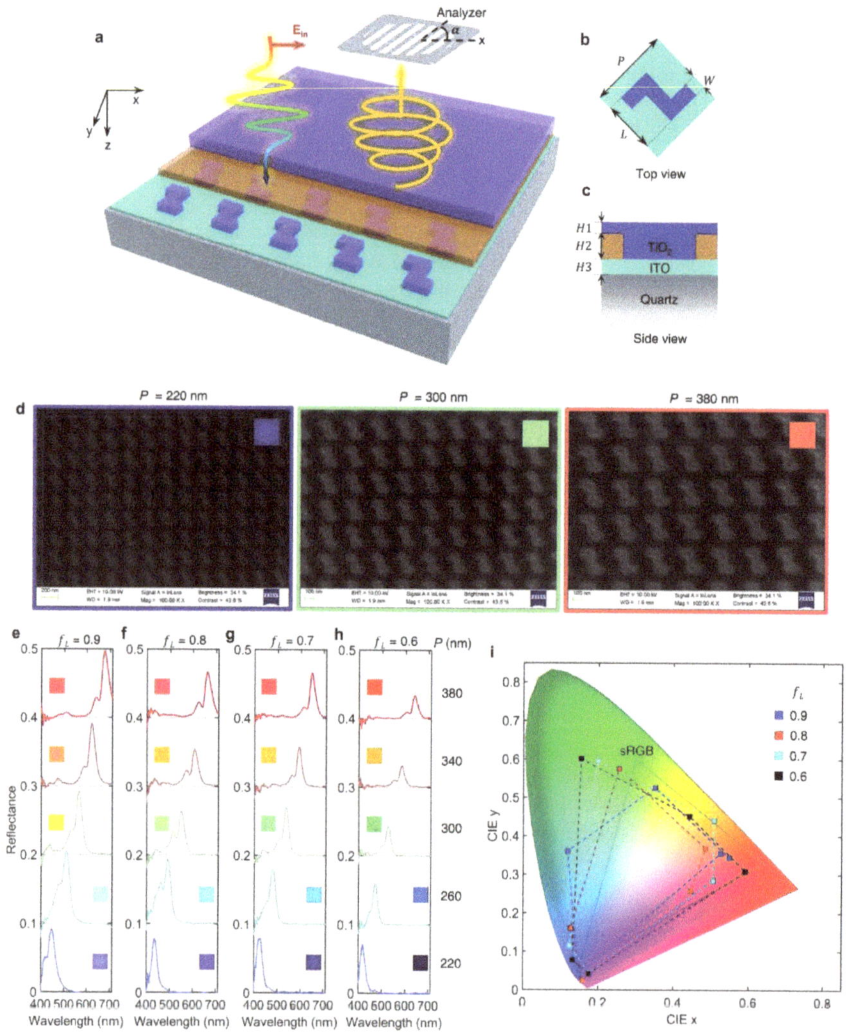

Figure 1. Full-color gamut of color-rendering performance of the embedded chiral metasurfaces. (**a**) Perspective view of the embedded chiral metasurface. (**b**) Top view and (**c**) side view of a single unit cell of the embedded chiral metasurface. (**d**) The top view SEM images of the embedded chiral metasurfaces with P of 220 nm, 300 nm, and 380 nm. The cross-polarization reflectance spectra and corresponding micrographs with f_L of (**e**) 0.9, (**f**) 0.8, (**g**) 0.7, and (**h**) 0.6 at different p from 220 nm to 380 nm. (**i**) The corresponding CIE 1931 chromaticity diagram of (**e**–**h**).

Figure 1e–h show the full-gamut color performance of the chiral metasurface with f_L from 0.9 to 0.6 and p from 220 nm to 380 nm at 40 nm intervals. The micrographs show that we successfully created full-color swatches in blue, cyan, yellow, green, orange, and red. The cross-polarization reflectance spectra are measured by a microscope and a fiber-coupled spectrometer. Here, the measured height of the top TiO_2 layer, PMMA layer, and ITO layer by ellipsometer are, respectively, about 40 nm, 100 nm, and 50 nm. The electric field polarization of the normally incident light is along the x-axis (E_{in}, red arrow in the oblique view schematic in Figure 1a), and there is a 45° angle between E_{in} and the arm of the half-gammadion-shaped resonator. In addition, an analyzer on the reflection direction has an angle α of 90° relative to the x-axis, and the reflectance spectra E_{out} along the analyzer direction are recorded. Figure 1e–h present that the resonance peak of the measured cross-polarization reflectance spectrum is significantly red-shifted with increasing periods for each fixed f_L, indicating the color hue mainly depending on the period. All resonance peaks have an accompanying small resonance peak to the left of the main resonance peak. For each fixed period, the resonance peak is slightly blue-shifted with f_L decreasing, and the reflection efficiency decreases. Figure 1i shows the calculated chromaticity coordinates by the reflectance spectra in Figure 1e–h, which cover the full-color gamut beyond the sRGB range. The blue, red, cyan, and black square markers are, respectively, for the nanostructures with an f_L of 0.9, 0.8, 0.7, and 0.6.

We also numerically calculate the cross-polarization reflectance spectra of the chiral metasurface and the electromagnetic field at the resonant wavelength. The simulation is performed by the finite-difference time-domain (FDTD) solver, EastWave. The simulation domain is a 3D unit cell consisting of a TiO_2 chiral half-gammadion-shaped resonator embedded in PMMA and a top TiO_2 layer on quartz coated with an ITO layer. The periodic boundary conditions are applied to the x- and y-directions and the PML boundary conditions are applied to the z-dimension to absorb the outgoing waves. Quartz's refractive index was acquired from the software, PMMA's was set to 1.5, and the refractive indices of the ITO and TiO_2 were measured by an ellipsometer (see details about optical constants of ITO and TiO_2 in Figure S2 of Figure S1 Materials). The incident light is along the x-axis, and the normalized cross-polarization reflectance power was calculated by the integration of the Poynting vector at the y-direction: $R_{cross} = \frac{1}{2}\int abs\left(\vec{P_y}\right)\cdot ds/P_s$, where P_s is the source power. Figure 2a–d show the simulated cross-polarization reflectance spectra with an f_L from 0.9 to 0.6 and p from 220 nm to 380 nm. All the simulated cross-polarization reflectance spectra show prominent sharp resonance peaks in the visible range and have very minimal background noise, which is essential for obtaining extremely saturated colors. More importantly, the resonance peak significantly redshifts with the period increasing and the reflectance efficiency decreases with the f_L decreasing. These are consistent with the experimental results, and the comparison of the simulated and measured results are detailed in Figure S3 in Supplementary Materials. The disparity in the simulation and measurement spectra is mainly from the difference in the source incident angle. Unlike the light source conditions of vertical incidence during the simulation, in the actual measurement, the light source is of a certain aperture angle, which leads to the broadening of the resonance peak, and at the same time, the absorption introduced by the fabrication will broaden the resonant peak. The calculated cross-polarization reflectance spectra are converted to chromaticity coordinates in the CIE 1931 chromaticity diagram in Figure 2e, which demonstrates that the colors can cover the full-color gamut far beyond that of sRGB. The blue, red, cyan, and black square markers are, respectively, for the nanostructures with an f_L of 0.9, 0.8, 0.7, and 0.6. The formation of resonance peaks is the hybridized mode from the embedded nanoresonator and the array effect [19], and we calculated the field distribution of the resonance peak at the red dot marker in Figure 2b, as shown in Figure 2f. The chiral metasurface is proposed due to its excellent polarization conversion properties, which makes it have good spectral properties under cross-polarization conditions to achieve high-saturation structural colors, and at the same time, our proposed embedded metasurface is less susceptible to damage and has the advantage that it can be used for a long time compared to other structural color work.

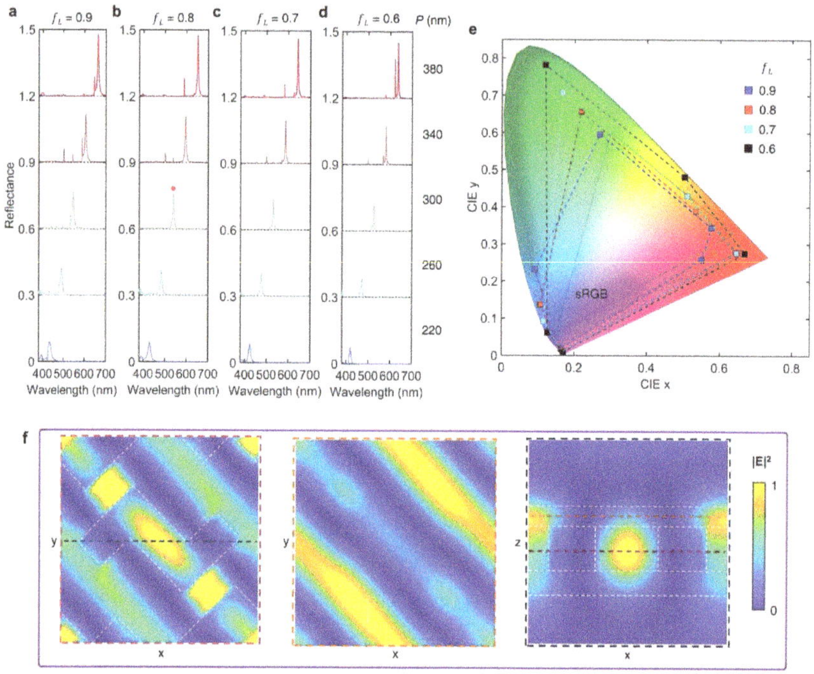

Figure 2. The color performance of simulated embedded chiral metasurface with different f_L and P. The cross-polarization reflectance spectra and corresponding micrographs with f_L of (**a**) 0.9, (**b**) 0.8, (**c**) 0.7, and (**d**) 0.6 at different p from 220 nm to 380 nm. (**e**) The corresponding CIE 1931 chromaticity diagram of (**a**–**d**). (**f**) The square of the absolute of the electric field distributions $|E|^2$ at the xy- and xz-plane of resonant peaks in (**b**) marked by a magenta dot.

The proposed embedded chiral metasurfaces have different modes under different observation conditions, such as those illustrated in Figure 3a,b. Figure 3a has a vibrant structural color covering the full-color gamut in Figure 3c under cross-polarization conditions with different periods from 220 nm to 400 nm, which we call the color mode. As a comparison, we observe the corresponding structural color at the angle of 135 degrees between the polarizer and the analyzer, which loses its bright color covering the full gamut. The change in the above conditions (the period and rotation angle of a single structure) does not change the hue but presents a grayish-yellow color, which we call the gray mode. The corresponding chromaticity diagram in Figure 3c also shows that its gamut changes very little and all the chromaticity coordinates are close to the white point, which is difficult to recognize with our naked eye. This feature allows us to use it to switch between the color and gray modes. The fabricated "rainbow" shows bright colors under cross-polarization conditions, as in Figure 3d, and dull colors in gray mode, as in Figure 3e. The stitching traces in Figure 3e are caused by two photographs due to the large area of "rainbow".

The proposed embedded chiral metasurface is anti-counterfeit due to its different color performance under different observation conditions (e.g., no polarization, single-polarization, and cross-polarization). As a proof of concept, we have designed and fabricated cryptographic nanoprints of a star encrypted with the letter "A" and the Chinese word " 李龙杰" which can only be decoded in the designated cross-polarization. The period parameters of the star, the Chinese word " 李龙杰", and the letter "A" are, respectively, 540 nm, 580 nm, and 620 nm. The other parameters f_L, f_W, $H1$, $H2$, and $H3$ are, respectively, 0.7, 0.6, 110 nm, 100 nm, and 50 nm. In the absence of polarization, which is the observation condition in our daily life, we can only observe a pink star, in which we can barely observe the Chinese word " 李龙杰" and letter "A" with the naked eye, as shown in Figure 4a. When the polarization direction is 0 or 90 degrees from the horizontal direction of the star, we can only see the internal information dimly, as shown in Figure 4b. When we observe using cross-polarization, the information will be decoded and we can visibly observe it with the naked eye, as shown in Figure 4c. The star changes colors and the embedded information becomes more or less clear as we rotate the sample. The letter "A" is easier to make out at any rotation angle, while the Chinese word " 李龙杰" is less distinct at 0 and 72 degrees, while it is very clear at 36, 108, and 144 degrees. In the case of cross-polarization, where the incident polarization and the reflected polarization are perpendicular to one another, only a few specific incident and reflected polarization tools are capable of fully decoding the encrypted information recorded in the embedded chiral metasurface. These phenomena offer platforms for use in cryptography and other secure applications. Figure 4d presents the SEM images of the full view of the entire star and one of the corners, where each pixel is made of the proposed embedded chiral metasurface.

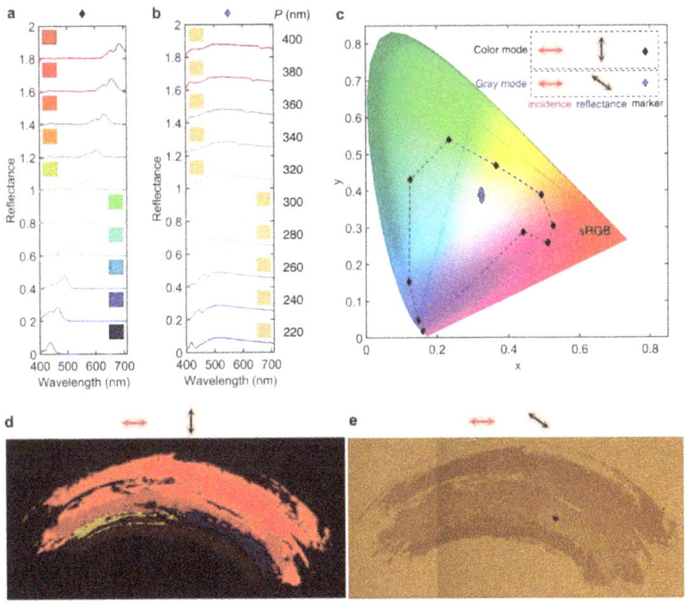

Figure 3. The color performance of fabricated embedded chiral metasurface under different observation conditions. (**a**) Reflectance spectra and micrographs of the embedded chiral metasurface under cross-polarization observation conditions (0 degree incidence, 90 degree reflectance). (**b**) Reflectance spectra and micrographs of the embedded chiral metasurface under observation conditions of 0 degree incidence and 135 degree reflectance. (**c**) The corresponding CIE 1931 chromaticity diagram of (**a**,**b**). Rainbow images in (**d**) color mode and (**e**) gray mode.

As an auxiliary validation, we calculated the reflectance spectra of the structures with periods of 540 nm, 580 nm, and 620 nm under different polarization light conditions, as shown in Figure 5a–c. The parameters of this design are the same as those of the previous section, with the exception that f_L and $H1$ are 0.7 and 110 nm, respectively. The polarization conditions for the three reflectance spectra are the polarized light parallel to the x-axis, calculating the single-polarized reflectance spectrum, as shown in Figure 5a; the polarized light perpendicular to x-axis, calculating the single-polarized reflectance spectrum, as shown in Figure 5b; and the polarized parallel to light x-axis, calculating the cross-polarized reflectance spectrum, as shown in Figure 5c. The cross-polarized reflectance spectra become more noticeable than that of the first two single-polarization shifts when the period is changed from 540 nm to 620 nm. In the 1931 CIE chromaticity diagram, it is obvious that, with the period increasing, the color from the cross-polarization reflectance changes (the blue square dot), while the colors from the single-polarized reflectance spectrum hardly change (red and cyan dots), as shown in Figure 5d. The anti-counterfeiting function is aided by such a spectral performance.

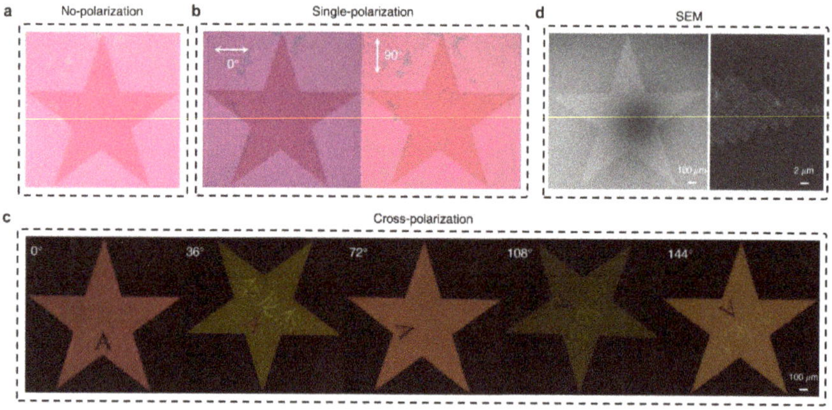

Figure 4. Anti-counterfeiting feature of the embedded chiral metasurfaces. (**a**) Micrograph of a star with encrypted information under no polarization. (**b**) Micrograph of a star with encrypted information under single-polarization (0 degrees or 90 degrees). (**c**) Micrograph of a star with encrypted information under cross-polarization with the sample rotated at an angle of 0, 36, 72, 108, and 144 degrees. (**d**) The top view SEM images of the entire star and one of the corners consist of embedded chiral metasurfaces.

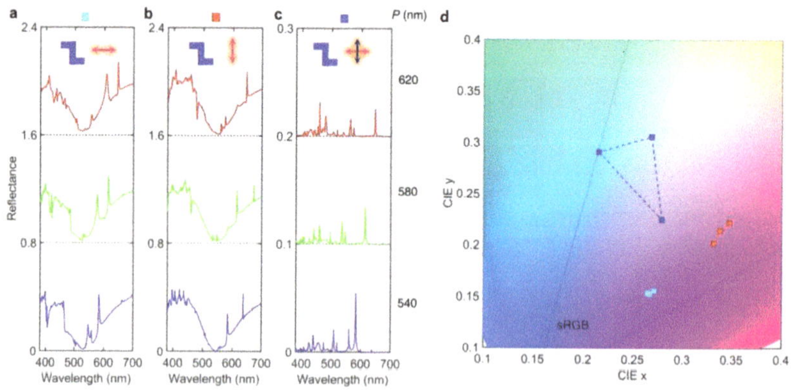

Figure 5. The color performance of embedded chiral metasurface with different *P* from 540 nm to 620 nm. The reflectance spectra under (**a**) single-polarization with 0 degrees to the x-axis, (**b**) single-polarization with 90 degrees with 90 degrees to the x-axis, (**c**) cross-polarization at different *p* from 540 nm to 620 nm. (**d**) The corresponding CIE 1931 chromaticity diagram of (**a**–**c**).

3. Conclusions

In summary, we have demonstrated the full-color gamut of the structural colors and anti-counterfeit functions of an all-dielectric chiral metasurface consisting of half-gammadion-shaped resonators embedded in PMMA and a top TiO_2 layer on quartz coated with an ITO layer. The excellent polarization conversion characteristics of the chiral structure are exploited, while the embedded structure is less susceptible to damage and thus better protected for the long-term use of the metasurface. Cross-polarization resonance peaks cross the visible range with the period p (from 220 nm to 380 nm) and f_L (from 0.6 to 0.9), resulting in highly saturated structured colors covering the full-color gamut considerably beyond sRGB. Moreover, we experimentally verify that the proposed embedded chiral metasurface can switch modes under different observation conditions: color mode and gray mode. Further, we designed and fabricated a star with internal hidden information including the letter "A" and the Chinese word " 李龙杰", achieving complete information decoding under cross-polarization, which is thus the anti-counterfeiting effect. This research has significant implications in micro-display, nano-printing, anti-counterfeiting, and information encryption.

Supplementary Materials: The following supporting information can be downloaded at: https://www.mdpi.com/article/10.3390/photonics10040401/s1, Figure S1: Comparision of reflectance with and without ITO layer; Figure S2: Optical constants of ITO and TiO_2; Figure S3: Comparison of simulated and measured cross-polarization reflectance spectra.

Author Contributions: Conceptualization, L.S.; methodology, L.L. and L.S.; software, L.L., H.L., X.S. and J.H.; validation, L.L., H.H., H.X., C.L., S.Z. and J.N. investigation, L.L.; resources, L.S.; writing—original draft preparation, L.L., L.S. and J.N.; writing—review and editing, L.L., L.S. and J.N.; visualization, L.L. and H.L.; funding acquisition, L.S. All authors have read and agreed to the published version of the manuscript.

Funding: The research was funded by the National Natural Science Foundation of China (Grant Nos. 61875225, 61888102).

Data Availability Statement: Not applicable.

Conflicts of Interest: The authors declare no conflict of interest.

References

1. Dean, N. Colouring at the nanoscale. *Nat. Nanotechnol.* **2015**, *10*, 15–16. [CrossRef] [PubMed]
2. Baranov, D.G.; Zuev, D.A.; Lepeshov, S.I.; Kotov, O.V.; Krasnok, A.E.; Evlyukhin, A.B.; Chichkov, B.N. All-dielectric nanophotonics: The quest for better materials and fabrication techniques. *Optica* **2017**, *4*, 814. [CrossRef]
3. Kumar, K.; Duan, H.; Hegde, R.S.; Koh, S.C.W.; Wei, J.N.; Yang, J.K.W. Printing colour at the optical diffraction limit. *Nat. Nanotechnol.* **2012**, *7*, 557–561. [CrossRef] [PubMed]
4. Lu, B.-R.; Xu, C.; Liao, J.; Liu, J.; Chen, Y. High-resolution plasmonic structural colors from nanohole arrays with bottom metal disks. *Opt. Lett.* **2016**, *41*, 1400. [CrossRef]
5. Wang, X.; Xu, D.; Jaquet, B.; Yang, Y.; Wang, J.; Huang, H.; Chen, Y.; Gerhard, C.; Zhang, K. Structural Colors by Synergistic Birefringence and Surface Plasmon Resonance. *ACS Nano* **2020**, *14*, 16832–16839. [CrossRef]
6. Shaltout, A.M.; Kim, J.; Boltasseva, A.; Shalaev, V.M.; Kildishev, A.V. Ultrathin and multicolour optical cavities with embedded metasurfaces. *Nat. Commun.* **2018**, *9*, 2673. [CrossRef]
7. Jiang, M.; Siew, S.Y.; Chan, J.Y.E.; Deng, J.; Wu, Q.Y.S.; Jin, L.; Yang, J.K.W.; Teng, J.; Danner, A.; Qiu, C.-W. Patterned resist on flat silver achieving saturated plasmonic colors with sub-20-nm spectral linewidth. *Mater. Today* **2020**, *35*, 99–105. [CrossRef]
8. Kim, Y.; Jung, K.; Cho, J.; Hyun, J.K. Realizing Vibrant and High-Contrast Reflective Structural Colors from Lossy Metals Supporting Dielectric Gratings. *ACS Nano* **2019**, *13*, 10717–10726. [CrossRef]
9. Jung, Y.; Jung, H.; Choi, H.; Lee, H. Polarization Selective Color Filter Based on Plasmonic Nanograting Embedded Etalon Structures. *Nano Lett.* **2020**, *20*, 6344–6350. [CrossRef]
10. Hail, C.U.; Schnoering, G.; Damak, M.; Poulikakos, D.; Eghlidi, H. A Plasmonic Painter's Method of Color Mixing for a Continuous Red–Green–Blue Palette. *ACS Nano* **2020**, *14*, 1783–1791. [CrossRef]
11. Shrestha, V.R.; Lee, S.-S.; Kim, E.-S.; Choi, D.-Y. Aluminum Plasmonics Based Highly Transmissive Polarization-Independent Subtractive Color Filters Exploiting a Nanopatch Array. *Nano Lett.* **2014**, *14*, 6672–6678. [CrossRef]
12. Tan, S.J.; Zhang, L.; Zhu, D.; Goh, X.M.; Wang, Y.M.; Kumar, K.; Qiu, C.-W.; Yang, J.K.W. Plasmonic Color Palettes for Photorealistic Printing with Aluminum Nanostructures. *Nano Lett.* **2014**, *14*, 4023–4029. [CrossRef]
13. Zhu, X.; Vannahme, C.; Højlund-Nielsen, E.; Mortensen, N.A.; Kristensen, A. Plasmonic colour laser printing. *Nat. Nanotechnol.* **2016**, *11*, 325–329. [CrossRef]
14. Song, M.; Wang, D.; Kudyshev, Z.A.; Xuan, Y.; Wang, Z.; Boltasseva, A.; Shalaev, V.M.; Kildishev, A.V. Enabling Optical Steganography, Data Storage, and Encryption with Plasmonic Colors. *Laser Photonics Rev.* **2021**, *15*, 2000343. [CrossRef]

15. Shi, L.; Niu, J.; Li, L.; Wang, C.; Shang, X.; Zhang, P.; Liu, Y.; Zhang, Y. Deep Subwavelength Wide-Angle Structural Colors at the Single Pixel Level. *Adv. Opt. Mater.* **2022**, *10*, 2200552. [CrossRef]
16. Yang, B.; Cheng, H.; Chen, S.; Tian, J. Structural colors in metasurfaces: Principle, design and applications. *Mater. Chem. Front.* **2019**, *3*, 750–761. [CrossRef]
17. Liu, X.; Huang, Z.; Zang, J. All-Dielectric Silicon Nanoring Metasurface for Full-Color Printing. *Nano Lett.* **2020**, *20*, 8739–8744. [CrossRef]
18. Yang, W.; Xiao, S.; Song, Q.; Liu, Y.; Wu, Y.; Wang, S.; Yu, J.; Han, J.; Tsai, D.-P. All-dielectric metasurface for high-performance structural color. *Nat. Commun.* **2020**, *11*, 1864. [CrossRef]
19. Jang, J.; Badloe, T.; Yang, Y.; Lee, T.; Mun, J.; Rho, J. Spectral Modulation through the Hybridization of Mie-Scatterers and Quasi-Guided Mode Resonances: Realizing Full and Gradients of Structural Color. *ACS Nano* **2020**, *14*, 15317–15326. [CrossRef]
20. Xu, Z.; Xu, Z.; Li, N.; Dong, Y.; Fu, Y.H.; Hu, T.; Zhong, Q.; Zhou, Y.; Li, D.; Zhu, S.; et al. Metasurface-based subtractive color filter fabricated on a 12-inch glass wafer using a CMOS platform. *Photon. Res. PRJ* **2021**, *9*, 13–20. [CrossRef]
21. Li, L.; Niu, J.; Shang, X.; Chen, S.; Lu, C.; Zhang, Y.; Shi, L. Bright Field Structural Colors in Silicon-on-Insulator Nanostructures. *ACS Appl. Mater. Interfaces* **2021**, *13*, 4364–4373. [CrossRef] [PubMed]
22. Liang, C.; Liang, C.; Deng, L.; Deng, L.; Dai, Q.; Li, Z.; Zheng, G.; Zheng, G.; Zheng, G.; Zheng, G.; et al. Single-celled multifunctional metasurfaces merging structural-color nanoprinting and holography. *Opt. Express* **2021**, *29*, 10737–10748. [CrossRef] [PubMed]
23. Lee, T.; Kim, J.; Koirala, I.; Yang, Y.; Badloe, T.; Jang, J.; Rho, J. Nearly Perfect Transmissive Subtractive Coloration through the Spectral Amplification of Mie Scattering and Lattice Resonance. *ACS Appl. Mater. Interfaces* **2021**, *13*, 26299–26307. [CrossRef] [PubMed]
24. Dong, Z.; Jin, L.; Rezaei, S.D.; Wang, H.; Chen, Y.; Tjiptoharsono, F.; Ho, J.; Gorelik, S.; Ng, R.J.H.; Ruan, Q.; et al. Schrödinger's red pixel by quasi-bound-states-in-the-continuum. *Sci. Adv.* **2022**, *8*, eabm4512. [CrossRef]
25. Badloe, T.; Kim, J.; Kim, I.; Kim, W.-S.; Kim, W.S.; Kim, Y.-K.; Rho, J. Liquid crystal-powered Mie resonators for electrically tunable photorealistic color gradients and dark blacks. *Light Sci. Appl.* **2022**, *11*, 118. [CrossRef]
26. Shang, X.; Niu, J.; Wang, C.; Li, L.; Lu, C.; Zhang, Y.; Shi, L. Mie Resonances Enabled Subtractive Structural Colors with Low-Index-Contrast Silicon Metasurfaces. *ACS Appl. Mater. Interfaces* **2022**, *14*, 55933–55943. [CrossRef]
27. Hentschel, M.; Koshelev, K.; Sterl, F.; Both, S.; Karst, J.; Shamsafar, L.; Weiss, T.; Kivshar, Y.; Giessen, H. Dielectric Mie voids: Confining light in air. *Light Sci. Appl.* **2023**, *12*, 3. [CrossRef]
28. Sun, S.; Zhou, Z.; Zhang, C.; Gao, Y.; Duan, Z.; Xiao, S.; Song, Q. All-Dielectric Full-Color Printing with TiO_2 Metasurfaces. *ACS Nano* **2017**, *11*, 4445–4452. [CrossRef]
29. Sun, S.; Yang, W.; Zhang, C.; Jing, J.; Gao, Y.; Yu, X.; Song, Q.; Xiao, S. Real-Time Tunable Colors from Microfluidic Reconfigurable All-Dielectric Metasurfaces. *ACS Nano* **2018**, *12*, 2151–2159. [CrossRef]
30. Koirala, I.; Lee, S.-S.; Choi, D.-Y. Highly transmissive subtractive color filters based on an all-dielectric metasurface incorporating TiO_2 nanopillars. *Opt. Express* **2018**, *26*, 18320. [CrossRef]
31. Yang, B.; Liu, W.; Li, Z.; Cheng, H.; Choi, D.-Y.; Chen, S.; Tian, J. Ultrahighly Saturated Structural Colors Enhanced by Multipolar-Modulated Metasurfaces. *Nano Lett.* **2019**, *19*, 4221–4228. [CrossRef]
32. Zhang, C.; Jing, J.; Wu, Y.; Fan, Y.; Yang, W.; Wang, S.; Song, Q.; Xiao, S. Stretchable All-Dielectric Metasurfaces with Polarization-Insensitive and Full-Spectrum Response. *ACS Nano* **2020**, *14*, 1418–1426. [CrossRef]
33. Huo, P.; Huo, P.; Song, M.; Song, M.; Zhu, W.; Zhu, W.; Zhang, C.; Chen, L.; Chen, L.; Lezec, H.J.; et al. Photorealistic full-color nanopainting enabled by a low-loss metasurface. *Optica* **2020**, *7*, 1171–1172. [CrossRef]
34. Eaves-Rathert, J.; Kovalik, E.; Ugwu, C.F.; Rogers, B.R.; Pint, C.L.; Valentine, J.G. Dynamic Color Tuning with Electrochemically Actuated TiO2 Metasurfaces. *Nano Lett.* **2022**, *22*, 1626–1632. [CrossRef]
35. Li, H.; Li, H.; Xu, Y.; Zhang, X.; Xiao, X.; Zhou, F.; Zhang, Z. All-dielectric high saturation structural colors enhanced by multipolar modulated metasurfaces. *Opt. Express* **2022**, *30*, 28954–28965. [CrossRef]
36. Yang, B.; Ma, D.; Liu, W.; Choi, D.-Y.; Li, Z.; Cheng, H.; Cheng, H.; Tian, J.; Chen, S.; Chen, S.; et al. Deep-learning-based colorimetric polarization-angle detection with metasurfaces. *Optica* **2022**, *9*, 217–220. [CrossRef]
37. Li, L.; Niu, J.; Wang, C.; Shang, X.; Xue, H.; Hu, J.; Li, H.; Lu, C.; Zhao, S.; Zhang, Y.; et al. High-Saturation Full-Color Printing with All-Dielectric Chiral Metasurfaces. *ACS Appl. Mater. Interfaces* **2023**.
38. Park, C.-S.; Koirala, I.; Gao, S.; Shrestha, V.R.; Lee, S.-S.; Choi, D.-Y. Structural color filters based on an all-dielectric metasurface exploiting silicon-rich silicon nitride nanodisks. *Opt. Express* **2019**, *27*, 667. [CrossRef]
39. Yang, J.-H.; Babicheva, V.E.; Yu, M.-W.; Lu, T.-C.; Lin, T.-R.; Chen, K.-P. Structural Colors Enabled by Lattice Resonance on Silicon Nitride Metasurfaces. *ACS Nano* **2020**, *14*, 5678–5685. [CrossRef]
40. Daqiqeh Rezaei, S.; Dong, Z.; You En Chan, J.; Trisno, J.; Ng, R.J.H.; Ruan, Q.; Qiu, C.-W.; Mortensen, N.A.; Yang, J.K.W. Nanophotonic Structural Colors. *ACS Photonics* **2021**, *8*, 18–33. [CrossRef]
41. Ko, J.H.; Yoo, Y.J.; Lee, Y.; Jeong, H.-H.; Song, Y.M. A review of tunable photonics: Optically active materials and applications from visible to terahertz. *iScience* **2022**, *25*, 104727. [CrossRef] [PubMed]
42. Kim, I.; Jang, J.; Kim, G.; Lee, J.; Badloe, T.; Mun, J.; Rho, J. Pixelated bifunctional metasurface-driven dynamic vectorial holographic color prints for photonic security platform. *Nat. Commun.* **2021**, *12*, 3614. [CrossRef] [PubMed]

43. Chen, L.; Ma, Q.; Luo, S.S.; Ye, F.J.; Cui, H.Y.; Cui, T.J. Touch-Programmable Metasurface for Various Electromagnetic Manipulations and Encryptions. *Small* **2022**, *18*, 2203871. [CrossRef] [PubMed]
44. Wang, H.; Qin, Z.; Huang, L.; Li, Y.; Zhao, R.; Zhou, H.; He, H.; Zhang, J.; Qu, S. Metasurface with dynamic chiral meta-atoms for spin multiplexing hologram and low observable reflection. *PhotoniX* **2022**, *3*, 10. [CrossRef]
45. Ma, Q.; Gao, W.; Xiao, Q.; Ding, L.; Gao, T.; Zhou, Y.; Gao, X.; Yan, T.; Liu, C.; Gu, Z.; et al. Directly wireless communication of human minds via non-invasive brain-computer-metasurface platform. *eLight* **2022**, *2*, 11. [CrossRef]
46. Mehmood, M.Q.; Seong, J.; Naveed, M.A.; Kim, J.; Zubair, M.; Riaz, K.; Massoud, Y.; Rho, J. Single-Cell-Driven Tri-Channel Encryption Meta-Displays. *Adv. Sci.* **2022**, *9*, 2203962. [CrossRef]
47. Bao, Y.; Yu, Y.; Xu, H.; Guo, C.; Li, J.; Sun, S.; Zhou, Z.-K.; Qiu, C.-W.; Wang, X.-H. Full-colour nanoprint-hologram synchronous metasurface with arbitrary hue-saturation-brightness control. *Light Sci.Appl.* **2019**, *8*, 95. [CrossRef]
48. Javed Satti, A.; Ashar Naveed, M.; Javed, I.; Mahmood, N.; Zubair, M.; Qasim Mehmood, M.; Massoud, Y. A highly efficient broadband multi-functional metaplate. *Nanoscale Adv.* **2023**, *5*, 2010–2016. [CrossRef]
49. Zhang, Y.; Shi, L.; Zhang, R.-Y.; Duan, J.; Ng, J.; Chan, C.T.; Fung, K.H. Metric-Torsion Duality of Optically Chiral Structures. *Phys. Rev. Lett.* **2019**, *122*, 200201. [CrossRef]
50. Choi, S.; Son, H.; Lee, B. Chirality-selective all-dielectric metasurface structural color display. *Opt. Express* **2021**, *29*, 41258–41267. [CrossRef]
51. Shu, X.; Li, A.; Hu, G.; Wang, J.; Alù, A.; Chen, L. Fast encirclement of an exceptional point for highly efficient and compact chiral mode converters. *Nat. Commun.* **2022**, *13*, 2123. [CrossRef]

photonics

MDPI

Article

Active Polarization Imaging for Cross-Linear Image Histogram Equalization and Noise Suppression in Highly Turbid Water

Huajun Zhang [1], Jianrui Gong [1], Mingyuan Ren [1], Ning Zhou [1], Hantao Wang [1], Qingguo Meng [2] and Yu Zhang [1,*]

[1] School of Physics, Harbin Institute of Technology, Harbin 150001, China
[2] School of Physics and Electronic Engineering, Harbin Normal University, Harbin 150025, China
* Correspondence: zhangyunn@hit.edu.cn

Abstract: The absorption and scattering of impurity particles in turbid water cause the target signal light to be attenuated and to produce backscattered light, resulting in the reduced quality of underwater polarimetric imaging. As water turbidity increases, the effect of backscattered light becomes greater, making polarization imaging in highly turbid water a challenge. Theory and experiment show that the increase in the intensity of backscattered light leads to high noise gain in the underwater active polarization imaging model. In order to enhance image contrast and suppress noise gain in highly turbid water, we propose an underwater imaging enhancement method that appropriately combines the non-physical and physical models. The method uses contrast limited adaptive histogram equalization (CLAHE) for a certain number of cross-linear images ($I_{\min}$) before calculating their polarization enhancement images, and it constructs joint filtering (multi-frame averaging and bilateral filtering) to suppress the high noise gain introduced by the imaging model and CLAHE. The experimental results in highly turbid water validate the rationality and feasibility of the proposed method, and the comparative processing results (52.7~98.6 NTU) outperform those of the conventional non-physical and physical model methods. The method maintains the complexity of the system and facilitates the application of conventional polarimetric imaging in harsher underwater environments.

Keywords: underwater imaging; polarimetric imaging; high-turbidity water; histogram equalization

Citation: Zhang, H.; Gong, J.; Ren, M.; Zhou, N.; Wang, H.; Meng, Q.; Zhang, Y. Active Polarization Imaging for Cross-Linear Image Histogram Equalization and Noise Suppression in Highly Turbid Water. *Photonics* **2023**, *10*, 145. https://doi.org/10.3390/photonics10020145

Received: 3 January 2023
Revised: 23 January 2023
Accepted: 26 January 2023
Published: 30 January 2023

1. Introduction

The underwater environment contains a large number of impurity particles, and light will be scattered and absorbed by the particles when propagating underwater, resulting in a reduction in the quality of underwater imaging, mainly in the form of reduced image brightness and loss of contrast. The turbidity in natural waters typically varies in the range of a few dozen to a few tens of nephelometric turbidity units (NTU) [1,2], and water is considered highly turbid when it is above 50 NTU [1]. Some underwater tasks [3–5] (e.g., submarine microbial imaging, pipeline maintenance, and shipwreck rescue) can rapidly increase the turbidity of the water and maintain it for a period of time due to the particles that are stirred up when operating close to the bottom area. This significantly reduces the effective distance of underwater imaging and severely hampers the execution of underwater missions. In recent years, in order to improve the quality of underwater imaging by reducing the effects of underwater backscattered light, different de-scattering imaging methods have been pioneered and are mainly divided into non-physical and physical model-based methods [6–9]. They provide strategies to address the different needs of underwater imaging accordingly.

Among the non-physical methods, the contrast limited adaptive histogram equalization (CLAHE) method is a common operation for underwater de-scattering [3,10]. Compared to using conventional histogram equalization (HE) to improve the operation of the dynamic range of the greyscales in the raw image, CLAHE operates on small regions of the image that are tiled, and it uses bilinear differences to reduce the effect of boundaries between the small regions, resulting in a more stable histogram equalization [11]. Thus, CLAHE can effectively enhance the contrast of the target signal light in the raw image. Among the physical model-based methods, the underwater polarimetric imaging model proposed by Schechner et al. is widely used for underwater imaging due to its compactness, ease of piggybacking, and operation [5,6,12]. It also belongs to the category of a time-sharing rotating polarization imaging system [3,13]. In order to obtain a better de-scattering effect, many optimization measures have been developed around the Schechner model, using, for example, optical correlation [14–16] and extrapolation fitting [17] and finding feasibility intervals [18] to optimize the selection of relevant parameters, as well as combining Mueller matrices [19,20] and computer algorithms [21–23] to suppress the backscattered light. In highly turbid water, Liu et al. found that red light was more effective than other wavelengths in suppressing scattering, and they combined this with active polarization imaging to make the leap from "undetectable" to "detectable" [1]. At the same time, Hu et al. cleverly applied the memory effect of circularly polarized light to underwater polarization imaging and realized target detection in highly turbid water through the combined action of linearly and circularly polarized light [24]. In contrast to Liu's and Hu's methods of modulating the light source and increasing the complexity of the imaging system, Li et al. used a histogram stretching operation with greyscale normalized for a cross-linear image (I_{min}) and combined it with polarimetric imaging to achieve experimental results in highly turbid water that were superior to those of the traditional methods [25]. Considering the high backscattered light intensity [26] and high noise gain characteristics in highly turbid water [6], the raw image will have overexposed regions due to backscattered light enhancement, and CLAHE will be more limited in its processing. In addition, normalized histogram stretching can weaken the contrast enhancement effect of the target signal light under high noise conditions. Therefore, it is necessary to design a method for polarization enhancement in highly turbid water without increasing the complexity of the system.

In this paper, the problem of strong backscattering and the high noise gain of the imaging model caused by it under high-turbidity conditions is discussed theoretically and experimentally. Based on the presence of two influencing factors, we appropriately combine the non-physical and the physical models and propose an active polarization imaging method based on the CLAHE histogram equalization of the cross-linear image and the joint noise suppression. CLAHE is used to equalize the intensity distribution of the target signal light in I_{min}. Meanwhile, the polarization enhancement results are processed using a combination of multi-frame averaging [27,28] and bilateral filtering [29] to counteract the high noise gain generated by the imaging model and CLAHE. Only a certain number of I_{min} need to be acquired in our method to achieve the multi-frame averaging effect, and experimentally, we discuss the dependence of the noise suppression effect on the number of averaged frames. Finally, real-world experiments in different water samples with high turbidity are set up to verify the effectiveness and advantages of our method.

2. Methodology

2.1. Underwater Polarization Imaging Model

Turbid water bodies contain a large number of impurities, including soluble substances as well as insoluble particulate matter and microorganisms. The light shining on the object in the turbid water is absorbed and scattered by the water molecules and impurities, resulting in the received image degradation. The attenuation of the target signal light intensity and the backscattered light covering the object result in a loss of contrast in the image. According to the underwater imaging model, the received image is the incoherent sum of the target signal light and the backscattered light [6].

$$I(x,y) = S(x,y) + B(x,y) \quad (1)$$

where $S(x,y)$ denotes the target signal light, derived from the irradiance of the target object. $B(x,y)$ denotes the backscattered light. When $S(x,y)$ propagates in the water, it is affected by two interfering factors; one is the absorption of water molecules and impurities, resulting in its intensity attenuation; the other is caused by the scattering of impurities by the backscattered light $B(x,y)$ superimposed on $S(x,y)$, resulting in a reduction in the image contrast, ultimately leading to a decline in imaging quality. Backscattered light has the characteristics of partial polarization; so, rotating the analyzer before the receiver will obtain the darkest (cross-linear) and brightest (co-linear) two orthogonal polarization images, noted as $I_{\min}$ and $I_{\max}$. According to Equation (1), $I_{\min}$ and $I_{\max}$ can be expressed as:

$$\begin{cases} I_{\min}(x,y) = S_{\min}(x,y) + B_{\min}(x,y) \\ I_{\max}(x,y) = S_{\max}(x,y) + B_{\max}(x,y) \end{cases} \quad (2)$$

$S_{\min}(x,y)$ and $S_{\max}(x,y)$ denote the darkest and brightest target signal light in the orthogonal polarization images, and $B_{\min}(x,y)$ and $B_{\max}(x,y)$ denote the darkest and brightest backscattered light in the orthogonal polarization images. As partial polarization light can be decomposed in two orthogonal directions, the total light intensity $I(x,y)$ can be expressed as:

$$I(x,y) = I_{\min}(x,y) + I_{\max}(x,y) \quad (3)$$

According to the definition of the degree of polarization (DOP), the DOP of the target signal light p_{targ} and the backscattered light p_{scat} can be calculated from the corresponding components in $I_{\min}(x,y)$ and $I_{\max}(x,y)$,

$$\begin{cases} p_{targ} = \frac{S_{\max} - S_{\min}}{S_{\max} + S_{\min}} \\ p_{scat} = \frac{B_{\max} - B_{\min}}{B_{\max} + B_{\min}} \end{cases} \quad (4)$$

The expressions for the target signal light and the backscattered light can be obtained by combining Equations (1)–(4),

$$\begin{cases} S(x,y) = \frac{1}{p_{scat} - p_{targ}} [I_{\min}(1 + p_{scat}) - I_{\max}(1 - p_{scat})] \\ B(x,y) = \frac{1}{p_{scat} - p_{targ}} [I_{\max}(1 - p_{targ}) - I_{\min}(1 + p_{targ})] \end{cases} \quad (5)$$

2.2. Noise Analysis of Polarization Imaging in High-Turbidity Water

There are three main components of the noise sources in the imaging process; these are readout noise, dark current noise, and photon noise. Define g_{electr} as the number of photogenerated electrons in the imaging system that cause a change per unit of gray level. Therefore, the noise variance of the readout noise can be represented by ρ^2/g_{electr}^2, where ρ is the standard deviation (STD) of the electronic readout noise of the imaging system. Dark current noise is related to the exposure time of the imaging system, and its noise variance is expressed by Dt/g_{electr}^2, where D represents the magnitude of the dark current, and t is the exposure time. Photon noise can be denoted as $I(x,y)/g_{electr}$, which is the signal-related noise independent of the parameters and quality of the camera. Thus, for an image, the noise variance of its pixel gray level is [6,30],

$$\sigma_I^2 = \rho^2/g_{electr}^2 + Dt/g_{electr}^2 + I(x,y)/g_{electr} \quad (6)$$

The underwater polarization imaging model relies on two polarization orthogonal sub-images, $I_{\min}$ and $I_{\max}$, which are considered as two statistically independent measurements and are linearly related to $S(x,y)$ and $B(x,y)$ in Equation (5). Assuming that the noise variances of $I_{\min}$ and $I_{\max}$ are $\sigma_{\min}^2$ and $\sigma_{\max}^2$, respectively, and combining Equations (5) and

(6), we can obtain the following relationship between the noise variance of the target signal light and the two polarization orthogonal sub-images,

$$\begin{aligned}\sigma_S^2 &= \left(\frac{1+p_{scat}}{p_{targ}-p_{scat}}\right)^2 (\rho^2/g_{electr}^2 + Dt/g_{electr}^2 + I_{min}(x,y)/g_{electr}) \\ &+ \left(\frac{1-p_{scat}}{p_{targ}-p_{scat}}\right)^2 (\rho^2/g_{electr}^2 + Dt/g_{electr}^2 + I_{max}(x,y)/g_{electr}) \\ &= \left(\frac{1+p_{scat}}{p_{targ}-p_{scat}}\right)^2 \sigma_{min}^2 + \left(\frac{1-p_{scat}}{p_{targ}-p_{scat}}\right)^2 \sigma_{max}^2\end{aligned} \tag{7}$$

For Equation (7), we consider it in two cases; first, we assume that the non-signal-related noise is dominant in the imaging process; that is, when $\sigma_{min}^2 = \sigma_{max}^2 = \sigma_0^2$, the noise variance of the target signal light becomes of the form,

$$\begin{aligned}\sigma_S^2 &= \left(\frac{1+p_{scat}}{p_{targ}-p_{scat}}\right)^2 \sigma_0^2 + \left(\frac{1-p_{scat}}{p_{targ}-p_{scat}}\right)^2 \sigma_0^2 \\ &= 2\sigma_0^2\left[\frac{1+p_{scat}^2}{(p_{targ}-p_{scat})^2}\right]\end{aligned} \tag{8}$$

From Equation (8), it can be seen that the noise in the target signal light gains constantly, and the magnitude of the noise variance is related to p_{scat} and Δp ($\Delta p = p_{scat} - p_{targ}$, the difference between p_{scat} and p_{targ}). The variation of the noise amplification σ_S/σ_0 in the target signal light with p_{scat} and p_{targ} is shown in Figure 1a. It can be seen that when the DOP of the backscattered light is $p_{scat} = 0$ and the DOP of the target signal light is $p_{targ} = 1$, the noise amplification of the target signal light reaches the minimum value of $\sqrt{2}$. However, when the two DOPs gradually approach, that is, $\Delta p \to 0$, the noise is continuously increased. Figure 1b,c show the changes in p_{scat}, p_{targ}, and Δp measured in the experiment with the increasing water turbidity and the increasing imaging distance, respectively. p_{scat} and p_{targ} are calculated according to the DOP definition after pre-labeling the target and the background regions in the experiment. It can be seen that in the imaging process of the increasing turbidity and distance of the water, the value of p_{targ} has a tendency to gradually approach p_{scat}; so, Δp gradually decreases, which leads to the continuous increase in the noise amplification rate in the recovered target signal light.

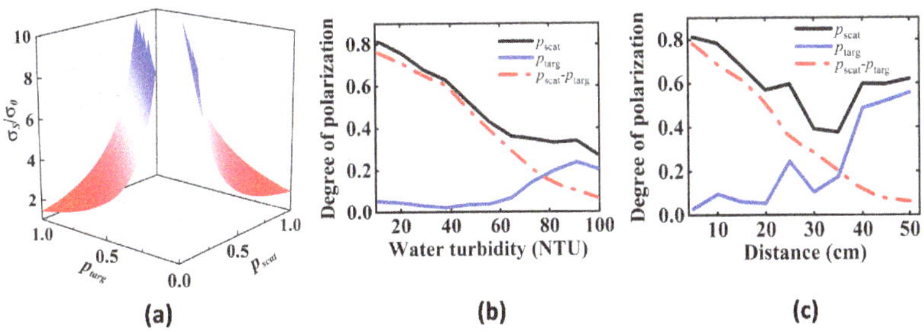

Figure 1. (**a**) Variation of noise amplification of target signal light with p_{scat} and p_{targ}; (**b**) variation trend of p_{scat}, p_{targ}, and Δp with increasing turbidity of the water; (**c**) variation trend of p_{scat}, p_{targ}, and Δp with increasing imaging distance in the water with turbidity of 34.0 NTU.

Next, we consider the practical case where the photon noise is not negligible, i.e., $\sigma_{min} \neq \sigma_{max}$. From the expression for $S(x,y)$ in Equation (5), it is clear that the contribution of p_{targ} to $S(x,y)$ is a scaling factor that adjusts the global intensity. Here, we assume that $p_{targ} = 0$ [6], in which case Equation (5) can be simplified to,

$$\begin{cases} S(x,y) = \frac{1}{p_{scat}}[I_{min}(x,y)(1+p_{scat}) - I_{max}(x,y)(1-p_{scat})] \\ B(x,y) = \frac{1}{p_{scat}}[I_{max}(x,y) - I_{min}(x,y)] \end{cases} \tag{9}$$

According to Equation (9), it is obtained that

$$\begin{cases} I_{\min}(x,y) = \frac{1}{2}[B(x,y)(1-p_{\text{scat}}) + S(x,y)] \\ I_{\max}(x,y) = \frac{1}{2}[B(x,y)(1+p_{\text{scat}}) + S(x,y)] \end{cases} \tag{10}$$

Through Equations (10) and (7), the noise variance expression of $S(x,y)$ with respect to the target signal light and backscattered light can be obtained,

$$\sigma_S^2 = \frac{1}{p_{\text{scat}}^2}\left\{\left[2\left(\rho^2/g_{\text{electr}}^2 + Dt/g_{\text{electr}}^2\right) + \frac{S(x,y)}{g_{\text{electr}}}\right]\left(1+p_{\text{scat}}^2\right) + \frac{B(x,y)}{g_{\text{electr}}}\left(1-p_{\text{scat}}^2\right)\right\} \tag{11}$$

From Equation (11), we can see that the noise variance of the target signal light $S(x,y)$ is related to the intensity of the backscattered light $B(x,y)$. Therefore, it is beneficial to reduce the intensity of the backscattered light appropriately in the actual scene to improve the signal-to-noise ratio (SNR) of the target signal light. Figure 2 shows the raw images, $I_{\min}$ and $I_{\max}$, acquired under different high-turbidity environments and the DOP distribution maps calculated using Equation (4). In fact, as the turbidity of the water increases, the content of impurity particles in the medium continues to rise, resulting in an increased probability of the photons interacting with the impurity particles. It can be seen from the raw images and the $I_{\max}$ in Figure 2 that the backscattered light saturates the receiver response to light intensity when a certain turbidity is reached. In addition, as the turbidity increases, the large-sized particles in the water will also increase, resulting in an increase in the scattering coefficient of the medium, and the photons gradually lose their initial polarization state [26]. Consequently, the intensity of the backscattered light in $I_{\min}$ increases, as shown in $I_{\min}$ in Figure 2. At the same time, the DOP of the backscattered light decreases, as shown in the DOP distribution maps in Figure 2.

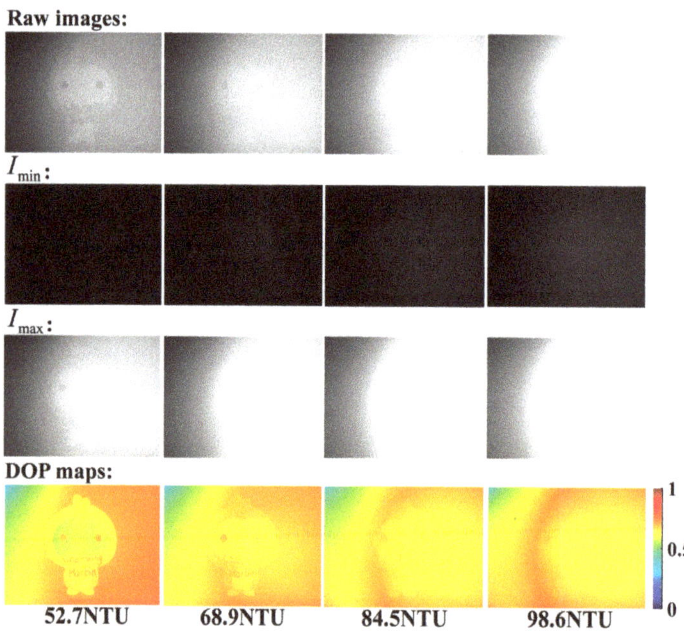

Figure 2. Raw images (first row), $I_{\min}$ (second row) and $I_{\max}$ (third row), and the calculated DOP distribution maps (last row) acquired in four high-turbidity environments.

The noise analysis of the underwater polarization imaging model shows that there are two main sources of noise in highly turbid water; one is the change in the DOP, and the other is the increase in the intensity of the backscattered light. Together, they cause a reduction in the SNR of the recovered image in a highly turbid environment. In addition, it can be seen from I_{min} in Figure 2 that the depolarization phenomenon of the backscattered light in the highly turbid water makes the target signal in I_{min} gradually submerge in the backscattered light. Therefore, in order to improve the quality of polarization differential imaging in highly turbid water, two aspects need to be addressed; one is contrast enhancement, and the other is noise suppression.

2.3. CLAHE-Based Cross-Linear Image Histogram Equalization and Joint Noise Suppression

According to the analysis in the previous section, the polarization enhancement effect of the target signal light in highly turbid water is affected by two aspects; one is the increased backscattered light intensity, and the other is the high noise gain with the increase in turbidity. Aiming at these two effects, we propose an underwater polarization differential imaging enhancement method based on CLAHE-based cross-linear image histogram equalization and joint noise suppression to improve the imaging quality. Figure 3 shows the processing flowchart of the entire method.

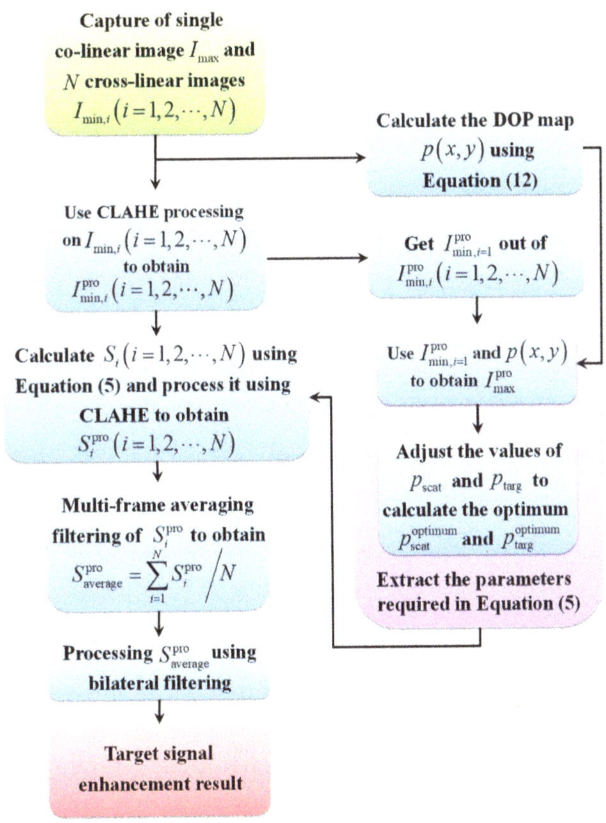

Figure 3. The flowchart of underwater polarization differential imaging enhancement based on CLAHE cross-linear image histogram equalization and joint noise suppression.

First, adjust the analyzer in front of the receiver to a co-linear state and capture a co-linear image I_{max}. Then, adjust the analyzer to the cross-linear state and continuously capture N cross-linear images $I_{\min,i}(i = 1, 2, \ldots, N)$. Using the captured $I_{\max}$ and $I_{\min,i=1}$, and according to the definition of the DOP, the DOP distribution map can be obtained,

$$p(x, y) = \frac{I_{\max}(x, y) - I_{\min,i=1}(x, y)}{I_{\max}(x, y) + I_{\min,i=1}(x, y)} \tag{12}$$

Meanwhile, the p_{scat} and p_{targ} required for the polarization differential processing can be obtained according to Equation (4).

As the grayscale of the image in the turbid water is compressed to a very narrow range, especially when the water turbidity increases, the intensity of the backscattered light on the target signal light increases. This results in a grayscale overexposed area in the $I_{\max}$ image and the attenuation of the target signal light intensity in the $I_{\min}$ image, ultimately causing further compression of the gray space of the captured image. To mitigate the extent to which the image grayscale is compressed, a grayscale histogram equalization of $I_{\min,i}(i = 1, 2, \ldots, N)$ is performed by using CLAHE to obtain the adjusted grayscale $I^{\text{pro}}_{\min,i}(i = 1, 2, \ldots, N)$,

$$I^{\text{pro}}_{\min,i}(i = 1, 2, \ldots, N) = \text{CLAHE}[I_{\min,i}(i = 1, 2, \ldots, N)] \tag{13}$$

It is worth noting that the $I_{\max}$ image acquired in highly turbid water contains a large number of grayscale overexposed areas, which is not conducive to the grayscale histogram equalization operation. Therefore, only $I_{\min,i}(i = 1, 2, \ldots, N)$ needs to be processed during the actual processing. The DOP distribution map $p(x, y)$ reflects the polarization relationship between two polarization orthogonal images and contains the information on the target and the backscattered light. To preserve the polarization relationship [25], the processed $I^{\text{pro}}_{\max}$ can be obtained using Equation (12),

$$I^{\text{pro}}_{\max}(x, y) = \frac{1 + p(x, y)}{1 - p(x, y)} I^{\text{pro}}_{\min,i=1}(x, y) \tag{14}$$

Based on the already obtained $I^{\text{pro}}_{\min,1}$, $I^{\text{pro}}_{\max}$, p_{scat}, and p_{targ}, the recovered target signal light can be obtained by using Equation (5). However, due to the high-turbidity conditions, the difference between the two DOPs gradually decreases, which reduces the effect of the common-mode suppression and leads to the poor recovery of the target signal light. To improve the image enhancement result, the values of p_{scat} and p_{targ} are adjusted by setting $\varepsilon_1 \in \left[0, \frac{1}{p_{\text{scat}}}\right]$ and $\varepsilon_2 \in \left[0, \frac{1}{p_{\text{targ}}}\right]$. The values of the two DOPs are optimal at this time when the values of ε_1 and ε_2 are taken so that the grayscale histogram of the recovered image $S(x, y)$ is the most uniform, i.e., $p^{\text{optimum}}_{\text{scat}} = \varepsilon^{\text{optimum}}_{\text{scat}} p_{\text{scat}}$ and $p^{\text{optimum}}_{\text{targ}} = \varepsilon^{\text{optimum}}_{\text{targ}} p_{\text{targ}}$.

With the known results given above, Equation (5) can be used to calculate the N processed target signal light results obtained from $I^{\text{pro}}_{\min,i}(i = 1, 2, \ldots, N)$, as follows:

$$\begin{aligned} S_i(i = 1, 2, \ldots, N) &= \frac{1}{p^{\text{optimum}}_{\text{scat}} - p^{\text{optimum}}_{\text{targ}}} \left[I^{\text{pro}}_{\min,i}(i = 1, 2, \ldots, N)\left(1 + p^{\text{optimum}}_{\text{scat}}\right) \right. \\ &\quad \left. - I^{\text{pro}}_{\max}\left(1 - p^{\text{optimum}}_{\text{scat}}\right)\right] \end{aligned} \tag{15}$$

The N $S_i(i = 1, 2, \ldots, N)$ obtained by using Equation (15) are then used with CLAHE to perform histogram equalization; so, there are,

$$S^{\text{pro}}_i(i = 1, 2, \ldots, N) = \text{CLAHE}[S_i(i = 1, 2, \ldots, N)] \tag{16}$$

Next, in order to reduce the gained noise in the target signal light, the N histogram equalized $S_i^{\text{pro}}(i = 1, 2, \ldots, N)$ is subjected to multi-frame averaging processing, and then, bilateral filtering is used,

$$\text{BF}\left(S_{\text{average}}^{\text{pro}}\right) = \text{BF}\left(\sum_{i=1}^{N} S_i^{\text{pro}} \Big/ N\right) \tag{17}$$

Here, $\text{BF}(\cdot)$ denotes the bilateral filtering operation, and $\text{BF}\left(S_{\text{average}}^{\text{pro}}\right)$ is the result of polarization enhancement by equalizing the histogram of the cross-linear polarization image and combining the noise suppression processing.

3. Real-World Experiment and Results

3.1. Experimental Setup

The experimental setup is shown in Figure 4, showing the schematic diagram and a photograph of the experimental setup, respectively. The emitting end of the imaging system is composed of a white LED light source (Supfire, X90, China, wavelength in the visible range) and a rotatable linear polarizer (Sigma, USP-30C0.4–38, Japan). The illumination beam is transformed into a stable output of linearly polarized light after passing through the polarizer. The experimental environment is a glass water tank of 60 cm (length) × 35 cm (width) × 40 cm (height), and the liquid level during the experiment is 25 cm. Milk is mainly divided into skimmed milk, semi-skimmed milk, and whole milk, which contain casein molecules (size 0.04 ~ 0.3 μm) and fat globules (size 1 ~ 20 μm) of different sizes [14,31]. The scattering types of the milk are mainly divided into Mie scattering and Rayleigh scattering, which is similar to the scattering type of seawater. As the scattering coefficient of the same volume of skimmed milk is smaller [31], which facilitates the preparation of water with different turbidities, the experiment uses skimmed milk to simulate different underwater turbid environments. The illumination beam passes through the medium to the target object (the target is 20 cm away from the wall of the water tank at the transmitting end) and is reflected and finally reaches the receiving end of the imaging system. The receiving end consists of a rotatable analyzer (Sigma, USP-30C0.4–38, Japan) and a monochrome charge-coupled device (CCD) camera (Daheng Optics, R-125–30UM, China, wavelength response range 400 ~ 1000 nm). During the acquisition process, we rotate the analyzer to acquire a co-linear polarization image and multiple cross-linear images, respectively. The acquisition time interval of the multi-frame images is 10 ms, and the image size is 964 (width) × 1292 (height) pixels.

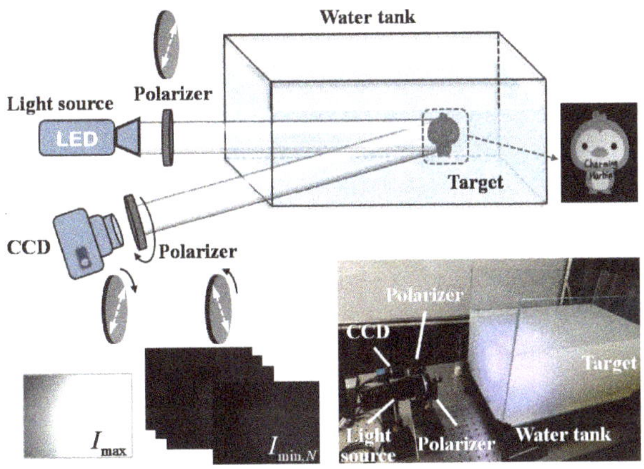

Figure 4. Underwater polarization imaging experimental setup.

3.2. Rationality and Feasibility Analysis of the Processing Flow

To verify the rationality and feasibility of the processing process in the proposed method, we added 200 mL of skimmed milk to the water tank; the turbidity of the water was 84.5 NTU. The analysis in Section 2.2 shows that as the turbidity of the water increases the backscattered light is continuously enhanced, which makes a large overexposure area in I_{max}. With the existing histogram equalization operation, it would be difficult to achieve a better processing effect. Therefore, our method chooses to perform the histogram equalization operation on I_{min}. At the same time, the continuous acquisition of I_{min} in a short period of time ensures the statistical characteristics of the noise distribution. Additionally, the DOP distribution map $p(x, y)$ in the whole scene can be considered as stable; so, we acquired 1 frame of I_{max} and 200 frames of I_{min} (with an acquisition interval of 10 ms).

According to the processing flow, the CLAHE operation is first performed on the captured I_{min}. CLAHE itself belongs to the classical approach to dealing with de-scattering in the non-physical model [3,10], and here, we combine it appropriately with the classical physical de-scattering model. CLAEH is a variant of the adaptive histogram equalization (AHE) algorithm. Compared with AHE, its contrast amplification is limited, which can better preserve the overall contrast distribution of the raw image and effectively reduce the noise amplification problem [11]. However, the additional noise introduced by CLAHE still needs to be considered. Figure 5a,d show the raw intensity image and the normalized grayscale histogram of I_{min}. Figure 5b,e show the results and the normalized grayscale histogram of I_{min} after processing using Li's method [25], and Figure 5c,f show the results of the I_{min} processing using CLAHE [11]. From Figure 5, it can be seen that both Li's method and the CLAHE method improve the contrast of the target signal in the image compared to the raw intensity of I_{min}. The grayscale distribution ranges of the two methods are basically the same, but the grayscale distribution after CLAHE processing is more uniform.

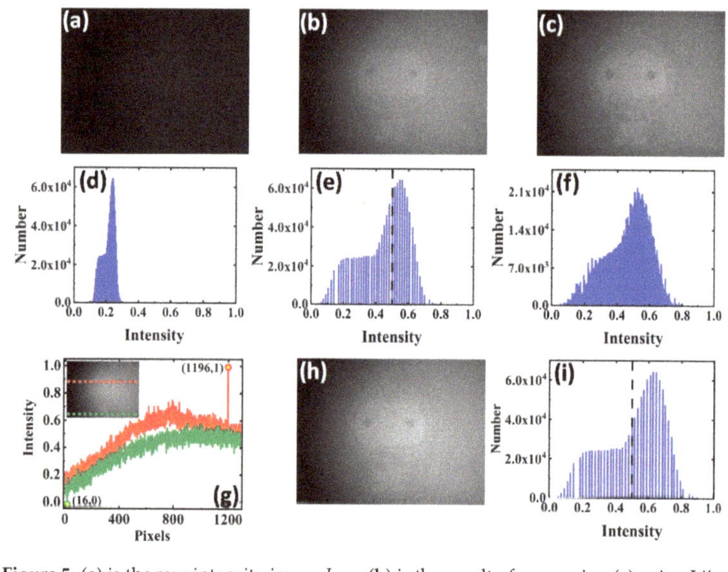

Figure 5. (**a**) is the raw intensity image I_{min}; (**b**) is the result of processing (**a**) using Li's method [25]; (**c**) is the result of processing (**a**) using CLAHE [11]; (**d**) is the normalized grayscale histogram of (**a**); (**e**) is the normalized grayscale histogram of (**b**); (**f**) is the normalized grayscale histogram of (**c**); (**g**) is the grayscale value row indexes of the extreme points in (**b**); (**h**) is (**b**) the grayscale normalization result after removing the extreme point; (**i**) is the normalized grayscale histogram of (**h**).

It is worth noting that the image acquired in the highly turbid water contains a large amount of noise, which may cause some grayscale extreme points in the image. Therefore, the histogram stretching operation for the grayscale normalization of I_{min} using Li's method will be disturbed. Figure 5g shows the row indexes of the grayscale values in Figure 5b that are through the minimum (green line) and the maximum (red line), and the presence of these extreme value points will interfere with the grayscale normalization effect of Li's method. Therefore, we remove the extreme value points and then normalize to obtain Figure 5h, and the grayscale histogram is shown in Figure 5i. It can be seen that the grayscale distribution of the region to the right of the black dashed line in Figure 5i is higher than that in Figure 5d, which proves that removing the extreme points is effective for enhancing the image brightness. However, the removal of the extreme points is relatively troublesome and uncertain in the actual processing, which is not conducive to the automation of the processing process. Therefore, we use CLAHE to improve the contrast of the target signal light in I_{min} in the highly turbid water scene. It should be emphasized that the original scene of the DOP distribution map $p(x,y)$, which contains the polarization information of the water, was calculated before processing I_{min}; so, it will not change the polarization relationship between I_{min}^{pro} and I_{max}^{pro} [25].

The reason for using the multi-frame averaging method as the first step of joint noise reduction is twofold: the first reason is that after the CLAHE process not only is the contrast of the target signal light improved, but also the noise in the image is gained, causing the overall peak signal-to-noise ratio (PSNR) of the image to decrease. Figure 6b,d show the results obtained using Li's method and the S_1^{pro} obtained using our method, respectively, where Figure 6a shows the target image in clear water. Figure 6c,e show the spatial frequency spectrum of Figure 6b,d, respectively. It can be seen from the PSNR and the spectrogram that the noise amplification in the results containing the CLAHE processing step is significantly higher than that of Li's method.

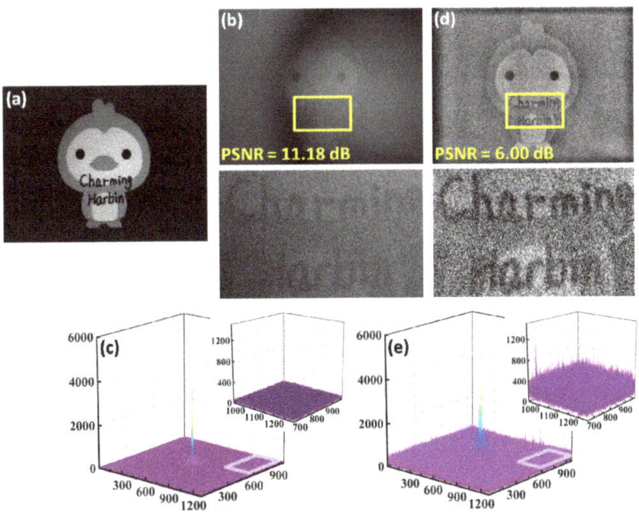

Figure 6. (**a**) is the target image in clear water; (**b**) is the result of Li's method [25]; (**c**) is the spatial frequency spectrum of (**b**); (**d**) is the S_1^{pro} obtained by our method; (**e**) is the spatial frequency spectrum of (**d**).

The second reason is that the polarization differential imaging itself belongs to the time-sharing rotating polarization imaging system [3,13], and the polarization images in the different directions are obtained by rotating the analyzer placed in front of the receiver. Hence, the imaging process itself does not exclude the noise suppression operation of the multi-frame averaging. At the same time, the method of multi-frame averaging has been widely used in reducing imaging noise [27,28]. In addition, multi-frame averaging preserves the high-frequency details of the image well and maintains the sharpness of the image compared to the methods that directly use low-pass filtering (such as Gaussian low-pass filtering [32]). It is worth noting that in our method, only a certain number of images $I_{\min}$ need to be continuously acquired to achieve the multi-frame averaging effect. We therefore choose multi-frame averaging as the first step in joint noise suppression. However, multi-frame averaging has certain limitations in the improvement of the image SNR. The SNR of N images after multi-frame averaging is,

$$I_{\mathrm{SNR}}^{N} = \sqrt{N}\frac{I \cdot \eta_{\mathrm{QE}}}{\sqrt{I \cdot \eta_{\mathrm{QE}} + I_{\mathrm{RON}}{}^{2} + D \cdot t}} \tag{18}$$

Here, I denotes the image signal, η_{QE} denotes the quantum efficiency of the optical signal, I_{RON} denotes the readout noise, and D is the dark current, and the dark current noise is time-dependent. According to Equation (18), it can be known that the SNR can be improved by $\sqrt{N}$ times when the multi-frame averaging operation is performed on N images. However, as N increases, the SNR increases more and more slowly. Figure 7a,b show the results of the images acquired in clear water and with different noises artificially added (a) Gaussian random noise with mean $\mu = 0$ and variance $\sigma^2 = 0.02$ and (b) Gaussian random noise with mean $\mu = 0$ and variance $\sigma^2 = 0.2$). Figure 7c shows the PSNR curves after averaging over different numbers of frames, with the maximum number of frames averaged at 200. It can be seen that the PSNR rises slowly after N reaches the value of 50. Figure 7d,e show the multi-frame averaging results of two images with different noise intensities added for $N = 50$. Figure 7f–h show the results obtained from the actual experiment, which are consistent with the simulation results, and the improvement of the PSNR tends to be stable after the value of N reaches 50. Comparing the enlarged part (green dashed area) in Figure 7f,h, it can be seen that the noise of the image is significantly suppressed. Therefore, the result of $N = 50$ is used in the multi-frame averaging operation in the following experiments.

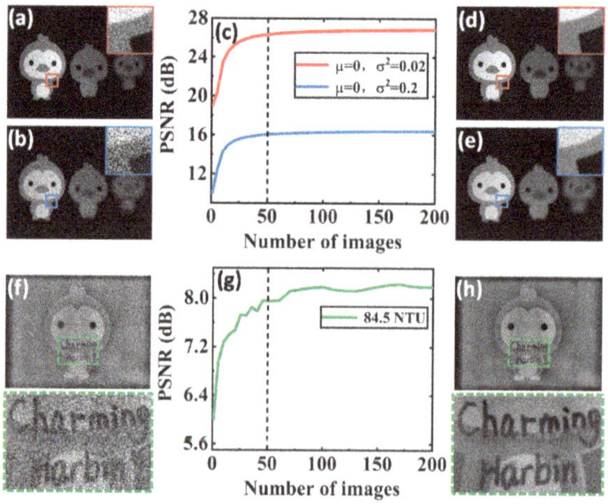

Figure 7. (**a**) and (**b**) are the results of artificially adding different random noises to the target image captured in clear water; (**c**) PSNR curves of the average results of different numbers of multiple frames in two cases; (**d**) and (**e**) are the multi-frame averaged results when N = 50 in two cases; (**f**) S_1^{pro} is obtained by processing the images captured in 84.5NTU turbid water; (**g**) PSNR curve after multi-frame averaging of different numbers of S_i^{pro}; (**h**) the results of S_i^{pro} after multi-frame averaging when N = 50.

For the SNR analysis of multi-frame averaging, it is known that the ability of multi-frame averaging to suppress noise is limited. Bilateral filtering has the effect of double filtering, which can suppress noise on the one hand and preserve the edge strength of the object well on the other hand [29,33]. Therefore, the bilateral filtering operation is used as the last step of joint noise suppression. Figure 8 shows the polarization enhancement results after comparing the different processing methods; Figure 8a is the raw image intensity; Figure 8b is the processing result of Li's method [25]; and Figure 8c is the processing result using Liu's method [32]. Figure 8d–f are the results of S_1^{pro}, 50 frames of S_i^{pro} multi-frame averaging $S_{\mathrm{average}}^{\mathrm{pro}}$, and the result $\mathrm{BF}\left(S_{\mathrm{average}}^{\mathrm{pro}}\right)$ after bilateral filtering on $S_{\mathrm{average}}^{\mathrm{pro}}$, respectively. It can be seen from the enlarged part in Figure 8f that the joint noise suppression method can effectively reduce the interference of noise and improve the clarity of the target signal. Figure 8g shows the pixel grayscale row index curves of the raw image intensity and the different processing methods, and the indexed pixel range is the yellow dashed line in Figure 8b. It can be seen from Figure 8g that our method better maintains the intensity distribution of the target signal and restores the details of the target signal compared to other methods.

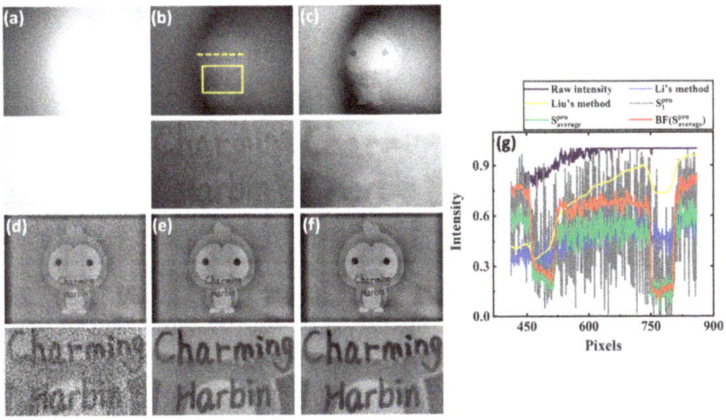

Figure 8. (**a**) is the raw image intensity; (**b**) the result of using Li's method [25]; (**c**) the result of using Liu's method [32]; (**d**) is S_1^{pro}; (**e**) is $S_{\mathrm{average}}^{\mathrm{pro}}$, results after averaging 50 frames of S_i^{pro}; (**f**) is $\mathrm{BF}\left(S_{\mathrm{average}}^{\mathrm{pro}}\right)$; (**g**) the pixel grayscale index curves of the raw image intensity and the processing results of different methods.

3.3. Results in Different High-Turbidity Water Samples

In the experiment, the raw target image in clear water was first acquired as a reference image for calculating the PSNR. Then, 120 mL of skimmed milk was added to the water tank, at which time the initial turbidity of the medium in the water tank was 52.7 NTU, and the images were acquired. Then, the volume of skimmed milk was continuously increased, and the corresponding images were acquired until 240 mL of skimmed milk was added and the water turbidity was 98.6 NTU (120 mL, 160 mL, 200 mL, 220 mL, and 240 mL, corresponding to the turbidity of 52.7 NTU, 68.9 NTU, 84.5 NTU, 92.5NTU, and 98.6 NTU, respectively). The first column in Figure 9 shows the intensity of the raw image, and it can be found that as the turbidity of the water increases the intensity of the backscattered light will rise rapidly and generate a large area of grayscale overexposure area. The target object information in the raw image gradually decreases with the increase in turbidity until it is completely lost. This also reflects the necessity of improving the imaging quality of the target objects in high-turbidity underwater scenes. The second and fourth columns in Figure 9 show the processing results using Schechner's method, Li's method, and the CLAHE method, respectively, and the fifth column is the processing result of our method.

Under the high-turbidity conditions, the signal light in $I_{\min}$ decreases rapidly, and the results of Schechner's method are low in grayscale value, and the image is dark overall. Li's method and the CLAHE method have adjusted the image grayscale histogram so that the grayscale is more uniform. However, as the intensity of the backscattered light increases and the noise of the imaging process continues to gain, the intensity of the target signal recovered by Li's method decreases significantly with the increase in turbidity. CLAHE is a histogram equalization operation for the raw intensity image, and the CLAHE method will gradually fail when the backscattered light intensity exceeds the dynamic range of the CCD. Our method takes into account the high backscattered light intensity and the low SNR characteristics of the imaging process in highly turbid water and uses CLAHE for cross-linear image processing combined with joint noise suppression for polarization enhancement of the target signal. In contrast, as can be seen from the results in the fifth column of Figure 9, although the recovery of the target signal decreases with the increasing turbidity, our processing results outperform the conventional method under the same turbidity conditions.

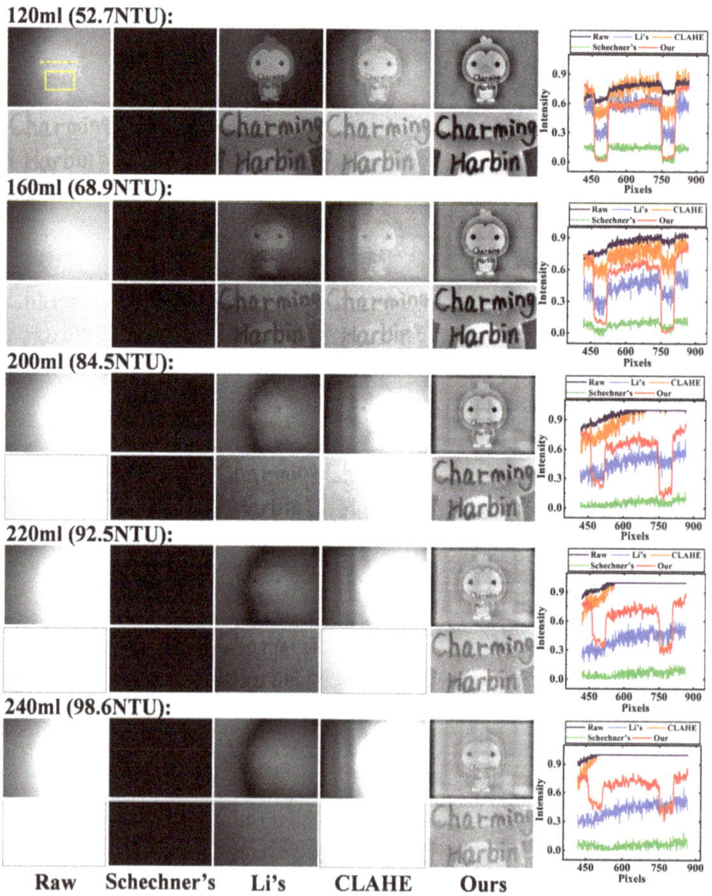

Figure 9. The first column is the raw image intensity; the second column is the processing result of Schechner's method [6]; the third column is the processing result of Li's method [25]; the fourth column is the CLAHE processing result [11]; and the fifth column is ours; the sixth column compares the pixel grayscale indexing results of the raw image intensity and the different methods (along the yellow dashed line).

To quantify the effect of the proposed method on the improvement of the recovered image contrast and noise suppression, the pixel coordinates of the yellow dashed line in Figure 9 are used as the indexing range to compare the differences in grayscale values between the raw image intensity and the results of the different methods of processing, as shown in the last column of Figure 9. The results of the image grayscale indexing under the different turbidity conditions are also shown. It can be seen from the grayscale indexing results that the backscattered light has a significant effect on the intensity of the raw image (purple line) and the CLAHE processing method (orange line). In addition, the lower signal light intensity makes the results recovered by Schechner's method (green line) have a low contrast. Where the blue line indicates the results of Li's method and the red line indicates our results, it can be seen that the red line has the best depiction effect on the details of the object under different turbidity conditions. Moreover, it can be verified from the degree of fluctuation of the curves that noise in the imaging results of the highly turbid water has a significant impact on the imaging quality. In order to objectively evaluate the effect of the image quality improvement, we use image entropy and PSNR to evaluate the effect of the different methods of processing, and the corresponding results are shown in Table 1.

Table 1. Quantitative comparisons of the recovered images in Figure 9.

		Raw	Schechner's	Li's	CLAHE	Ours
52.7NTU	PSNR(dB)	8.90	6.49	13.73 [a]	10.43	**17.33**
	Entropy	1.69	1.62	2.04	2.05	**2.27**
68.9NTU	PSNR(dB)	6.61	5.95	11.41	8.87	**16.14**
	Entropy	1.68	1.24	1.34	1.96	**2.21**
84.5NTU	PSNR(dB)	5.05	5.64	11.18	5.63	**13.05**
	Entropy	1.09	1.05	1.18	1.17	**2.00**
92.5NTU	PSNR(dB)	4.73	5.58	10.66	4.99	**11.12**
	Entropy	0.48	1.05	1.21	0.66	**1.87**
98.6NTU	PSNR(dB)	4.63	5.56	**10.71**	4.68	10.22
	Entropy	0.12	0.99	1.24	0.19	**1.75**

[a] For each evaluation metric, the best results are marked in bold red, and the second best results are marked in blue with underlining.

It can be seen in Table 1 that the image entropy value obtained by the proposed method is the highest under the different turbidity conditions; the PSNR is not as good as Li's method under the condition of 98.6 NTU, but it is also at the second best value and the difference is not significant. The effectiveness of our method is verified by the quantized results; the PSNR can be improved by a maximum of about 2.6 times (84.5 NTU), and the image entropy can be improved by a maximum of about 14.6 times (98.6 NTU) compared to the raw image.

The limitations of image histogram equalization operating on a large number of overexposed regions can be seen from the CLAHE processing results in Figure 9. Our method applies it to the cross-linear image I_{min} and combines it with joint noise suppression to obtain better results than using CLAHE directly. However, the interaction probability between the photons and particles increases due to the rise in large-size particles in the highly turbid water, causing the photons to gradually lose their initial polarization state and become randomly polarized [26]. This eventually manifests itself as an increase in the intensity of the backscattered light in I_{min}, which is also detrimental to the operation of CLAHE. To illustrate the applicability of the proposed method, peak-to-correlation energy (PCE) was used to evaluate the correlation between the polarization orthogonal sub-images as the turbidity increased [15,16,34]. Figure 10 shows the PCE curves calculated by I_{min} and I_{max} under the different turbidity conditions. It can be found that the intensity of the backscattered light in I_{min} and I_{max} gradually increases as the turbidity increases, leading to a gradual decrease in the correlation between them. The effect of the histogram equalization of I_{min} using CLAHE is thus weakened because the target signal light is gradually submerged in the background. In addition, the effect of CLAHE is also related to its setting of the number of tiles (NT) and the clip limit (CL). Therefore, in order to improve the applicability of this method in more complex environments, the optimization of NT and CL [35] and the means of signal enhancement under low-light conditions [36] can be combined to increase the signal proportion in I_{min} so as to achieve better imaging quality.

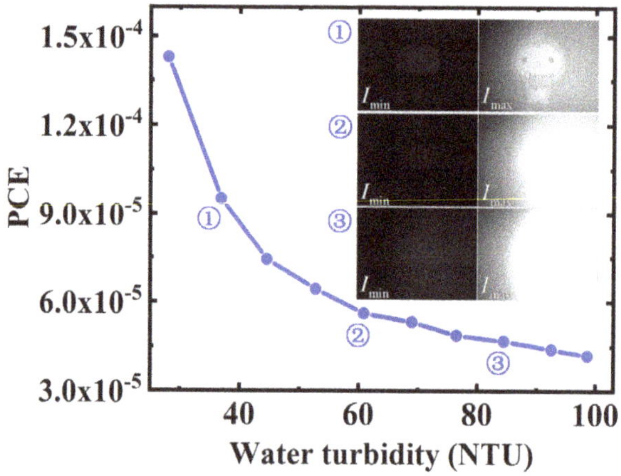

Figure 10. PCE curves of polarization orthogonal sub-images under different turbidity.

4. Discussion

Underwater imaging enhancement methods based on non-physical and physical models have been widely used in dealing with turbid media. However, as the turbidity increases, especially in highly turbid water, the processing effect of the traditional methods becomes limited. Therefore, it is necessary to optimize the imaging system [1,24] or combine different algorithms [5,25,37] to improve the imaging quality. We theoretically and experimentally analyzed the influence of highly turbid water on the underwater active polarization imaging model and showed that the increase in the backscattered light intensity and the high noise gain of the imaging model mainly affect the imaging quality. Inspired by the normalized histogram stretching of I_{min} by Li's method, we proposed an active polarization imaging processing method combining non-physical and physical models. Our method uses CLAHE to process a certain number of I_{min} and then performs polarization enhancement and uses joint filtering noise suppression (multi-frame averaging and bilateral filtering) on the polarization enhancement results to reduce the high noise gain generated by the imaging model and CLAHE. The quantitative results of the experiments verify that our method is suitable for polarization imaging in highly turbid water.

Underwater active polarization imaging takes into account the DOP of target signal light p_{targ}; so, we compared the experimental results of a high-DOP (HDOP) object. Figure 11 shows the experimental results of two metal blades glued to a metal plate. From the results, it is clear that our method is capable of handling the HDOP object. However, the HDOP object and the backscattered light both have a certain degree of polarization-maintaining ability; so, the overexposed area produced by the backscattered light leads to a distortion of the restoration result of the corresponding area, as shown by the red dotted line in Figure 11.

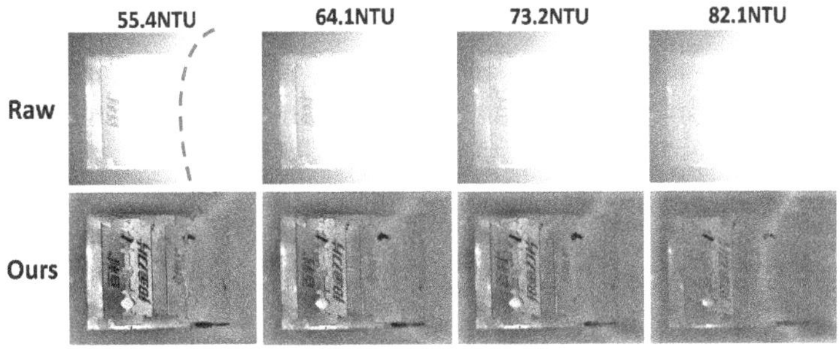

Figure 11. Processing results of HDOP object at different turbidities.

Our method is implemented on the basis of the non-physical model (CLAHE) and the physical model (underwater active polarization imaging) and borrows the processing idea from Li's method. Therefore, we compare their processing results in Figure 9. To compare our result more fully, we show the results of the classic and the effective dark channel prior (DCP) [7,38] and retinex [39,40] methods in computer vision in Figure 12. Due to the active illumination and with the increasing turbidity, the backscattered light produces inhomogeneity and causes a large number of overexposed areas, which affects the processing effectiveness of the DCP and retinex.

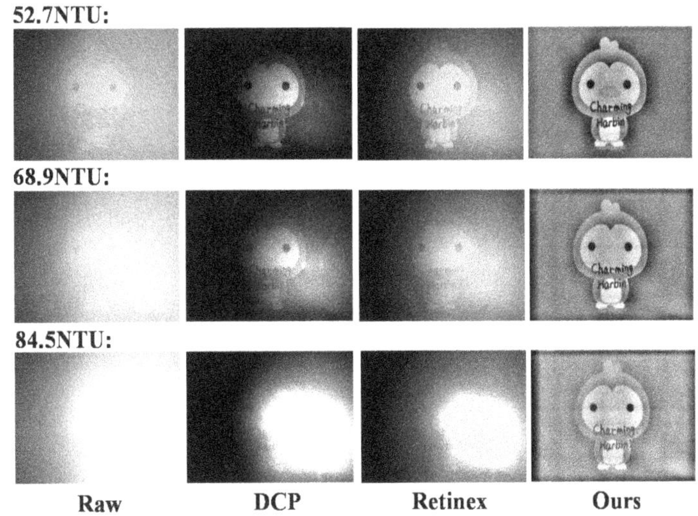

Figure 12. Comparison of the results of DCP [7,38] and retinex [39,40] at different turbidities.

While our method uses only classical processing methods from non-physical and physical models, it is effective to cross them reasonably to form a new processing method to achieve target imaging in harsh underwater environments. In future work, the development from "detectable" to "better detectable" can be achieved by setting up more suitable light sources (e.g., red light [1] or circularly polarized light illumination [24], etc.) and combining more advanced computational methods (e.g., computer vision [7,38–40] or machine learning [41,42], etc.).

5. Conclusions

In this paper, it is recognized through theoretical and experimental analysis that the influence of backscattered light on underwater active polarization imaging becomes more and more obvious as the turbidity increases, especially in the case of highly turbid water (>50 NTU). It is mainly reflected in two aspects: one is the increase in backscattered light intensity, and the other is the decrease in Δp. Together, they produce a processing result with a significantly lower contrast and high noise gain. We proposed an active polarization imaging method based on the CLAHE of cross-linear images and joint noise suppression (multi-frame averaging and bilateral filtering). The experimental results in the different high-turbidity water samples (52.7 $\sim$ 98.6 NTU) show that our recovery is better than the conventional methods. The PSNR and entropy of our method are improvements of up to 2.6 times and 14.6 times compared to the raw image. This method, which appropriately combines non-physical and physical methods [25], facilitates the application of underwater active polarization imaging in harsher environments without changing the complexity of the system.

Author Contributions: Conceptualization, H.Z.; methodology, H.Z.; software, H.Z. and Q.M.; validation, H.Z.; formal analysis, H.Z., J.G. and N.Z.; investigation, H.Z.; resources, Y.Z.; data curation, M.R. and N.Z.; writing—original draft, H.Z.; writing—review and editing, J.G., M.R., H.W., Q.M. and Y.Z.; visualization, H.Z.; supervision, Y.Z.; project administration, Y.Z. All authors have read and agreed to the published version of the manuscript.

Funding: This research received no external funding.

Institutional Review Board Statement: Not applicable.

Informed Consent Statement: Not applicable.

Data Availability Statement: Data underlying the results presented in this paper are not publicly available at this time but may be obtained from the authors upon reasonable request.

Conflicts of Interest: The authors declare no conflict of interest.

References

1. Liu, F.; Han, P.; Wei, Y.; Yang, K.; Huang, S.; Li, X.; Zhang, G.; Bai, L.; Shao, X. Deeply seeing through highly turbid water by active polarization imaging. *Opt. Lett.* **2018**, *43*, 4903–4906. [CrossRef] [PubMed]
2. Olmanson, L.G.; Brezonik, P.L.; Bauer, M.E. Airborne hyperspectral remote sensing to assess spatial distribution of water quality characteristics in large rivers: The Mississippi River and its tributaries in Minnesota. *Remote Sens. Environ.* **2013**, *130*, 254–265. [CrossRef]
3. Li, X.; Han, Y.; Wang, H.; Liu, T.; Chen, S.; Hu, H. Polarimetric Imaging Through Scattering Media: A Review. *Front. Phys.* **2022**, *10*, 153. [CrossRef]
4. Mullen, A.D.; Treibitz, T.; Roberts, P.L.D.; Kelly, E.L.A.; Horwitz, R.; Smith, J.E.; Jaffe, J.S. Underwater microscopy for in situ studies of benthic ecosystems. *Nat. Commun.* **2016**, *7*, 12093. [CrossRef]
5. Amer, K.O.; Elbouz, M.; Alfalou, A.; Brosseau, C.; Hajjami, J. Enhancing underwater optical imaging by using a low-pass polarization filter. *Opt. Express* **2019**, *27*, 621–643. [CrossRef]
6. Treibitz, T.; Schechner, Y.Y. Active Polarization Descattering. *IEEE Trans. Pattern Anal. Mach. Intell.* **2009**, *31*, 385–399. [CrossRef]
7. He, K.; Sun, J.; Tang, X. Single Image Haze Removal Using Dark Channel Prior. *IEEE Trans. Pattern Anal. Mach. Intell.* **2011**, *33*, 2341–2353. [CrossRef]
8. Liang, J.; Ren, L.; Qu, E.; Hu, B.; Wang, Y. Method for enhancing visibility of hazy images based on polarimetric imaging. *Photonics Res.* **2014**, *2*, 38–44. [CrossRef]
9. Zhou, J.; Liu, D.; Xie, X.; Zhang, W. Underwater image restoration by red channel compensation and underwater median dark channel prior. *Appl. Opt.* **2022**, *61*, 2915–2922. [CrossRef]
10. Li, X.; Zhang, L.; Qi, P.; Zhu, Z.; Xu, J.; Liu, T.; Zhai, J.; Hu, H. Are Indices of Polarimetric Purity Excellent Metrics for Object Identification in Scattering Media? *Remote Sens.* **2022**, *14*, 4148. [CrossRef]
11. Reza, A.M. Realization of the Contrast Limited Adaptive Histogram Equalization (CLAHE) for real-time image enhancement. *J. VLSI Signal Process. Syst. Signal Image Video Technol.* **2004**, *38*, 35–44. [CrossRef]
12. Schechner, Y.Y.; Karpel, N. Recovery of underwater visibility and structure by polarization analysis. *IEEE J. Ocean. Eng.* **2005**, *30*, 570–587. [CrossRef]

13. Zhou, J.; Zhang, D.; Zhang, W. Classical and state-of-the-art approaches for underwater image defogging: A comprehensive survey. *Front. Inf. Technol. Electron. Eng.* **2020**, *21*, 1745–1769. [CrossRef]
14. Dubreuil, M.; Delrot, P.; Leonard, I.; Alfalou, A.; Brosseau, C.; Dogariu, A. Exploring underwater target detection by imaging polarimetry and correlation techniques. *Appl. Opt.* **2013**, *52*, 997–1005. [CrossRef]
15. Han, P.; Liu, F.; Wei, Y.; Shao, X. Optical correlation assists to enhance underwater polarization imaging performance. *Opt. Lasers Eng.* **2020**, *134*, 106256. [CrossRef]
16. Zhang, H.; Ren, M.; Wang, H.; Yao, J.; Zhang, Y. Fast processing of underwater polarization imaging based on optical correlation. *Appl. Opt.* **2021**, *60*, 4462–4468. [CrossRef]
17. Hu, H.; Zhao, L.; Li, X.; Wang, H.; Liu, T. Underwater Image Recovery Under the Nonuniform Optical Field Based on Polarimetric Imaging. *IEEE Photonics J.* **2018**, *10*, 6900309. [CrossRef]
18. Huang, B.; Liu, T.; Hu, H.; Han, J.; Yu, M. Underwater image recovery considering polarization effects of objects. *Opt. Express* **2016**, *24*, 9826–9838. [CrossRef]
19. Liu, F.; Zhang, S.; Han, P.; Chen, F.; Zhao, L.; Fan, Y.; Shao, X. Depolarization index from Mueller matrix descatters imaging in turbid water. *Chin. Opt. Lett.* **2022**, *20*, 022601. [CrossRef]
20. Guan, J.; Ma, M.; Sun, P. Optimization of rotating orthogonal polarization imaging in turbid media via the Mueller matrix. *Opt. Lasers Eng.* **2019**, *121*, 104–111. [CrossRef]
21. Jin, H.; Qian, L.; Gao, J.; Fan, Z.; Chen, J. Polarimetric Calculation Method of Global Pixel for Underwater Image Restoration. *IEEE Photonics J.* **2020**, *13*, 6800315. [CrossRef]
22. Zhao, Y.; He, W.; Ren, H.; Li, Y.; Fu, Y. Polarization descattering imaging through turbid water without prior knowledge. *Opt. Lasers Eng.* **2021**, *148*, 106777. [CrossRef]
23. Qi, P.; Li, X.; Han, Y.; Zhang, L.; Xu, J.; Cheng, Z.; Liu, T.; Zhai, J.; Hu, H. U2R-pGAN: Unpaired underwater-image recovery with polarimetric generative adversarial network. *Opt. Lasers Eng.* **2022**, *157*, 107112. [CrossRef]
24. Hu, H.; Zhao, L.; Li, X.; Wang, H.; Yang, J.; Li, K.; Liu, T. Polarimetric image recovery in turbid media employing circularly polarized light. *Opt. Express* **2018**, *26*, 25047–25059. [CrossRef]
25. Li, X.; Hu, H.; Zhao, L.; Wang, H.; Yu, Y.; Wu, L.; Liu, T. Polarimetric image recovery method combining histogram stretching for underwater imaging. *Sci. Rep.* **2018**, *8*, 12430. [CrossRef]
26. Wang, D.; Qi, J.; Huang, B.; Noble, E.; Stoyanov, D.; Gao, J.; Elson, D. Polarization-based smoke removal method for surgical images. *Biomed. Opt. Express* **2022**, *13*, 2364–2379. [CrossRef]
27. Rong, L.; Xiao, W.; Pan, F.; Liu, S.; Li, R. Speckle noise reduction in digital holography by use of multiple polarization holograms. *Chin. Opt. Lett.* **2010**, *8*, 653–655. [CrossRef]
28. Wang, J.; Wan, M.; Gu, G.; Qian, W.; Ren, K.; Huang, Q.; Chen, Q. Periodic integration-based polarization differential imaging for underwater image restoration. *Opt. Lasers Eng.* **2022**, *149*, 106785. [CrossRef]
29. Han, P.; Liu, F.; Zhang, G.; Tao, Y.; Shao, X. Multi-scale analysis method of underwater polarization imaging. *Acta Phys. Sin.* **2018**, *67*, 054202. [CrossRef]
30. Treibitz, T.; Schechner, Y.Y. Polarization: Beneficial for visibility enhancement? In Proceedings of the 2009 IEEE Conference on Computer Vision and Pattern Recognition, Miami, FL, USA, 20–25 June 2009; pp. 525–532. [CrossRef]
31. Piederriere, Y.; Boulvert, F.; Cariou, J.; Le Jeune, B.; Guern, Y.; Le Brun, G. Backscattered speckle size as a function of polarization: Influence of particle-size and -concentration. *Opt. Express* **2005**, *13*, 5030–5039. [CrossRef]
32. Liu, F.; Wei, Y.; Han, P.; Yang, K.; Bai, L.; Shao, X. Polarization-based exploration for clear underwater vision in natural illumination. *Opt. Express* **2019**, *27*, 3629–3641. [CrossRef]
33. Tomasi, C.; Manduchi, R. Bilateral filtering for gray and color images. In Proceedings of the Sixth International Conference on Computer Vision (IEEE Cat. No.98CH36271), Bombay, India, 7 January 1998; pp. 839–846. [CrossRef]
34. Zhang, H.; Zhou, N.; Meng, Q.; Ren, M.; Wang, H.; Zhang, Y. Local optimum underwater polarization imaging enhancement based on connected domain prior. *J. Opt.* **2022**, *24*, 105701. [CrossRef]
35. Campos, G.F.C.; Mastelini, S.M.; Aguiar, G.J.; Mantovani, R.G.; de Melo, L.F.; Barbon, S., Jr. Machine learning hyperparameter selection for Contrast Limited Adaptive Histogram Equalization. *EURASIP J. Image Video Process.* **2019**, *2019*, 59. [CrossRef]
36. Hu, H.; Lin, Y.; Li, X.; Qi, P.; Liu, T. IPLNet: A neural network for intensity-polarization imaging in low light. *Opt. Lett.* **2020**, *45*, 6162–6165. [CrossRef]
37. Liang, J.; Ren, L.; Liang, R. Low-pass filtering based polarimetric dehazing method for dense haze removal. *Opt. Express* **2021**, *29*, 28178–28189. [CrossRef]
38. Jiao, Q.; Liu, M.; Li, P.; Dong, L.; Hui, M.; Kong, L.; Zhao, Y. Underwater image restoration via Non-Convex Non-Smooth variation and thermal exchange optimization. *J. Mar. Sci. Eng.* **2021**, *9*, 570. [CrossRef]
39. Hassan, N.; Ullah, S.; Bhatti, N.; Mahmood, H.; Zia, M. The Retinex based improved underwater image enhancement. *Multimed. Tools Appl.* **2021**, *80*, 1839–1857. [CrossRef]
40. Fu, X.; Zhuang, P.; Huang, Y.; Liao, Y.; Zhang, X.-P.; Ding, X. A retinex-based enhancing approach for single underwater image. In Proceedings of the 2014 IEEE International Conference on Image Processing (ICIP), Paris, France, 27–30 October 2014; pp. 4572–4576. [CrossRef]

41. Qi, Q.; Li, K.; Zheng, H.; Gao, X.; Hou, G.; Sun, K. SGUIE-Net: Semantic Attention Guided Underwater Image Enhancement with Multi-Scale Perception. *arXiv* **2022**, arXiv:2201.02832. [CrossRef]
42. Han, P.; Li, X.; Liu, F.; Cai, Y.; Yang, K.; Yan, M.; Sun, S.; Liu, Y.; Shao, X. Accurate passive 3D polarization face reconstruction under complex conditions assisted with deep learning. *Photonics* **2022**, *9*, 924. [CrossRef]

photonics

MDPI

Article

An Algorithm to Extract the Boundary and Center of EUV Solar Image Based on Sobel Operator and FLICM

Shuai Li, Jianhua Zhang, Bei Liu *, Chengzhi Jiang, Lanxu Ren, Jingjing Xue and Yansong Song

China Academy of Space Technology (Xi'an), Xi'an 710100, China
* Correspondence: liu_bei_0504@163.com; Tel.: +86-15191430289

Abstract: An algorithm to extract the disk boundary and center of EUV solar image using the Sobel operator, Fuzzy Local Information C-Means Clustering algorithm (FLICM), and the least square circle fitting method is proposed in this paper. The Sobel operator can determine the solar disk boundary preliminarily, and then the image is processed further using the FLICM algorithm. After the background is removed based on the clustered image and the boundary points can be highlighted, these points are fitted using the least square circle fitting method as the final boundary circle. The solar data used in this paper was from the observation of the Solar Dynamics Observatory Atmospheric Imaging Assembly (SDO/AIA) instrument. The 2523 19.3 nm solar images covering solar minimum, moderate solar activity, and more active suns were calculated using the proposed algorithm to analyze the accuracy statistically. The statistical comparison results demonstrate that the method is accurate and effective. This method can support the processing of solar EUV images and serve the operational system of a space weather forecast.

Keywords: Sobel operator; FLICM; EUV solar boundary; solar center; AIA/SDO

Citation: Li, S.; Zhang, J.; Liu, B.; Jiang, C.; Ren, L.; Xue, J.; Song, Y. An Algorithm to Extract the Boundary and Center of EUV Solar Image Based on Sobel Operator and FLICM. *Photonics* **2022**, *9*, 889. https://doi.org/10.3390/photonics9120889

Received: 18 October 2022
Accepted: 21 November 2022
Published: 22 November 2022

1. Introduction

Imaging the sun in extreme ultraviolet (EUV) band from space is an important approach to monitoring a hot coronal plasma in active solar phenomena. It can improve the forecast of space weather and early warnings of possible impacts on the Earth's environment. Space-borne optical remote instruments for the Sun have been developed for more than 40 years, and many EUV imaging instruments have been launched to study solar atmospheric dynamics. The Extreme-ultraviolet Imaging Telescope (EIT) onboard the Solar and Heliospheric Observatory (SOHO) was launched in 1995 to observe the corona and transition region on the solar disk in 17.1 nm (Fe IX), 19.5 nm (Fe XII), 28.4 nm (Fe XV) and 30.4 nm (He II) [1–3]. In 1998, the Transition Region and Coronal Explorer (TRACE) mission with three EUV imaging channels was launched to image the solar corona at 17.1 nm, 19.5 nm, and 28.4 nm for diagnosis of coronal plasmas between 10^5 K and 10^6 K [4,5]. As the successor to TRACE, the National Aeronautics and Space Administration (NASA) launched the Solar Dynamics Observatory (SDO) mission in 2010, and the Atmospheric Imaging Assembly (AIA) onboard SDO has provided near-continuous monitoring of the Sun in 7 narrowband EUV channels [6,7]. The Extreme Ultraviolet Imager (EUI), equipped in the Solar Orbiter mission, was launched last year, which aims to provide full-disk EUV and high-resolution EUV and Lyman-α imaging of the solar atmosphere by imaging the three spectral lines of 17.4 nm, 30.4 nm and 121.6 nm [8,9]. In 2021, the solar X-ray and EUV telescope, on board the Fengyun-3E satellite, was launched in China, which imaged the sun at 19.5 nm [10]. These instruments accumulated a large number of solar EUV images for solar study.

Before the solar image is released, it needs to go through a lot of processes and corrections. Among them, image positioning is an extremely important step [11–13]. Finding the accurate center and radii is not a trivial problem for solar EUV images since the corona is complex. For solar images in visible and infrared bands, the boundary can be calculated using solar limb darkening [14–16]. However, this method is not suitable for EUV images. Also, finding accurate limb positions for EUV images was challenging for the AIA team. After some experimentation, they found techniques that work on all AIA channels except 30.4 nm (which has a particularly noisy limb). The EUV images were processed with Sobel transform after a 3-pixel Gaussian smooth. To isolate a limb signal better, they calculated a radial direction for the first derivative between $\pm 5^{\circ}$ around the rough boundary. The next step is to use these points to get an estimate of the limb and use it to eliminate bad points from the set. Finally, a least squares fit finds the optimal circle that intersects a maximum number of the remaining points [17,18].

This paper described a method to extract the center and boundary of solar EUV images based on the Sobel operator, FLICM algorithm, and the least square circle fitting. The EUV images are preprocessed with the Sobel operator to find the rough boundaries. Then, the FLICM algorithm is used to cluster images and remove the discrete background. After the two steps, the extreme points of the boundary will be fitted using the least square circle fitting method. The outline of this paper is as follows. The data sets used in this paper will be introduced in Section 2. The Sobel operator and FLICM algorithm will be described in detail in Sections 3 and 4, respectively. The fitting results will be displayed in Section 5, and the statistical analysis of extraction results will be discussed in Section 6. Finally, a conclusion will be given in Section 7.

2. Data Sets

The solar EUV images used in this study were obtained by the Solar Dynamics Observatory Atmospheric Imaging Assembly (SDO/AIA) instrument [6,7]. SDO/AIA is the follow-on to the extremely successful Transition Region and Coronal Explorer (TRACE) mission, which consists of 4 individual telescopes. Each telescope consists of a mirror system coated in halves, each of which responds to a different portion of the solar spectrum, resulting in a virtual system of 8 distinct telescopes. The field of view of the AIA telescopes is 41 arcmins, large enough to encompass the full sun. The AIA is to provide narrow-band imaging of seven extreme ultraviolet (EUV) band passes centered on specific lines: Fe XVIII (9.4 nm), Fe VIII, XXI (13.1 nm), Fe IX (17.1 nm), Fe XII, XXIV (19.3 nm), Fe XIV (21.1 nm), He II (30.4 nm), and Fe XVI (33.5 nm), at a resolution consistent with 0.6 arcsec detector pixels, once every 10 s. One telescope observes C IV (near 160 nm) and the nearby continuum (170 nm) and has a filter that observes in visible to enable alignment with images from other telescopes. The AIA was launched with an SDO mission on 11 February 2010. Operating in geosynchronous orbit, the AIA has observed and accumulated a large amount of solar data so far. The data used in this paper is the Lev 1 product of SDO/AIA from https://sdac.virtualsolar.org/cgi/search (21 November 2022). There are 1076 solar 19.3 nm images from 1 August 2021 to 31 October 2021, 718 images with a time interval of 12 h in all of 2014, and 729 images with a time interval of 12 h in all of 2019 were calculated in this paper. Figure 1A shows the 19.3 nm solar image observed by AIA at 09:39:40 UTC on 28 August 2021. Figure 1B is the local enlarged image in the rectangular box in Figure 1A, which is used to display the boundary information of the EUV image.

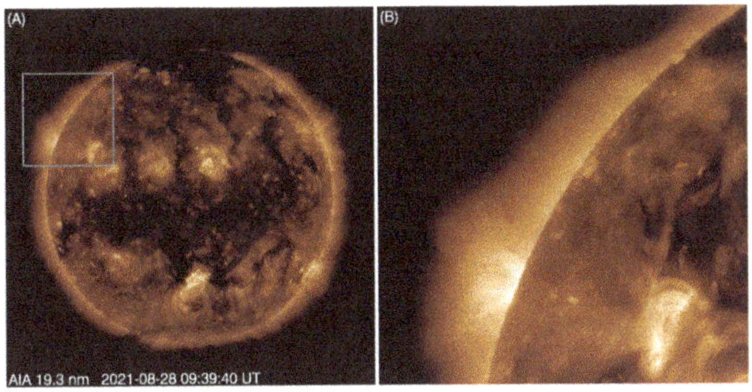

Figure 1. (**A**) is the 19.3 nm solar image observed by AIA at 09:39:40 UTC on 28 August 2021, and (**B**) is the local enlarged image in the rectangular box in A. The observation time is marked in the lower left corner of A.

3. Sobel Operator

To extract the disk boundary of the EUV solar image, the first step is to recalculate the image using the Sobel operator. The Sobel operator is an edge detection with the first derivative [19,20]. In the process of the algorithm, a 3 × 3 template is used as the kernel to perform convolution and operation with each pixel in the image. Then a suitable threshold is selected to extract the edge. Technically, it is a discrete differentiation operator, computing an approximation of the gradient of the image intensity function. At each point in the image, the result of the Sobel operator is either the corresponding gradient vector or the norm of this vector. Sobel operator is the partial derivative of $F_{x,y}$ as the central computing 3 × 3 neighborhood at x, y direction. To suppress the noise, a certain weight is correspondingly increased on the center point, and its digital gradient approximation equations may describe as follows:

$$\begin{aligned} G_x &= \{F_{x+1,y-1} + 2F_{x+1,y} + F_{x+1,y+1}\} - \{F_{x-1,y-1} + 2F_{x-1,y} + F_{x-1,y-1}\} \\ G_y &= \{F_{x-1,y+1} + 2F_{x,y+1} + F_{x+1,y+1}\} - \{F_{x-1,y-1} + 2F_{x,y-1} + F_{x+1,y-1}\} \end{aligned} \quad (1)$$

The size of its gradient is

$$g(x,y) = |G_x| + |G_y| \quad (2)$$

The convolution template operator is as follows:

$$T_x = \begin{bmatrix} -1 & 0 & 1 \\ -2 & 0 & 2 \\ -1 & 0 & 1 \end{bmatrix} \quad T_y = \begin{bmatrix} -1 & -2 & -1 \\ 0 & 0 & 0 \\ 1 & 2 & 1 \end{bmatrix} \quad (3)$$

When calculating the edge of the image, the horizontal template T_x and the vertical template T_y are used to convolute with the image. Then, two gradient matrices of the same size as the original image can be obtained. Finally, the total gradient value can be calculated by adding the two matrices, and the edge can be obtained by the threshold method. The result is shown in Figure 2, while Figure 2A is the original EUV image, and Figure 2B is the edge distribution detected by the Sobel operator. As shown in Figure 2, the location of the solar disk boundary is further enhanced.

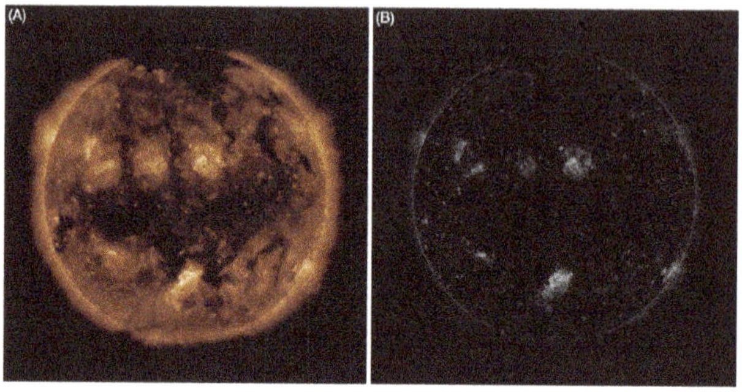

Figure 2. The original solar image (**A**) and the image calculated by Sobel operator (**B**).

4. FLICM Algorithm

In this section, a robust Fuzzy Local Information C-Means Clustering algorithm (FLICM) is used to further extract the basic positional information of solar disk boundary [21,22]. This algorithm incorporates local spatial and gray-level information in a fuzzy way to preserve robustness and noise insensitiveness by adding a novel fuzzy factor

$$G_{ki} = \sum_{\substack{j\in N_i \\ i\neq j}} \frac{1}{d_{ij}+1}(1-u_{kj})^m \|x_j - v_k\|^2 \tag{4}$$

to the squared error objective function

$$J_m = \sum_{i=1}^{N}\sum_{k=1}^{c}[u_{ki}^m \|x_i - v_k\|^2 + G_{ki}] \tag{5}$$

where N is the total number of pixels, c is the number of clusters with $2 \leq c < N$, x_i is the gray level of the ith pixel, which is the center of the local window (e.g., 3×3 pixels), v_k is the prototype of the center of cluster k, u_{ki} is the degree of the membership of x_i in the kth cluster, m is a weighting exponent on each fuzzy membership that determines the amount of fuzziness of the resulting classification and is usually set to be 2, N_i stands for the set of neighbors falling into a local window around pixel x_i, and d_{ij} is the spatial Euclidean distance between pixels i and j. The necessary conditions for J_m to be at its minimal local extreme, with respect to u_{ki} and v_k are obtained in the following formulas:

$$u_{ki} = \frac{1}{\sum\limits_{j=1}^{c}\left(\frac{\|x_i - v_k\|^2 + G_{ki}}{\|x_i - v_j\|^2 + G_{ji}}\right)^{1/m-1}} \tag{6}$$

$$v_k = \frac{\sum\limits_{i=1}^{N} u_{ki}^m x_i}{\sum\limits_{i=1}^{N} u_{ki}^m} \tag{7}$$

In this paper, the solar image after the Sobel operator calculation will be clustered into 10 categories using the FLICM algorithm (c was set to 10), and the pixels lower than the median gray value will be zeroed. As shown in Figure 3, Figure 3B is the result after the FLICM algorithm operation, and after background removal, the image in Figure 3C. Figure 3D–F is a partially enlarged view of the rectangular boxes in Figure 3A–C. The green line-1 and blue line-2 are selected to verify that the algorithm can effectively remove interference points near the boundary. The normalized gray values vs. pixels along the lines are given in Figure 4, and the red dotted line marks the location of the solar disk boundary along the line. Mass calculations have proved that this method can efficiently remove the interference points nearby the boundary, and Figure 3C will be further used to find the disk boundary.

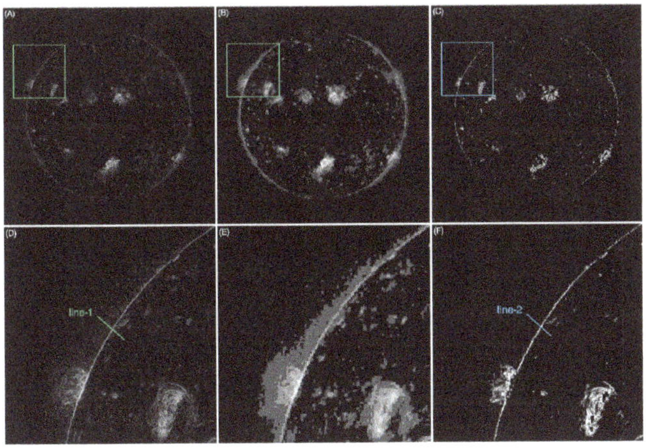

Figure 3. The images calculated by Sobel operator (**A**), the image calculated by FLICM algorithm (**B**), and the image after background removal (**C–F**) are the partially enlarged views corresponding to (**A–C**).

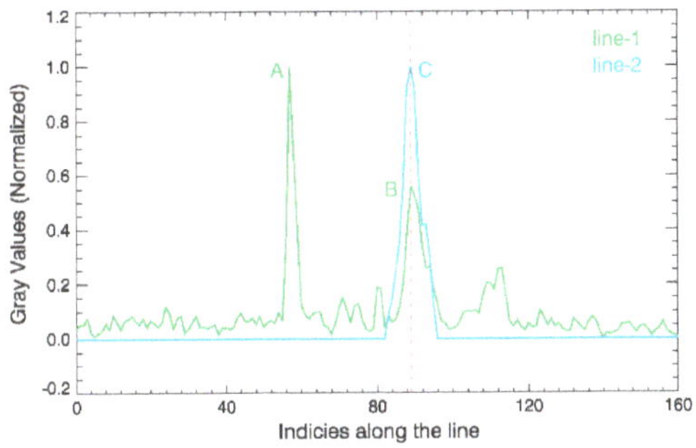

Figure 4. The normalized gray values vs. pixels along the lines in Figure 3D (green) and Figure 3F (blue). The red dotted line represents the boundary along the line.

5. Least Square Circle Fitting

Drawing lines from the image center from 0° to 360°, and the step size is set to 0.5°. Since the CCD size of AIA is 4096 × 4096, the image center can be roughly considered as (2048, 2048). Then, 721 data arrays can be obtained, and the coordinates corresponding to the outermost non-zero point of each array are the disk boundaries to be searched. Using these coordinates, the least square circle fitting method is used to fit the disk circle boundary [23–25]. As shown in Figure 5A, the fitted boundary is displayed by the red circle, Figure 5B gives a partially enlarged view, and Figure 5C is the calculated boundary in the original solar EUV image. Figure 6 gives the extraction results of the solar 19.3 nm, 17.1 nm, and 13.1 nm images observed by AIA using the proposed method. Visually, this method can accurately extract the boundary and center of the solar EUV image. The accuracy of this method will be analyzed in the next chapter.

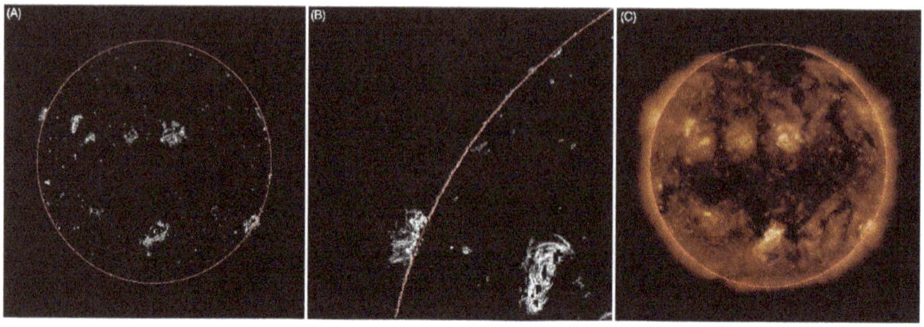

Figure 5. The fitted circle boundary using the least square circle fitting (**A**,**B**) is a partially enlarged view of (**A**,**C**) the fitted boundary in the original solar EUV image.

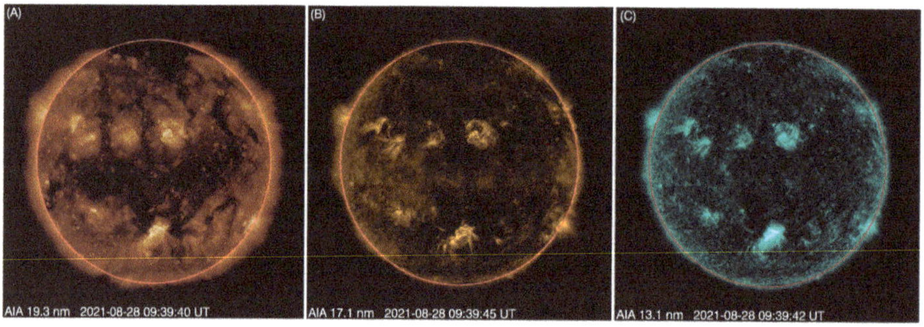

Figure 6. The calculated boundaries of the solar 19.3 nm (**A**), 17.1 nm (**B**), and 13.1 nm (**C**) images observed by AIA using the proposed method. The observation time is marked in the lower left corner of each image.

6. Statistical Analysis

To verify the effectiveness of the proposed algorithm and its applicability under different solar activities, the solar 19.5 nm images from three years were selected for statistical comparison analysis. There were 718 images with a time interval of 12 h in all of 2014, 729 images with a time interval of 12 h in all of 2019, and 1076 solar images from 1 August 2021 to 31 October 2021. The sun in 2014 was more active, while in 2019, it reached a minimum, and between August and October 2021, the solar activity was moderate. The solar disk centers and boundaries calculated by the proposed method were compared with the results released by the AIA team. The comparison results, given in Table 1, were classified according to four constraints with $|X_{calculation} - X_{AIA}| \leq 2$ pixels and $|Y_{calculation} - Y_{AIA}| \leq 2$ pixels, $|X_{calculation} - X_{AIA}| > 2$ pixels or $|Y_{calculation} - Y_{AIA}| > 2$ pixels, $|R_{calculation} - R_{AIA}| \leq 3$ pixels, and $|R_{calculation} - R_{AIA}| > 3$ pixels. From Table 1, we can conclude that the algorithm performs better during smaller solar activity than when the sun is more active. It is to be expected cause the solar boundary will be clearer after FLICM and background removal processes. The scatter charts for statistical comparison under the conditions of $|X_{calculation} - X_{AIA}| \leq 2$ pixels and $|Y_{calculation} - Y_{AIA}| \leq 2$ pixels, and $|R_{calculation} - R_{AIA}| \leq 3$ pixels are given in Figures 7–9. The image data in Figure 7 was from 2014, and the data in Figures 8 and 9 were obtained in 2019 and 1 August 2021 to 31 October 2021, respectively. The blue points in each panel are the calculation of the proposed algorithm, and the red points are the results of AIA. The ratios are marked in red font at the upper right corners. A are the comparisons of solar center coordinates, and B are the comparisons of solar radii in each figure.

Table 1. Statistical comparison of the calculated solar centers and radii with the results of AIA.

Date	2014	2019	2021
Solar level	active	quiet	moderate
Total images	718	729	1076
$\lvert X_{calculation} - X_{AIA} \rvert \leq 2$ pixels and $\lvert Y_{calculation} - Y_{AIA} \rvert \leq 2$ pixels	609 (84.82%)	665 (91.22%)	965 (89.68%)
$\lvert X_{calculation} - X_{AIA} \rvert > 2$ pixels or $\lvert Y_{calculation} - Y_{AIA} \rvert > 2$ pixels	109 (15.18%)	64 (8.78%)	111 (10.32%)
$\lvert R_{calculation} - R_{AIA} \rvert \leq 3$ pixels	643 (89.55%)	675 (92.59%)	970 (90.15%)
$\lvert R_{calculation} - R_{AIA} \rvert > 3$ pixels	75 (10.45%)	54 (7.41%)	106 (9.85%)

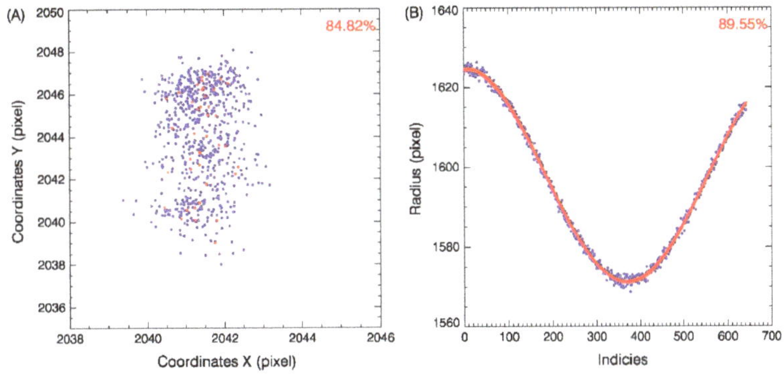

Figure 7. The image data was obtained in 2014, and the sun was more active. (**A**) is the comparison of the 609 samples of solar centers between our calculations (blue points) and the results of AIA (red points), and (**B**) is the comparison of the 643 samples of solar radii between our calculations (blue points) and the results of AIA (red points).

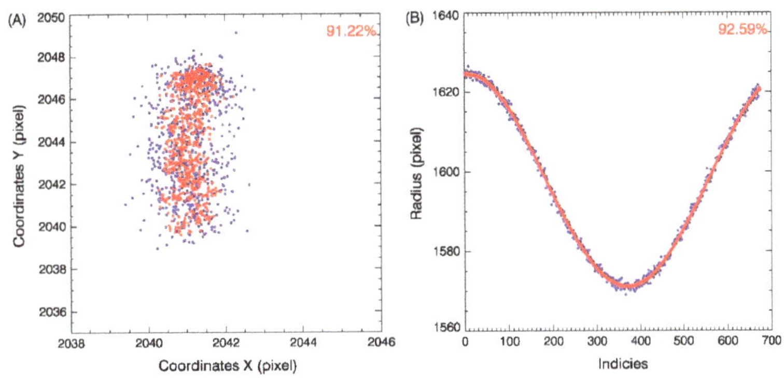

Figure 8. The image data was obtained in 2019, and the solar activity was minimum. (**A**) is the comparison of the 665 samples of solar centers between our calculations (blue points) and the results of AIA (red points), and (**B**) is the comparison of the 675 samples of solar radii between our calculations (blue points) and the results of AIA (red points).

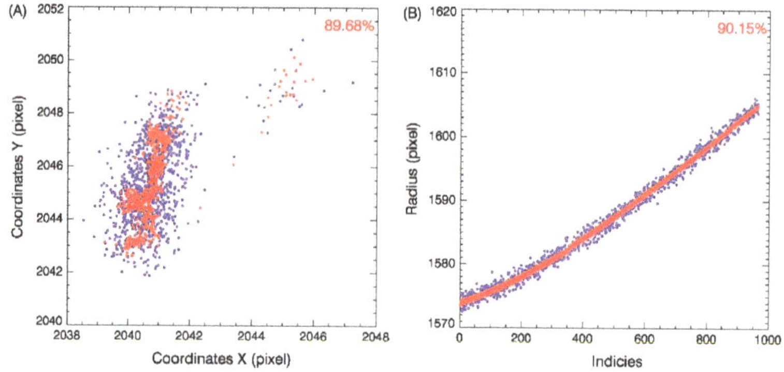

Figure 9. The image data was obtained from 1 August 2021 to 31 October 2021, and the solar activity was moderate. (**A**) is the comparison of the 965 samples of solar centers between our calculations (blue points) and the results of AIA (red points), and (**B**) is the comparison of the 970 samples of solar radii between our calculations (blue points) and the results of AIA (red points).

7. Conclusions

In this paper, an algorithm to extract the boundary and center of EUV solar image using the FLICM algorithm and Sobel operator was proposed. Based on the solar EUV images observed by SDO/AIA, the extraction results and accuracy were given and discussed. Using the Sobel operator, the image boundary can be found preliminarily, and the boundary image was further clustered by the FLICM algorithm. After the FLICM operation, the useful data points and the noise are further separated, and finally, the boundary points are determined using the least square circle fitting. The statistical comparison results of 1076 images from 1 August 2021 to 31 October 2021, 718 images in all of 2014, and 729 images in all of 2019 demonstrate that the algorithm is accurate and effective. The processes of the algorithm are summarized as follows:

(1) Calculating the preliminary boundary using the Sobel operator;
(2) Clustering the preliminary boundary image with the FLICM algorithm and the image is clustered into 10 categories;
(3) The background is generally removed based on the clustered image; the 5 categories with the smaller values are considered the background;
(4) Searching the peak value points from outside to inside around the image;
(5) Fitting these points as the final boundary circle using the least square circle fitting method.

Author Contributions: Conceptualization, S.L. and B.L.; methodology, S.L. and J.Z.; software, S.L.; validation, J.Z., B.L. and C.J.; formal analysis, S.L.; investigation, S.L., C.J. and Y.S.; resources, B.L.; data curation, L.R. and J.X.; writing—original draft preparation, S.L.; writing—review and editing, S.L.; visualization, S.L.; supervision, B.L.; project administration, J.Z.; funding acquisition, J.X. All authors have read and agreed to the published version of the manuscript.

Funding: This work was supported by the First Class Fund for Distinguished Young Scholars of the Xi'an Branch of China Academy of Space Technology (Y21-RCFYJQ1-09).

Institutional Review Board Statement: Not applicable.

Informed Consent Statement: Not applicable.

Data Availability Statement: The solar data used in this paper are available at https://sdac.virtualsolar.org/cgi/search.

Conflicts of Interest: The authors declare no conflict of interest.

References

1. Domingo, V.; Fleck, B.; Poland, A.I. SOHO: The Solar and Heliospheric Observatory. *Space Sci. Rev.* **1995**, *72*, 81–84. [CrossRef]
2. Martens, P.; Muglach, K. Scientific Highlights from the Solar and Heliospheric Observatory. In *Solar Polarization*; Springer: Dordrecht, The Netherlands, 1999.
3. Delaboudinière, J.P.; Artzner, G.E.; Brunaud, J.; Gabriel, A.H.; Hochedez, J.F.; Millier, F.; Song, X.Y.; Au, B.; Dere, K.P.; Howard, R.A.; et al. EIT: Extreme-ultraviolet Imaging Telescope for the SOHO mission. *Sol. Phys.* **1995**, *162*, 291–312. [CrossRef]
4. Strong, K.; Bruner, M.; Tarbell, T.; Wolfson, C.J. Trace—The transition region and coronal explorer. *Space Sci. Rev.* **1994**, *70*, 119–122. [CrossRef]
5. Handy, B.; Bruner, M.; Tarbell, T.; Title, A.; Wolfson, C.; LaForge, M.; Oliver, J. UV Observations with the Transition Region and Coronal Explorer. *Sol. Phys.* **1998**, *183*, 29–43. [CrossRef]
6. Cheimets, P.; Caldwell, D.C.; Chou, C.; Gates, R.; Lemen, J.; Podgorski, W.A.; Wolfson, C.J.; Wuelser, J.P. SDO-AIA telescope design. In Proceedings of the SPIE Optical Engineering + Applications, San Diego, CA, USA, 2–6 August 2009; Volume 7438, p. 74380G.
7. Lemen, J.R.; Akin, D.J.; Boerner, P.F.; Chou, C.; Drake, J.F.; Duncan, D.W.; Edwards, C.G.; Friedlaender, F.M.; Heyman, G.F.; Hurlburt, N.E.; et al. The Atmospheric Imaging Assembly (AIA) on the Solar Dynamics Observatory (SDO). In *The Solar Dynamics Observatory*; Springer: New York, NY, USA, 2011.
8. Marsch, E.; Fleck, B.; Schwenn, R. Solar Orbiter—A High Resolution Mission to the Sun and Inner Helisophere. *COSPAR Colloq. Ser.* **2001**, *11*, 445.
9. Rochus, P.; Auchère, F.; Berghmans, D.; Harra, L.; Schmutz, W.; Schühle, U.; Addison, P.; Appourchaux, T.; Cuadrado, R.A.; Baker, D.; et al. The Solar Orbiter EUI instrument: The Extreme Ultraviolet Imager. *Astron. Astrophys.* **2020**, *642*, A8. [CrossRef]
10. Chen, B.; Ding, G.-X.; He, L.-P. Solar X-ray and Extreme Ultraviolet Imager (X-EUVI) loaded onto China's Fengyun-3E Satellite. *Light. Sci. Appl.* **2022**, *11*, 29. [CrossRef] [PubMed]
11. Shimizu, T.; Katsukawa, Y.; Matsuzaki, K.; Ichimoto, K.; Kano, R.; DeLuca, E.E.; Lundquist, L.L.; Weber, M.; Tarbell, T.D.; Shine, R.A.; et al. Hinode Calibration for Precise Image Co-alignment between SOT and XRT (November 2006–April 2007). *Publ. Astron. Soc. Jpn.* **2007**, *59* (Suppl. S3), S845–S852. [CrossRef]
12. Couvidat, S.; Schou, J.; Hoeksema, J.T.; Bogart, R.S.; Bush, R.I.; Duvall, T.L.; Liu, Y.; Norton, A.A.; Scherrer, P.H. Observables Processing for the Helioseismic and Magnetic Imager Instrument on the Solar Dynamics Observatory. *Sol. Phys.* **2016**, *291*, 1887–1938. [CrossRef]
13. Denker, C.; Johannesson, A.; Marquette, W.; Goode, P.R.; Wang, H.; Zirin, H. Synoptic Hα Full-Disk Observations of the Sun from Big Bear Solar Observatory—I. Instrumentation, Image Processing, Data Products, and First Results. *Sol. Phys.* **1999**, *184*, 87–102. [CrossRef]
14. Wang, Y.; Liu, S.; Deng, Y.; Bai, X.; Mao, X. The measurement of flat fields and polarization offset from the routine observation data of a solar rotation. *Chin. Sci. Bull.* **2017**, *63*, 301–310. [CrossRef]
15. Wang, Y.; Bai, X.; Liu, S.; Deng, Y.; SUN, Y. Flat-field measuring and correction method for full-disk solar image based on ground glass. *Chin. Sci. Bull.* **2017**, *62*, 3057–3066. [CrossRef]
16. Wang, Y.; Bai, X.; Liu, S.; Deng, Y.; Zhang, Z.; Sun, Y. Flat-fielding of Full-disk Solar Images with a Gaussian-type Diffuser. *Sol. Phys.* **2019**, *294*, 127. [CrossRef]
17. Shine, R.A.; Nightingale, R.W.; Boerner, P.; Tarbell, T.D.; Wolfson, C.J. Flat Fielding and Image Alignments for AIA/SDO Data Images. In Proceedings of the AGU Meeting, San Francisco, CA, USA, 13–17 December 2010; p. SH23C-1872.
18. Shine, R.A.; Wolfson, C.; Boerner, P.F.; Tarbell, T.D.; Nightingale, R.W. Monitoring Image Alignments and Flat Fields for AIA/SDO Data Images. In Proceedings of the SPD Meeting #42, Las Cruces, NM, USA, 12–16 June 2011; pp. 21–26.
19. Zhang, J.Y.; Yan, C.; Huang, X.X. Edge detection of images based on improved Sobel operator and genetic algorithms. In Proceedings of the 2009 International Conference on Image Analysis and Signal Processing, Linhai, China, 11–12 April 2009.
20. Kanopoulos, N.; Vasanthavada, N.; Baker, R. Design of an image edge detection filter using the Sobel operator. *IEEE J. Solid-State Circuits* **1988**, *23*, 358–367. [CrossRef]
21. Krinidis, S.; Chatzis, V. A Robust Fuzzy Local Information C-Means Clustering Algorithm. *IEEE Trans. Image Process.* **2010**, *19*, 1328–1337. [CrossRef] [PubMed]
22. Ding, G.; He, F.; Zhang, X.; Chen, B. A new auroral boundary determination algorithm based on observations from TIMED/GUVI and DMSP/SSUSI. *J. Geophys. Res. Space Phys.* **2017**, *122*, 2162–2173. [CrossRef]
23. Liu, F.; Han, P.; Wei, Y.; Yang, K.; Huang, S.; Li, X.; Zhang, G.; Bai, L.; Shao, X. Deeply seeing through highly turbid water by active polarization imaging. *Opt. Lett.* **2018**, *43*, 4903–4906. [CrossRef] [PubMed]
24. Liu, F.; Wei, Y.; Han, P.; Yang, K.; Bai, L.; Shao, X. Polarization-based exploration for clear underwater vision in natural illumination. *Opt. Express* **2019**, *27*, 3629–3641. [CrossRef] [PubMed]
25. Liu, F.; Zhang, S.; Han, P.; Chen, F.; Zhao, L.; Fan, Y.; Shao, X. Depolarization index from Mueller matrix descatters imaging in turbid water. *Chin. Opt. Lett.* **2022**, *20*, 022601. [CrossRef]

photonics

MDPI

Communication

Preparation of an Integrated Polarization Navigation Sensor via a Nanoimprint Photolithography Process

Ze Liu [1,2], Jinkui Chu [1,2,*], Ran Zhang [1,2], Chuanlong Guan [1,2] and Yuanyi Fan [1,2]

1 Key Laboratory of Precision and Non-Traditional Machining Technology of Ministry of Education, Dalian University of Technology, Dalian 116024, China
2 Key Laboratory of Micro/Nano Technology and System of Liaoning Province, Dalian University of Technology, Dalian 116024, China
* Correspondence: chujk@dlut.edu.cn; Tel.: +86-411-8470-6108

Abstract: Based on the navigation strategy of insects utilizing the polarized skylight, an integrated polarization sensor for autonomous navigation is presented. The polarization sensor is fabricated using the proposed nanoimprint photolithography (NIPL) process by integrating a nanograting polarizer and an image chip. The NIPL process uses a UV-transparent variant template with nanoscale patterns and a microscale metal light-blocking layer. During the NIPL process, part of the resist material is pressed to fill into the nanofeatures of the variant template and is cured under UV exposure. At the same time, the other parts of the resist material create micropatterns according to the light-blocking layer. Polymer-based variant templates can be used for conformal contacts on non-flat substrates with excellent pattern transfer fidelity. The NIPL process is suitable for cross-scale micro–nano fabrication in wide applications. The measurement error of the polarization angle of the integrated polarization sensor is ±0.2°; thus, it will have a good application prospect in the polarization navigation application.

Keywords: nanoimprint photolithography process; cross-scale micro–nano process; polarization sensor; navigation

Citation: Liu, Z.; Chu, J.; Zhang, R.; Guan, C.; Fan, Y. Preparation of an Integrated Polarization Navigation Sensor via a Nanoimprint Photolithography Process. *Photonics* **2022**, *9*, 806. https://doi.org/10.3390/photonics9110806

Received: 16 September 2022
Accepted: 21 October 2022
Published: 27 October 2022

1. Introduction

Navigation technique is an essential ability to survive and develop for human beings and becomes increasingly important in a wide range of applications in both military and civil fields [1,2]. Consequent to several billion years of evolution, many animals, especially insects, have formed ingenious navigational capabilities. Some of them are sensitive to the polarization of light and utilize polarization information for communication and navigation [3–6]. The bionic polarization navigation inspired by the biological celestial polarization orientation method has advantages such as autonomy, immunity to interference, and no error accumulation [7]. Biomimetic navigation devices based on the polarization of the skylight have been developed [8–12]. Compared with photodiode-based polarization sensors, image-based polarization sensors are less susceptible to the surrounding environment, occlusion, or other external factors. In our previous paper [10], a multidirectional nanowire grid polarizer was integrated into a complementary metal oxide semiconductor (CMOS) sensor through nanoimprint lithography. A gap exists between the nanowire polarizer and the image plane due to the glass package cover of the CMOS sensor, and the level of integration has further room for improvement.

With respect to the integration of the nanowire polarizer and the image chip, it involves a micro–nano cross-scale process for the nanowires and other patterns such as the electrodes of the image chip. Among the fabrication techniques of micro- and nanoscale-integrated structures, the multistep shaping method [13–15] combines the commonly used microprocesses such as MEMS with nanofabrication technology to perform multiple pattern transfers on the same substrate, and finally, the micro- and nanoscale-integrated structures are obtained. It is flexible and can realize multiscale-integrated structures with complex patterns. However, the multistep shaping method usually requires multiple "lithography–etching" processes to process patterns for different sizes, in which the steps need to be precisely aligned with each other, and the accuracy of pattern transfer is easily reduced, resulting in a complex, time-consuming, and costly process. As commonly used nanofabrication techniques, electron-beam lithography (EBL) and focused ion beam (FIB) have high resolution but costly equipment and low efficiency and, therefore, are especially not suitable for micronlevel or larger patterns [16,17]. Nanoimprint lithography (NIL) is highly adaptable in both research and industry because of its high throughput and low cost, which can prepare micro- and nanostructures on large areas with high efficiency [18–21]. NIL replicates nanostructures on a template onto a polymer film by driving polymer flow through mechanical extrusion. With a core of polymer flow, NIL thus has some potential deficiency; for example, the amount of polymer flow required is different for different sizes of feature structures at different locations, and larger structure features on the template will require more polymer flow transferred at a far distance. Filling efficiency is different, and pattern replication may result in pattern defects or incomplete pattern filling. This problem is more obvious when producing cross-scale structures with non-uniform distributions. Such limitations of NIL are inherent and are related to the fabricated patterns, which cannot be solved by methods such as optimizing the process parameters of nanoimprinting technology. To enable the fabrication of micro–nano cross-scale structures, especially when it comes to the fabrication of large-area structures, NIL needs to break through its limitations. In addition, the lithography process is the most central bottleneck process in the micro- and nanomanufacturing industries such as integrated circuit manufacturing. In addition to the further development of new lithography techniques suitable for smaller patterns, improving the accuracy and flexibility of the existing lithography process is also of extraordinary significance for micro- and nanodevice fabrication. Some methods based on nanoimprinting and standard photolithography are reported. Cheng et al. proposed the combined nanoimprint and photolithography (CNP) using the hybrid mask–mold (HMM) and with the CNP method, resist patterns with a higher aspect ratio and with no residual layer can be obtained in a single step [22]. The utilized HMM was made from fused silica by appropriate lithography and RIE process. Christiansen et al. demonstrated the fabrication of polymer optics with nm to mm features by CNP [23,24]. The CNP stamp was fabricated through EBL on a quartz wafer. Lohse et al. demonstrated a scalable method via CNP for the preparation of microfluidic channels with nanopatterns. The excellent compatibilities of the polymer master to the UV-PDMS stamp and further between the UV-PDMS stamp and the imprint resist mr-NIL are the key factors for replication on large scale with high accuracy [25].

This paper demonstrates a nanoimprint photolithography (NIPL) process that integrates conventional optical lithography into nanoimprint lithography, sharing the advantages and compensating for each other's limitations, using the mechanical deformation mechanism of nanoimprinting to replicate nanoscale patterns and the selective curing mechanism of optical lithography to make micronscale patterns, by a freely definable micro–nano variant template. The variant template is made from a common UV transparent soft stamp via the T-NIL and lift-off process, which can achieve conformal contacts on non-flat substrates with excellent pattern transfer fidelity. Through the NIPL process, the polarization navigation sensor is fabricated by integrating an image chip with a multiorientation nanowire polarizer. Finally, a performance test is implemented, and the measurement results are presented.

2. Nanoimprint Photolithography (NIPL) Process for Micro–Nano Cross-Scale Integration Process

Depending on the curing principle of the utilized resist, NIL can be divided into thermal NIL (T-NIL) and UV NIL, with the use of thermally or UV-cured resist, respectively. For the UV NIL process [26], when exposed to UV light, the small molecules of the resist polymer undergo photochemical reactions to form macromolecular networks, with a dramatic increase in mechanical strength and thermal and chemical stability, which are not easily soluble. Some photolithography negative resists also have similar photochemical reaction characteristics, such as SU-8 photoresist, so these photosensitive photolithography negative resists can be used in the UV NIL. Another similarity between UV NIL and conventional optical lithography is that the required templates in both are made of UV-transparent materials. While the masks commonly used in conventional optical lithography are graphite printed on a PET film or chromium deposited on glass, the molds commonly used in the UV NIL are quartz glass or polymeric soft molds. Quartz glass mold is hard and brittle and tends to exert permanent damage such as fractures when in use. A flexible polymer mold, by contrast, can avoid damages, where the pattern of the master mold is first copied 1:1 to the polymer mold and then transferred to the polymer film. The polymer mold not only avoids the "hard–hard" contact between the master mold and the substrate, thus prolonging the service life of the expensive master mold, but also ensures high fidelity by conformal contacts on uneven substrates due to the flexible characteristics of the soft mold, which extends the application of NIL for complex functional devices. In this paper, a nanoimprint photolithography (NIPL) process is proposed based on the respective characteristics of UV NIL and photolithography, correlating the similarities between them by using a UV-curable, imprinting resist material and a freely definable micro–nano variant template according to the requirements. A cross-scale micro–nano-integrated fabrication is realized.

2.1. Fabrication of the Variant Template

Variant templates are made of UV-transparent polymer films that include nanoscale surface structures as well as larger metal light-blocking layers that can be used as both imprint molds for defining nanopatterns and lithographic masks for defining the patterns at the micronscale and larger. The nanoscale features of the variant template are fabricated via the T-NIL process, followed by deposition and lift-off processes to fabricate the light-blocking layers on the variant template as a lithographic mask. The variant templates can be made from the same NIL master mold to apply in different applications, and the light-blocking layer of the variant template can be designed according to the application requirements using conventional MEMS processes, which greatly enhances the flexibility and adaptability of the NIPL process and dispenses with the needs for additional expensive equipment.

The process flow of fabricating the variant template is shown in Figure 1.

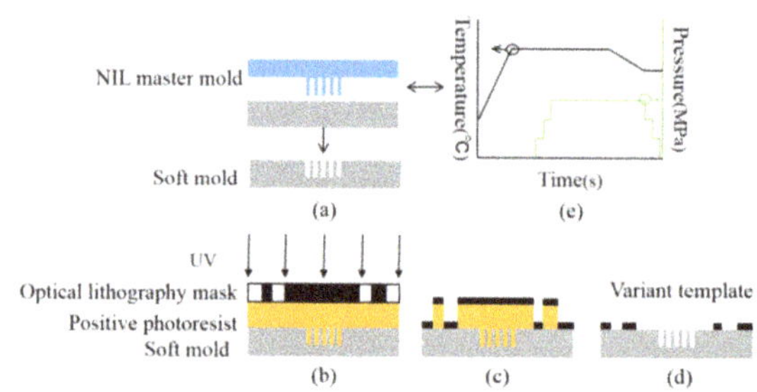

Figure 1. Process flow diagram of variant template fabrication: (**a**) replication of nanostructures of the variant template by T-NIL; (**b**) fabrication of micropatterns of the variant template via optical lithography; (**c**) deposited metal as a light-blocking layer of the variant template; (**d**) the variant template after lift-off process; (**e**) parameters for T-NIL.

Firstly, the nanostructures of the variant template are replicated from the master mold (generally silicon or nickel mold) using T-NIL (Figure 1a). The parameters for the T-NIL are shown in Figure 1e. The soft mold is placed on top of the master mold with sufficient surface contact to remove the air bubbles between the soft mold and the master mold. The imprinting temperature is generally up to 20–50 °C above the glass transition temperature (Tg) of the polymer mold, and therefore the polymer mold has good fluidity. High temperature and high pressure are maintained for a period to ensure the polymer mold sufficiently flows and fills into the concave structure of the master mold. Subsequently, the temperature is reduced to the demolding temperature, which is generally 20–40 °C below the Tg of the polymer mold.

Secondly, the metal film is deposited as a light-blocking layer with the desired micropattern in combination with optical lithography (Figure 1b,c).

Finally, after the lift-off process, the variant template is obtained with the nanoscale patterns and micronscale light-blocking layer (Figure 1d).

2.2. NIPL Process

The steps of the NIPL process are very simple, as shown in Figure 2. The variant template is first pressed into the resist film while being UV-exposed (Figure 2a,b). The part of the resist under UV exposure is cured, and the other part under the metal light-blocking layer remains. After the variant template and substrate are separated, the substrate is immersed in the developing solution to remove the unexposed resist. Both nanoscale and larger-scale patterns can be produced in the resist film in one single step. The NIPL process requires no additional equipment and can be performed with existing NIL machines; it can produce complex, cross-scale structures including nanoscale patterns in a single step without the need for multiple alignments and lithography processes, and the production accuracy is comparable to that of conventional NIL.

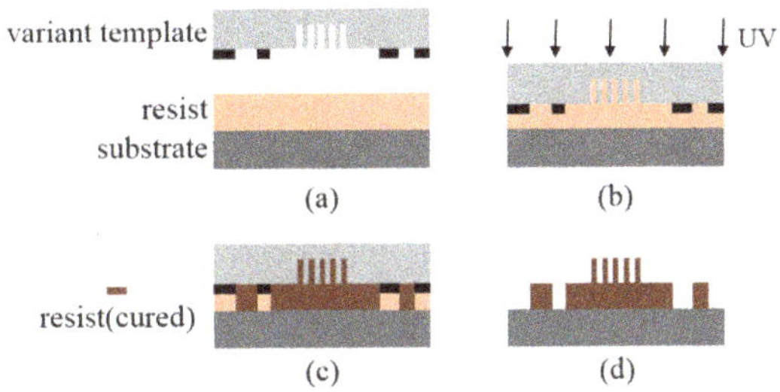

Figure 2. NIPL process flow chart: (**a**) resist is spin-coated on the substrate; (**b**) UV exposure; (**c**) cured resist under UV exposure; (**d**) the cross-scale structures with nano- and larger-scale patterns after demolding and development.

3. Polarization Navigation Sensor with the Integration of Multidirectional Nanogratings and the Image Chip

3.1. Integrated Fabrication of Multidirectional Nanogratings and the Image Chip

The area of the image chip is 6300 um × 5748 um, with 55 pin electrodes distributed around the edges, and the electrode size is 80 um × 63 um. The effective pixel array of the image chip is 732(H) × 492(V), with a pixel area of 6.5 um × 6.5 um, and the effective image area is 4760 um × 3200 um, shown as "Pixels" in Figure 3.

In the integrated fabrication process, two key issues need to be solved: (1) nanogratings should be integrated into the pixel area of the image chip; (2) to ensure that the edge-pin electrodes of the image chip can be connected to the external data acquisition processing circuit, the metal layer of the chip electrodes should be kept intact and undamaged, and the electrode surface should be clean and kept free from contamination by reagents such as the photoresist material during the integration process; both of these factors need to be considered.

For the integration process, the conventional process is first to fabricate the nanogratings on the pixel area via EBL, NIL, or other nanotechnology; subsequently, the metal film is deposited to form a metal nanowire polarizer, and the premise is that the electrodes are covered by the resist film to avoid short circuit to each other; finally, the residual resist and metal films on the electrodes are removed through, for example, dry etching or RIE. The conventional process is complicated and particularly difficult due to the edge bead effect. The edge bead effect is inevitable when coating the resist due to the flow characteristics of the resist, which leads to a thicker resist film at the edge of the substrate. Small-size substrates and non-circular substrates will aggravate the edge bead phenomenon [27]. When using full-chip-exposed lithography techniques such as UV NIL, the inhomogeneity of the resist film thickness can cause the over-etching of the pattern at the center or under-etching at the edges. To reduce the effect of edge beads, one of the common methods is chemical edge bead removal, which employs a chemical solvent dispensed on the edge of the substrate through a nozzle to eliminate the accumulated resist material on the substrate edge [28–30]. Other approaches, such as covering the edge of the substrate with heat-resistant tape, using a polymer holder to increase the planarity of the photoresist film, using a bell-shaped cover to minimize air turbulence while rotating, and the constant volume injection method, have been reported [31–34].

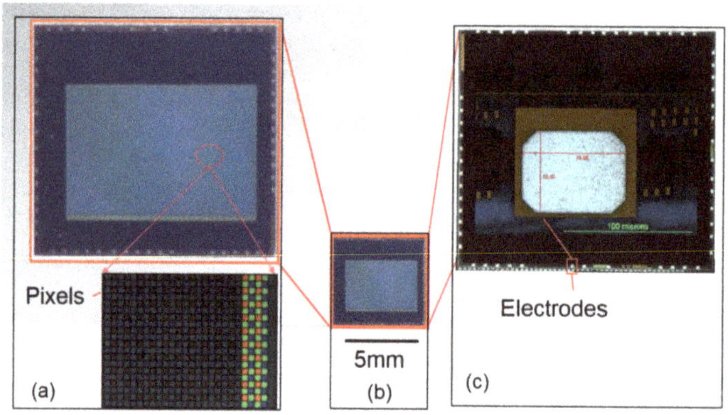

Figure 3. Optical and microscope photos of the image chip: (**a**) the photos of the enlarged pixels; (**b**) the photo of the full image chip; (**c**) the photos of the enlarged electrodes.

In this paper, NIPL is used to integrate the nanogratings and the image chip. With the proper variant template, nanogratings can be fabricated on the pixels, while the electrodes at the edges can remain clean and intact (Figure 4).

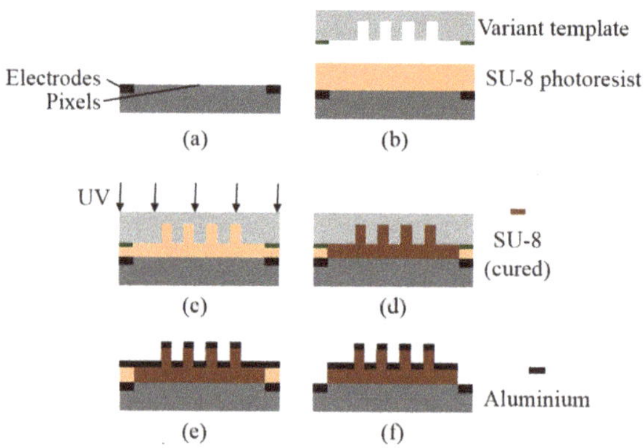

Figure 4. Integration process based on NIPL process: (**a**) cleaning the image chip; (**b**) spin-coating SU-8 photoresist on the image chip; (**c**,**d**) NIPL and the exposed SU-8 is cured; (**e**) metal Al deposited via thermal evaporation process; (**f**) the residual SU-8 resist and Al film on the electrodes are removed via a lift-off process.

Firstly, the image chip is cleaned to remove dust and organic pollutants. The image chip is immersed in acetone in a water bath (90 °C) for 10 min, followed by ultrasonic cleaning for 10 min with residual heat. The chip is moved into anhydrous ethanol, similar to the previous step, in a water bath (90 °C) for 10 min and ultrasonic cleaning for 10 min. After rinsing with deionized water, the image chip is ultrasonically cleaned in deionized water for 10 min and put into an oven to dry.

Secondly, SU-8 2015 is mixed with a diluent SU-8 2000 thinner at a ratio of 5:4 and spin-coated on the image chip surface. The spin-coating process begins at a low speed of 500 rpm for 6 s to spread the SU-8 resist evenly on the chip surface. Then, the excess solution is shaken out at the high speed of 4000 rpm for 30 s to make the SU-8 resist film more uniform. The image chip is placed on the horizontal heating plate for prebaking to remove the solvent in the SU-8 resist and to improve the adhesion between the SU-8 resist and the image chip. To eliminate the stress inside the SU-8 resist during spin-coating, the prebaking temperature is stepped up from 65 °C to 95 °C, increasing by 5 °C every 10 min, and remaining at 95 °C for 30 min, and then cooling naturally in air, to room temperature. This stepped changing temperature method can reduce the thermal stress caused by the drastic change in temperature and avoid the defect of SU-8 resist film wrinkle.

Then, the variant template for NIPL is fabricated. IPS soft mold is replicated via the T-NIL method for nanopatterns of the variant template, which contain nanogratings in four directions of 0°, 90°, 60°, and 150°, each with a period of 200 nm, a duty cycle of 0.5 and an area of 1.3 mm × 1.3 mm. The imprint parameters are 4 MPa at 155 °C, and the demolding temperature is 115 °C. Subsequently, the metal light-blocking layer is deposited via the magnetron sputtering process, including a 10 nm chromium (Cr) film layer and a 200 nm nickel (Ni) film layer, successively. The thin Cr film can improve the adhesion of Ni to the IPS soft mold, and the 200 nm Ni film has a UV transmission of about 0. The metal light-blocking layer covers the variant template except for the edges corresponding to the electrodes of the image chip.

Next, the NIPL process is performed using the SU-8 resist with a temperature of 85 °C and pressure of 4 MPa. After UV radiation, the exposed SU-8 resist is cured.

Then, 80 nm of metallic Al is deposited via the thermal vapor deposition process on the chip. Double-layer metal nanogratings are made on the pixel area.

Finally, the chip is immersed into the SU-8 developer to remove the uncured SU-8 resist with the attached Al film. After the SU-8 photoresist is spin-coated on the whole chip, there is a stack at the corners of the chip due to the edge bead effect. The selective local exposure of the NIPL process can leave the photoresist at the edges unexposed to UV, and therefore it can be easily removed using the developer.

As shown in Figure 5, the polarization chip is fabricated by integrating the image chip with multidirectional metal nanogratings, with nanograting directions of 0°, 90°, 60°, and 150°, a period of 200 nm, and a duty cycle of 0.5. The photograph of the integrated polarization sensor is shown in Figure 5a. As shown in Figure 5b, the electrodes are clean and intact after the integrating process. Figure 5c illustrates the SEM images of the multidirectional nanogratings on the test piece of silicon by the same integration process.

3.2. Hardware System of Integrated Polarization Sensor

The output electrical signals of the integrated polarization chip contain polarization information that is positively correlated with the intensity of the incident light. In the control processing module, STM32 is selected as a microcontroller unit (MCU) for the polarization information solution and real-time output. STM32 series microcontrollers have low cost and power consumption and are easy in terms of hardware design and program development, with a maximum operating frequency of 72 MHz, flash memory, and 64 K SRAM memory. The STM32 microcontroller contains an internal RTC clock circuit that enables continuous timing using the ticking timer when an external power supply is available. The sensor data can be read through the serial interface while quickly being transferred to the upper machine through another serial interface. The control processing module also includes EEPROM memory, a crystal oscillator, a decoupling capacitor, a filter capacitor, etc. The EEPROM is used to store the registered configuration of the integrated polarization chip, and the crystal provides the clock required for the STM32 microcontroller to work. The photos of the integrated polarization sensor are shown in Figure 5.

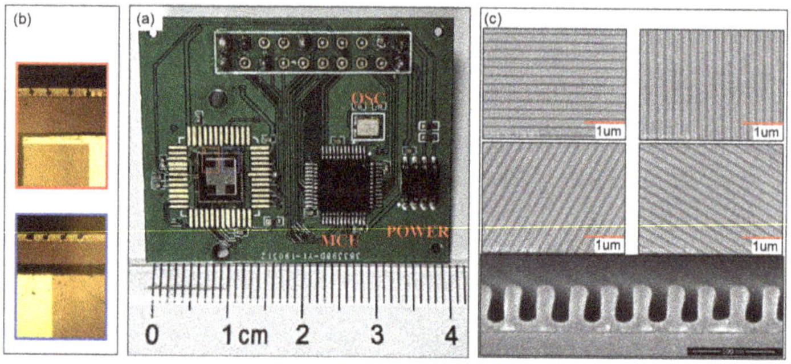

Figure 5. (**a**) The photograph of the integrated polarization sensor including the polarization chip, MCU and Power. (**b**) The photographs of partially enlarged details of nanogratings and adjacent electrodes of the polarization chip (red and blue boxes) in (**a**). (**c**) SEM images of multidirectional nanogratings on the test piece of silicon by the same integration process.

3.3. *Calculation of Polarization Angle of the Incident Light*

The integrated polarization chip contains four polarization units with nanogratings and covered pixels. The polarization units are sensitive to the polarization angle of 0°, 90°, 60°, and 150°, respectively. In theory, the intensities of the pixels in a polarization unit should be identical under irradiation. The output of a polarization unit is calculated as the mean value of the intensities of its pixels, as shown in the following equation:

$$\overline{p} = \frac{1}{XY}\sum_{y=1}^{Y}\sum_{x=1}^{X} p(x,y) \tag{1}$$

where $p(x,y)$ is the original intensity of a pixel, and x and y are the row and column numbers of the pixel in the pixel array. X and Y are the row and column dimensions. $\overline{p}$ is the mean value as the output of a polarization unit.

In practice, the variance between the pixels and the non-uniformity of the nanogratings may lead to inequality in pixel intensities in a polarization unit. The intensities of some pixels deviant from the majority of the intensities are labeled as abnormal values, which need to be eliminated to avoid reducing the reliability of the sensor and are determined as follows:

$$\sigma = \sqrt{\frac{1}{XY-1}\sum_{y=1}^{Y}\sum_{x=1}^{X}(p(x,y)-\overline{p})^2} \tag{2}$$

$$|p(x,y)-\overline{p}| > 3\sigma \tag{3}$$

where σ is the standard deviation. When the intensity of one pixel satisfies Equation (3), it can be regarded as an abnormal value and be eliminated. The above steps are repeated until there are no abnormal values. After eliminating the abnormal values, the mean intensities of pixels are calculated as the final intensity of a polarization unit. A large number of pixels would be sufficient for the calculation to compensate for unexpected situations such as some pixels breaking down, thus enhancing the robustness of the sensor.

When the sensor is illuminated with polarized light, the outputs of the polarization units can be described by the following equations:

$$\begin{aligned} s_1(\theta) &= KI(1 + d\cos 2\theta) \\ s_2(\theta) &= KI(1 - d\cos 2\theta) \\ s_3(\theta) &= KI\left[1 + d\cos\left(2\theta - \tfrac{2\pi}{3}\right)\right] \\ s_4(\theta) &= KI\left[1 - d\cos\left(2\theta - \tfrac{2\pi}{3}\right)\right] \end{aligned} \tag{4}$$

where I is the total intensity, d is the degree of polarization, θ is the polarization angle relative to the reference direction of the sensor, and K is a constant.

To eliminate the effect of the intensity of the incident light, the following calculation is implemented:

$$\begin{aligned} t_1 &= \frac{s_1(\theta)}{s_2(\theta)} = \frac{1+d\cos 2\theta}{1-d\cos 2\theta} \\ t_2 &= \frac{s_3(\theta)}{s_4(\theta)} = \frac{1+d\cos\left(2\theta-\frac{2\pi}{3}\right)}{1-d\cos\left(2\theta-\frac{2\pi}{3}\right)} \end{aligned} \tag{5}$$

With further elimination of the effect of the polarization degree, the calculated polarization angle is:

$$\begin{aligned} k_1 &= \frac{t_1-1}{t_1+1} \\ k_2 &= \frac{t_2-1}{t_2+1} \\ \theta &= \frac{1}{2}\tan^{-1}\left[\frac{1}{\sqrt{3}}\left(2\frac{k_2}{k_1}+1\right)\right] \end{aligned} \tag{6}$$

3.4. Performance Testing of the Integrated Polarization Sensor

A performance test of the integrated polarization navigation sensor was conducted on a test platform in the laboratory, as shown in Figure 6. A hollow precise rotary table was placed at the output port of the integrating sphere. A polaroid sheet was fastened on the rotary table. The output light of the integrating sphere was uniform and linearly polarized using the polaroid sheet. The polaroid sheet rotated, which was driven by the rotary table, thus resulting in a change in the polarization angle of the light. The performance of the integrated polarization sensor was evaluated by the accuracy of the measured change in the polarization angle.

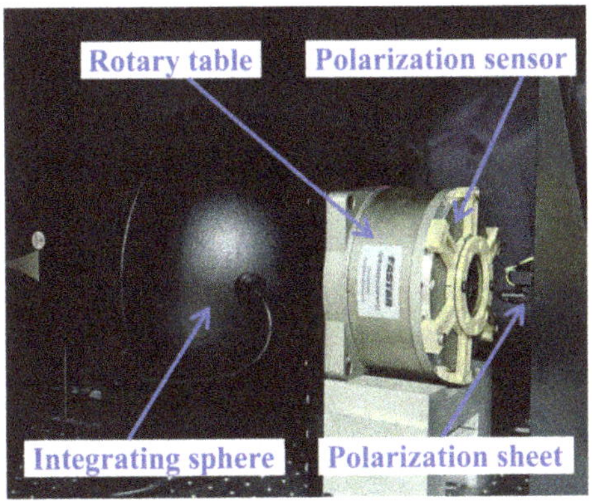

Figure 6. Measurement setup for polarization performance of the sensor.

4. Results and Discussion

The rotary table stopped every 4° of rotation for a total of 180°. The rotary table remained for 1 min at every stop, and at the same time, the polarization angles of the integrated polarization sensor were calculated. The rotation angle of the rotary table from the starting point was regarded as the theoretical rotation angle. The change in the angle of the sensor between the currently measured angle and the measured angle at the starting point was regarded as the measured rotation angle. The difference between the measured rotation angle and the theoretical rotation angle at every stop was the measured angle error of the polarization sensor. As shown in Figure 7, the error of the measured polarization angle was ±0.2°. The causes of this measurement error are possibly the presence of pixel noise, circuit noise, and the non-uniformity of the nanogratings. Through the performance test, it was shown that the integrated polarization sensor has the capability to detect the polarization angle of incident light. In a further study, we plan to investigate the potential of the integrated polarization sensor for application in autonomous navigation in combination with the distribution of a polarized skylight pattern.

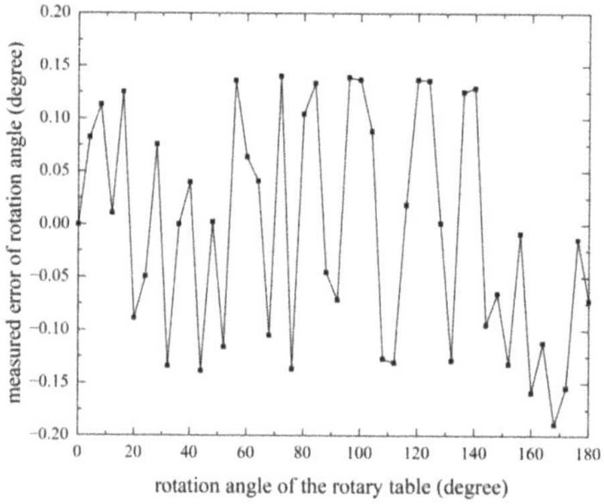

Figure 7. The error of the measured polarization angle of the integrated polarization sensor.

5. Conclusions

This paper presents a nanoimprint photolithography (NIPL) process by combining the respective characteristics of UV NIL and photolithography. In one step, the NIPL process can make cross-scale structures with nanoscale and microscale patterns simultaneously. A polarization chip was fabricated by integrating an image chip with nanogratings using the NIPL process. The polarization angle measurement error of the fabricated polarization navigation sensor was within ±0.2°.

The minimum dimension of the NIPL process demonstrated in this paper was 100 nm, but the limit of dimension for the NIPL process, dependent on the NIL master mold, can be smaller, even down to 10 nm theoretically. The NIPL process is widely applicable to cross-scale integration processes because the utilized variant templates are highly flexible for different application requirements.

Author Contributions: Z.L. conceived the idea of the fabrication of the integrated polarization navigation sensor. Y.F., C.G. and Z.L. performed the experiments. Z.L. analyzed the experimental data and wrote the initial draft of the manuscript. J.C. and R.Z. supervised the project. All authors have read and agreed to the published version of the manuscript.

Funding: This research was funded by the National Natural Science Foundation of China (No. 52175265) and Fundamental Research Funds for the Central Universities (Nos. DUT21ZD101, DUT21GF308).

Institutional Review Board Statement: Not applicable.

Informed Consent Statement: Not applicable.

Data Availability Statement: The study did not report any data.

Conflicts of Interest: The authors declare no conflict of interest. The funders had no role in the design of the study; in the collection, analyses, or interpretation of data; in the writing of the manuscript; or in the decision to publish the results.

References

1. Shakeri, R.; Al-Garadi, M.A.; Badawy, A.; Mohamed, A.; Khattab, T.; Al-Ali, A.K.; Harras, K.A.; Guizani, M. Design challenges of Multi-UAV systems in Cyber-Physical applications: A comprehensive survey and future directions. *IEEE Commun. Surv. Tut.* **2019**, *21*, 3340–3385. [CrossRef]
2. Olikhov, I.; Golchenko, A. Electron-beam-pumped semiconductor laser emitters in glideslope aircraft landing system. *Photonics* **2013**, *4*, 76–93.
3. Greif, S.; Borissov, I.; Yovel, Y.; Holland, R.A. A functional role of the sky's polarization pattern for orientation in the greater mouse-eared bat. *Nat. Commun.* **2014**, *5*, 4488. [CrossRef] [PubMed]
4. Hartmann, G.; Wehner, R. The ants path integration system-a neural architecture. *Biol. Cybern.* **1995**, *73*, 483–497. [CrossRef]
5. Wehner, R. Desert ant navigation: How miniature brains solve complex tasks. *J. Comp. Physiol. A* **2003**, *189*, 579–588. [CrossRef]
6. Froy, O.; Gotter, A.L.; Casselman, A.L.; Reppert, S.M. Illuminating the circadian clock in monarch butterfly migration. *Science* **2003**, *300*, 1303–1305. [CrossRef]
7. Powell, S.B.; Garnett, R.; Marshall, J.; Rizk, C.; Gruev, V. Bioinspired polarization vision enables underwater geolocalization. *Sci. Adv.* **2018**, *4*, eaao6841. [CrossRef]
8. Lambrinos, D.; Maris, M.; Kobayashi, H.; Labhart, T.; Pfeifer, R.; Wehner, R. An autonomous agent navigating with a polarized light compass. *Adapt. Behav.* **1997**, *6*, 131–161. [CrossRef]
9. Chu, J.K.; Wang, Z.W.; Guan, L.; Liu, Z.; Wang, Y.L.; Zhang, R. Integrated polarization dependent photodetector and its application for polarization navigation. *IEEE Photonic. Technol. Lett.* **2014**, *26*, 469–472.
10. Liu, Z.; Zhang, R.; Wang, Z.W.; Guan, L.; Li, B.; Chu, J.K. Integrated polarization-dependent sensor for autonomous navigation. *J. Micro Nanolith. Mem.* **2015**, *14*, 015001. [CrossRef]
11. Fan, C.; Hu, X.P.; Lian, J.X.; Zhang, L.L.; He, X.F. Design and calibration of a novel Camera-Based Bio-Inspired polarization navigation sensor. *IEEE Sens. J.* **2016**, *16*, 3640–3648. [CrossRef]
12. Lu, H.; Zhao, K.C.; Wang, X.C.; You, Z.; Huang, K.L. Real-time imaging orientation determination system to verify imaging polarization navigation algorithm. *Sensors* **2016**, *16*, 144. [CrossRef] [PubMed]
13. Bhushan, B.; Jung, Y.C. Natural and biomimetic artificial surfaces for superhydrophobicity, self-cleaning, low adhesion, and drag reduction. *Prog. Mater. Sci.* **2011**, *56*, 1–108. [CrossRef]
14. Kim, J.; Bae, W.; Choung, H.; Lim, K.T.; Seonwoo, H.; Jeong, H.E.; Suh, K.; Jeon, N.L.; Choung, P.; Chung, J.H. Multiscale patterned transplantable stem cell patches for bone tissue regeneration. *Biomaterials* **2014**, *35*, 9058–9067. [CrossRef] [PubMed]
15. He, Y.; Jiang, C.; Yin, H.; Chen, J.; Yuan, W. Superhydrophobic silicon surfaces with micro-nano hierarchical structures via deep reactive ion etching and galvanic etching. *J. Colloid Interf. Sci.* **2011**, *364*, 219–229. [CrossRef]
16. Ruffato, G.; Capaldo, P.; Massari, M.; Mezzadrelli, A.; Romanato, F. Pancharatnam-Berry optical elements for spin and orbital angular momentum division demultiplexing. *Photonics* **2018**, *5*, 46. [CrossRef]
17. Munshi, A.H.; Kephart, J.M.; Abbas, A.; Danielson, A.; Gelinas, G.; Beaudry, J.N.; Barth, K.L.; Walls, J.M.; Sampath, W.S. Effect of CdCl2 passivation treatment on microstructure and performance of CdSeTe/CdTe thin-film photovoltaic devices. *Sol. Energ. Mat. Sol. C* **2018**, *186*, 259–265. [CrossRef]
18. Toyoda, H.; Kimino, K.; Kawano, A.; Takahara, J. Incandescent Light Bulbs Based on a Refractory Metasurface. *Photonics* **2019**, *6*, 105. [CrossRef]
19. Hillmer, H.; Woidt, C.; Kobylinskiy, A.; Kraus, M.; Istock, A.; Iskhandar, M.S.Q.; Brunner, R.; Kusserow, T. Miniaturized interferometric sensors with spectral tunability for optical fiber Technology-A comparison of size requirements, performance, and new concepts. *Photonics* **2021**, *8*, 332. [CrossRef]
20. Dundar, A.F.; Kolewe, K.W.; Homyak, B.; Kurtz, I.S.; Schiffman, J.D.; Watkins, J.J. Bioinspired photocatalytic Shark-Skin surfaces with antibacterial and antifouling activity via nanoimprint lithography. *ACS Appl. Mater. Interfaces* **2018**, *10*, 20055–20063. [CrossRef]
21. Suresh, V.; Ding, L.; Chew, A.B.; Yap, F.L. Fabrication of Large-Area flexible SERS substrates by nanoimprint lithography. *ACS Appl. Nano Mater.* **2018**, *1*, 886–893. [CrossRef]
22. Cheng, X.; Guo, L.J. A combined-nanoimprint-and-photolithography patterning technique. *Microelectron. Eng.* **2004**, *71*, 277–282. [CrossRef]
23. Brokner-Christiansen, M.; Scholer, M.; Gersborg-Hansen, M.; Kristensen, A.; IEEE. Combined Nanoimprint and Photolithography of Integrated Polymer Optics. In Proceedings of the IEEE LEOS Summer Topical Meeting 2007, Portland, OR, USA, 23–25 July 2007.
24. Christiansen, M.B.; Scholer, M.; Kristensen, A. Integration of active and passive polymer optics. *Opt. Express* **2007**, *15*, 3931–3939. [CrossRef]
25. Lohse, M.; Heinrich, M.; Grutzner, S.; Haase, A.; Ramos, I.; Salado, C.; Thesen, M.W.; Grutzner, G. Versatile fabrication method for multiscale hierarchical structured polymer masters using a combination of photo- and nanoimprint lithography. *Micro. Nano. Eng.* **2021**, *10*, 100079. [CrossRef]
26. Haisma, J.; Verheijen, M.; Vandenheuvel, K.; Vandenberg, J. Mold-assisted nanolithography: A process for reliable pattern replication. *J. Vac. Sci. Technol. B* **1996**, *14*, 4124–4128. [CrossRef]
27. Nithi, A.; On-Uma, N.; Wutthinan, J.; Sitthisuntorn, S.; Charndet, H.; Amporn, P. Study of optimization condition for spin coating of the photoresist film on rectangular substrate by Taguchi design of an experiment. *Songklanakarin J. Sci. Technol.* **2009**, *31*, 331–335.

28. Hun, L.; Kangsun, L.; Byungwook, A.; Jing, X.; Linfeng, X.; Kwang, W.O. A new fabrication process for uniform SU-8 thick photoresist structures by simultaneously removing edge bead and air bubbles. *J. Micromech. Microeng.* **2011**, *21*, 125006.
29. Shaw, M.; Nawrocki, D.; Hurditch, R.; Johnson, D. Improving the process capability of SU-8. *Microsyst. Technol.* **2003**, *10*, 1–6. [CrossRef]
30. Delgadillo, P.; Gronheid, R.; Thode, C.J.; Wu, H.P.; Cao, Y.; Neisser, M.; Somervell, M.; Nafus, K.; Nealey, P.F. Implementation of a chemo-epitaxy flow for directed self-assembly on 300-mm wafer processing equipment. *J. Micro Nanolith. MEMS MOEMS* **2012**, *11*, 31302. [CrossRef]
31. Park, H.W.; Kim, H.J.; Roh, J.H.; Choi, J.K.; Cha, K.R. Simple and Cost-Effective method for edge bead removal by using a taping method. *J. Korean Phys. Soc.* **2018**, *73*, 1473–1478. [CrossRef]
32. Lin, C.H.; Lee, G.B.; Chang, B.W.; Chang, G.L. A new fabrication process for ultra-thick microfluidic microstructures utilizing SU-8 photoresist. *J. Micromech. Microeng.* **2002**, *12*, 590–597. [CrossRef]
33. Pannek, M.; Dunkel, T.; Schubert, D.W. Effect of a bell-shaped cover in spin coating process on final film thickness. *Mater. Res. Innov.* **2001**, *4*, 340–343. [CrossRef]
34. Torino, S.; Bhowmick, S.; Iodice, M.; Casalino, M.; Tufano, V.; Selvaggi, D.; Coppola, G.; Gioffrè, M. Edge beads removal in a photolithography process through a polymer mold. *Songklanakarin J. Sci. Technol.* **2021**, *43*, 485–491.

MDPI AG
Grosspeteranlage 5
4052 Basel
Switzerland
Tel.: +41 61 683 77 34

Photonics Editorial Office
E-mail: photonics@mdpi.com
www.mdpi.com/journal/photonics

www.ingramcontent.com/pod-product-compliance
Lightning Source LLC
Chambersburg PA
CBHW040748250926
51850CB00007BB/1438